NONSMOOTH DYNAMICS OF CONTACTING THERMOELASTIC BODIES

For other titles published in this series, go to
www.springer.com/series/5613

Advances in Mechanics and Mathematics

VOLUME 16

Aims and Scope

Mechanics and mathematics have been complementary partners since Newton's time, and the history of science shows much evidence of the beneficial influence of these disciplines on each other. The discipline of mechanics, for this series, includes relevant physical and biological phenomena such as: electromagnetic, thermal, quantum effects, biomechanics, nanomechanics, multiscale modeling, dynamical systems, optimization and control, and computational methods.

Driven by increasingly elaborate modern technological applications, the symbiotic relationship between mathematics and mechanics is continually growing. The increasingly large number of specialist journals has generated a complementarity gap between the partners, and this gap continues to widen. *Advances in Mechanics and Mathematics* is a series dedicated to the publication of the latest developments in the interaction between mechanics and mathematics and intends to bridge the gap by providing interdisciplinary publications in the form of monographs, graduate texts, edited volumes, and a special annual book consisting of invited survey articles.

J. Awrejcewicz
Yu. Pyryev

NONSMOOTH DYNAMICS OF CONTACTING THERMOELASTIC BODIES

Springer

Jan Awrejcewicz
Department of Automatics
and Biomechanics
Technical University of Lodz
Lodz, Poland
awrejcew@p.lodz.pl

Yu. Pyryev
Department of Automatics
and Biomechanics
Technical University of Lodz
Lodz, Poland
jupyrjew@p.lodz.pl

Series Editors:

David Y. Gao
Department of Mathematics
Virginia Tech
Blacksburg, VA 24061
gao@vt.edu

Ray W. Ogden
Department of Mathematics
University of Glasgow
Glasgow, Scotland, UK
rwo@maths.gla.ac.uk

ISBN: 978-1-4419-1880-2 e-ISBN: 978-0-387-09653-7
DOI: 10.1007/978-0-387-09653-7

Mathematics Subject Classification (2000): 70-xx, 34-xx, 35-xx, 37-xx, 65-xx

Printed on acid-free paper

springer.com

Contents

Preface

In this work, methods of analysis and models of contacting systems dynamics, including heat generation and wear exhibited by such systems, are presented. It should be emphasised that the methods and mathematical models of contacting systems exhibited by rigid/elastic bodies and heat/wear processes have been so far applied separately. Tribological processes occurring on a contact surface were not taken into consideration in the analysis of the dynamic rigid or elastic body models. On the other hand, most of the introduced models of bodies in contact that took tribological effects into consideration did not allow for their inertia analysis. This study contributes to the development of this field, as the models presented here yield prediction of the behaviour of contacting systems taking into account both mentioned aspects simultaneously.

When considered from the mathematical point of view, the method of analysis is reduced to the solution of the system of differential equations describing the velocities of contacting bodies and Volterra integral equation modelling contact pressure. The latter equation is obtained with the use of the Laplace integral transform.

Many aspects of the problem are considered in our analysis: the influence of various types of nonlinearities, damping, and excitations on the system behaviour are studied. Apart from Duffing elastic nonlinearity, the nonlinearity of frictional energy flux, the nonlinear relation between the friction factor and the velocity of slipping and contact temperature, and the nonlinearity of a wear factor are taken into consideration. The problem defined and discussed is solved with the use of perturbation theory, Melnikov's method, Laplace transforms, Volterra integral equation theory, and numerical analysis.

Numerical solutions of the problems are obtained, and the influence of heat generation and frictional wear on the motion of the bodies in contact is shown in many diagrams, because the authors intend to present the obtained results in a form useful for practical engineering.

The aim of this work is to define the dynamics of bodies in contact, the thermal field, the tensor of quasi-static stresses, displacements, and wear amount. The methods presented here apply to the problems defined using one or two coordinates of space and time coordinates. First, one-dimensional nonstationary contact models are considered, which enable an analytical solution. These simple models provide definitions of the basic dynamic characteristics of the analysed systems.

The subjects of this research are nonstationary dynamic phenomena and thermal conduction, as well as thermal stresses and wear in the bodies being in contact and subject to external load action.

For the solution of the problems concerning movable couplings, strains of the elements of kinematic pairs are taken into account, unlike in the research of many other authors. Energy dissipation is a result of the friction force that acts on the surfaces of bodies slipping against each other. In the case of movable couplings, self-excited vibrations may occur, depending on the properties of frictional contact. The self-excited vibrations may sometimes cause fast wear and destruction of the analysed kinematic pair.

The problem is formulated with the use of a classical linear theory of thermoelasticity. Special emphasis is put on the assumption that the material is homogeneous and isotropic as far as thermal and mechanical constants are concerned, and that its physical features are independent of temperature. Unlike other works, this study takes into account the inertial effects and the influence of thermoelastic coupling in the models of bodies in contact.

In Chapter 1 both the subject and aim of the study, as well as the current state-of-the-art of knowledge in the field of the dynamics of contacting bodies are presented. General rules of modelling contact problems are also outlined.

Chapter 2 is devoted to mathematical modelling of vibrations of a friction pair that consists of a rigid body (a bush) connected with a casing by means of springs and dampers and a rotating thermoelastic shaft.

Frictional heat generation, wear of a bush, and thermal expansion of a cylinder (shaft) are taken into account. The model of a rigid body and a cylinder is a system with two degrees-of-freedom. The model of a thermoelastic cylinder is described with the use of the theory of thermal stresses. The contact of a bush placed on a cylinder is governed by the Stribeck model of friction. In order to obtain the value of a contact pressure of friction, the Laplace method of transformation is applied. Eventually, the analysed problem is cast (Section 2.2) into the system of nonlinear differential equations and an integral equation governing the angular velocities of a cylinder and of a bush and contact pressure. The cases of kinematic (Section 2.3) and mechanical (Section 2.4) external excitations of a cylinder for various initial conditions of the bodies in contact are considered.

A new model is analysed, in which the changes of rotational speed of a bush and a cylinder, the changes of contact pressure, contact temperature and wear are interrelated. The analysis of the stability of solutions is also conducted. The analytical prediction of the occurrence of periodic and stick-slip motions is carried out, and then it is verified via numerical calculations. Possibilities of the occurrence of self-excited vibrations and chaos are analyzed in such real-like systems. The following have been shown, among others: (i) how stick-slip dynamics decreases and vanishes due to time and wear; and (ii) how stable regimes are extended due to shaft heating and wear.

It has been shown that in the considered system where an inertia of the moving body is taken into account, frictional stability loss cannot be achieved and the system cannot be overheated.

In the case when the motion of one of the contacting bodies has a constant velocity, contact pressure increases, yielding an increase either of frictional force or frictional torque. Frictional heat generation increases, and the analyzed system is overheated. Energy is supplied to the system owing to an increase of force or torque required for keeping constant velocity of the system.

In the case when the body inertia is taken into account, in spite of that the system heat extension is bounded and contact pressure may increase by increasing frictional force or torque, overheating cannot appear, because the system starts its braking process. Owing to heat exchange, a cooling process occurs; both contact pressure and friction force (torque) decrease which again causes an increase of the bodies' velocity and frictional heat generation. The mentioned process is repeated. In other words, the system controls its heating energy input to resist overheating.

A rigid body vibration model (bush–shaft) inserted into a shaft moving at nonconstant velocity is analysed. Melnikov's method is used to study chaotic dynamics of the externally driven bush. An analytical form of Melnikov's function is formulated and various nonlinear phenomena are analysed, including chaotic behaviour. The following conclusions are drawn: (i) Melnikov's function can be defined analytically, and hence one may control the system behaviour by realising either regular or chaotic dynamics; and (ii) there are zones of parameters where Melnikov's function can be significantly simplified to give an explicit formula for chaos occurrence.

Chaos thresholds, phase portraits, Poincaré cross-sections, Lapunov exponents, and the frequency spectra of vibrations are determined. Numerical calculations lead to many conclusions concerning the influence of frictional heat generation and wear on the character of system vibrations. Contact pressure and temperature are monitored numerically, and analysis of wear kinetics is conducted. In this work conditions of occurrence of periodic motion in systems with dry friction and impacts are investigated. We are aimed at estimation of parameters of the investigated system associated with occurrence of periodic motion. In particular, both the restitution coefficient and period of periodic dynamics are defined. For an arbitrary restitution coefficient two periodic orbits (stable and unstable) presented on the phase plane are studied.

In Chapter 3, vibrations of a cubicoid-shaped body and the phenomenon of frictional heat generation during contact with a surrounding medium are analysed. Two cases of thermoelastic contacts between two bodies have been studied. The bodies move vertically and they are situated between two walls. In the first considered model (Section 3.1) a rigid cubicoid (plate) is coupled with a massless spring and it can vibrate. Walls compress the body (plate) and move down at a constant speed. In the second model (Section 3.2) a contact between the plate and unmovable walls occurs owing to wall heating and thermal plate extension. The novelty of the proposed model includes inertia of the bodies, dependence of friction on the relative velocity, and the wear phenomenon. The proposed model consists of the system of both differential and integral equations that govern the plate velocity and contact pressure behaviours. Stability of stationary solutions has been analysed and conditions for stick-slip vibrations have been derived. It is expected that the proposed model may serve for estimation of the contacting characteristics of moving engineering tribological systems. A novel thermomechanical model of frictional self-excited stick-slip vibrations is proposed. A mechanical system consisting of two masses that are coupled by an elastic spring and moving vertically between two walls is considered. It is assumed that between masses and walls a Coulomb friction occurs, and stick-slip motion of the system is studied. The applied friction force depends on a relative velocity of the sliding bodies. Stability of stationary solutions is considered. A computation of contact parameters during heating of the bodies is performed. The possibility of existence of frictional autovibrations is illustrated and discussed. One degree-of-freedom system driven by either a force or a kinematic excitation is studied. An additional mass is added to the mentioned mechanical system via a special pressing device initiating dry friction occurrence on the contacting surfaces. We are focused on solution to the nonlinear problem of thermal stresses and strongly nonlinear equations governing dynamics of the investigated system. The carried-out analysis yields directions for a proper construction of the mechanical vibration dampers. It should be emphasised that the dynamic damper with dry friction may not achieve the expected properties. As we have shown, heat generation on the contacting surface between the damper and the oscillating body as well as heat expansion eliminate resonance phenomena. The real system is self-regulating, that is, it controls achievement of an optimal contacting pressure. The thermoelastic parallelepiped extends itself according to the conditions of both sliding velocity and heat transfer.

In Chapter 4 contact characteristics are studied (velocity of slipping, contact temperature, contact pressure, and wear) during the process of braking. A solution of the system of equations of heat dynamics of friction and wear is analysed (HDFW) for a tribomechanical system consisting of three different layers. Such a model gives not only the possibility of taking the thickness of frictional elements into account, but also enables determination of thermal regime and wear of a tribomechanical system consisting of the system of metal–ceramic weld and a metal plate. A solution of the thermal problem for the nonlinear relation between friction and wear coefficients and contact temperature is found. Contact characteristics of the relation of friction coefficients and wear on contact temperature are analysed numerically, and contact characteristics of a two-layer model of a brake during the braking process (the change of dimensionless velocity of slipping, contact temperature, and the kinetics of wear) are determined. Moreover, the maximum contact temperature, as well as braking time and the so-called effective thickness of a body are calculated. A particular case of a linear relation yielding the analytical form of wear is also studied.

The obtained solution can be used for monitoring heat regimes and wear of frictional surfaces of brakes working in the regime of a linear dependence of friction coefficient on temperature. Additionally, the case of harmonic variation of external load has been analysed. A numerical analysis has shown that the vibration of contact pressure causes a minor vibration of contact temperature, and both of them do not influence velocity and wear during the braking process.

The following conclusions have been obtained owing to the analysis of the steel–ceramic layer inserted on the steel basis of a frictional pad during braking: (i) for a constant thickness of the plate cover and disc an increase of the steel–ceramic layer induces an increase of contact temperature and decrease of wear; and (ii) there is thickness of the plate cover and disc such that a change of the steel–ceramic layer thickness does not influence contact characteristics.

Chapter 5 concerns analysis of the one-dimensional model of thermoelastic contact of two layers. The upper layer slips on the surface of the other one under the action of a force. Friction forces generate heat on the contact surface of friction pairs and result in wear. Heat exchange between external surfaces and their surroundings takes place, owing to Newton's law. Special emphasis is put on the research concerning frictional contact of two layers with different thicknesses.

The main novelty of the presented model lies in taking into account the inertia of one of the layers in contact, as well as the relation of the friction coefficient and wear on contact temperature, allowing us to provide the analysis for the case of close proximity of the layers. In the latter case, contact pressure is an unknown value and depends on heat expansion of the layers in contact.

It is illustrated how Frictional ThermoElastic Instability (FTEI) occurs in the case of constant velocity of motion of one layer. The effect of nonstationary heat generation and wear on the conditions of FTEI occurrence is also studied. FTEI is characterized by an exponential increase in time of contact characteristics, that is, temperature, pressure, and wear, when the relative velocity of slipping is higher than a certain critical velocity.

In the case of constant relative velocity of moving bodies, the mechanism of occurrence of FTEI of the bodies in contact modelling dynamics of seals, brake blocks, and other tribological systems is studied. An equation for critical speed determination is derived. It is illustrated how the inclusion of wear results in a higher critical speed, which means that FTEI occurs. In the case when wear is higher than heat expansion, FTEI does not occur. The obtained results can be applied in resistance calculations for movable elements of machines and mechanisms, as well as in introducing recommendations concerning improvement of the reliability and durability of various mechanisms and machines with friction, especially in formulating the criteria of an optimal choice of materials of rubbing pairs and in the analysis of the processes of heat generation and wear while grinding or braking of transport mechanisms.

The proposed theory and modelling can be applied in a numerical study of systems with many degrees-of-freedom or other more advanced models for wear and friction.

Owing to the introduction of a moment applied to a shaft (characteristic of the energy source) dependent on rotational velocity one may also obtain various

M_{fr} moment of friction force (per length unit) [N]

M_e moment of elasticity force (per length unit) [N]

$M(\tau_0)$ Melnikov function

$N(t)$ normal component of the reaction (normal load per length unit) [N m^{-1}]

$P(t)$ contact pressure [N m^{-2}]

P_* characteristic contact pressure value [N m^{-2}]

p dimensionless contact pressure

p^* perturbation of contact pressure

p_{st} dimensionless steady-state contact pressure

r dimensionless radius

R_1 radius of a cylinder [m]

R_2 external radius of the bush [m]

s Laplace transform parameter

s_m roots of a characteristic equation $\Delta(s) = 0$

t time [s]

t_D characteristic system time related to the bush oscillations period [s]

t_T characteristic time related to heat transfer [s]

t_* characteristic time of a system [s]

t_c period of time in which the bodies stay in contact [s]

T temperature of a contact surface [°C]

T_{sm} temperature of the surrounding medium [°C]

T_l body temperature ($l = 1, 2$) [°C]

T_p period [s]

u dimensionless displacement

u^w dimensionless wear

$U_0 h_U(t)$ compression [m]

U^w wear (displacement of the working surfaces along $\mathbf{N}_c$ due to wear) [m]

U displacement component along radial direction in the shaft [m]

V_r relative velocity of the contact bodies [m s^{-1}]

α_l coefficient of thermal expansion of the bodies ($l = 1, 2$) [$°C^{-1}$]

α_T heat transfer coefficient [W m^{-2} $°C^{-1}$]

γ dimensionless thermomechanical parameter

$\Delta(s)$ characteristic function

ζ dimensionless amplitude of external force excitation

ζ_k dimensionless amplitude of external kinematic excitation

θ dimensionless temperature

θ^* temperature perturbation

θ_{st} dimensionless steady-state contact temperature

λ^l, μ^l Lamé constants ($l = 1, 2$) [N m^{-2}]

λ_l thermal conductivity ($l = 1, 2$) [W m^{-1} $°C^{-1}$]

μ_m roots of a characteristic equation

ν_l Poisson's ratios ($l = 1, 2$)

σ_R stress component along radial direction in the shaft [N m^{-2}]

τ dimensionless time

ϕ dimensionless shaft position angle

$\dot{\phi} = \omega_1(\tau)$ dimensionless angular velocity of a cylinder

$\varphi(\tau)$ dimensionless bush position angle

φ^* perturbation of the angle of bush rotation

φ_{st} dimensionless static angle of bush rotation

$\varphi_l(t)$ angle of the shaft ($l = 1$) and the bush ($l = 2$) position

$\dot{\varphi}_l(t)$ angular velocities of a cylinder ($l = 1$) and the bush ($l = 2$) [s^{-1}]

$\omega_r(\tau)$ dimensionless relative velocity

$\tilde{\omega}$ relation of time characterizing the vibrations of a material system to the time characterizing the process of temperature equalizing

ω' frequency of external excitation [s^{-1}]

ω_0 dimensionless frequency of external perturbation

ω Fourier transformation parameter

Ω angular velocity [s^{-1}]

coefficient of thermal expansion of the bodies [illegible]

[illegible]

Chapter 1
Introduction

1.1. Object of the study

Friction, wear, heat generation accompanying friction, and heat expansion (or contraction) are all very complex phenomena that interact and form one complex multidimensional dynamic system analysed together with friction. For the nonstationary friction process, all of its time-dependant parameters are interrelated.

This study deals both with the phenomena occurring in material systems and with methods of analysing them. The problems discussed here concern the analysis of dynamic phenomena occurring in machine systems and are closely related to theoretical mechanics, friction theory, mechanics of materials, methods of the analysis of differential equations, analysis of thermal stresses, tribology (friction, wear), and to dynamic process control.

The authors are interested mainly in the search for approximate solutions of the obtained differential and integral equations and the analysis of their properties. For this purpose approximate analytical methods and numerical methods are applied. What is very important from the practical point of view is finding periodic stick-slip solutions. So-called deterministic chaos [Awrejcewicz (1989), (1996)] may occur in the studied systems, in which case the minimal change of initial conditions may lead to the exponential divergence of the neighbouring trajectories of the phase flow.

1.1.1 Equations of motion of bodies in contact

Let us consider the system of two bodies taking in the Euclidean space $\Re^{m_{\rm dim}}$, ($m_{\rm dim} = 1, 2, 3$) occupying volumes Ω^1, Ω^2 bounded by the surfaces $\Gamma^1 = \partial\Omega^1$, $\Gamma^2 = \partial\Omega^2$ (Fig. 1.1).

For a general case $\Gamma^l = \Gamma^l_u \cup \Gamma^l_\sigma \cup \Gamma^l_{u\sigma} \cup \Gamma^l_c$ or $\Gamma^l = \Gamma^l_T \cup \Gamma^l_q \cup \Gamma^l_{Tq} \cup \Gamma^l_c$, ($l = 1, 2$), where $\Gamma^l_c = \Gamma^l_s \cup \Gamma^l_{sl} \cup \Gamma_{st}$, Γ^l_u, Γ^l_σ, $\Gamma^l_{u\sigma}$ denote the boundaries for

J. Awrejcewicz and Yu. Pyryev, *Nonsmooth Dynamics of Contacting Thermoelastic Bodies*,
DOI: 10.1007/978-0-387-09653-7_1,

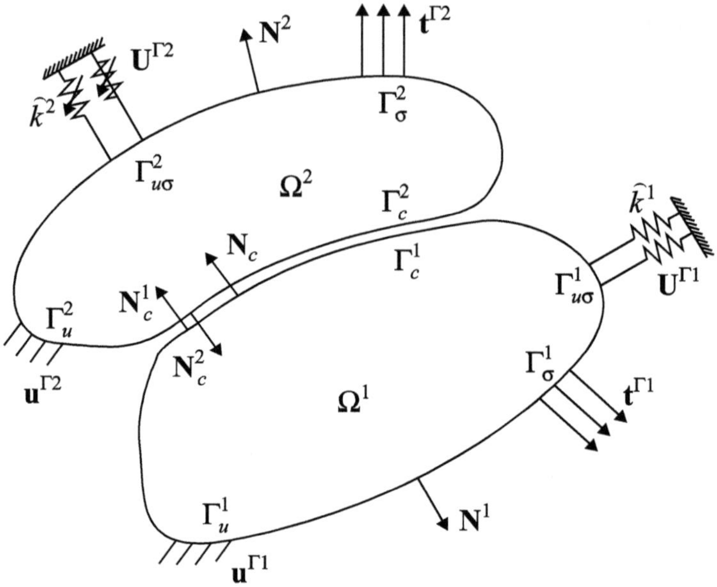

Fig. 1.1: Modelling the contact of two bodies.

each point, where displacements, stresses, or Winkler conditions are given. Γ^l_T, Γ^l_q, Γ^l_{Tq}, Γ^l_c denote the boundaries where temperature together with its gradient, and Newton condition or the condition of thermal contact, are determined. Γ^l_s, Γ_{st}, Γ^l_{sl} are the boundaries where clearance, stick, or slip occurs.

Let Ω denote the sum of the volumes Ω^1 and Ω^2 ($\Omega = \Omega^1 \cup \Omega^2$), whereas $\Gamma = \Gamma^1 \cup \Gamma^2$. The contact in the system under the influence of loads (both mechanical and nonmechanical) is accompanied by friction. The models (systems of equations) describing both mechanical and nonmechanical processes can be cast in the form:

$$\Im^l(\partial/\partial x, \partial/\partial t)U^l(x,t) = -\Phi^l(x,t), \quad l = 1,2, \quad (x,t) \in \mathcal{T}_\infty, \tag{1.1}$$

where $\Im^l(\partial/\partial x, \partial/\partial t)$ is the operational matrix of the form

$$\Im^l(\partial/\partial x, \partial/\partial t) = \|\Im^l_{nm}(\partial/\partial x, \partial/\partial t)\|_{N\times N}. \tag{1.2}$$

In the above $U^l = \|u^l_1, u^l_2, \ldots, u^l_n\|^T$ is an N-dimensional single-column matrix that governs the processes taking place in the bodies; the symbol $\|\cdot\|^T$ denotes a column matrix; the operators $\Im^l_{nm}(\partial/\partial x, \partial/\partial t)$ for homogeneous bodies are the polynomials obtained from the partial differentials $\partial/\partial x_m$ ($m = 1,2,3$), $\partial/\partial t$ with constant coefficients; $\mathcal{T}_\infty\{(x,t) : x \in \Omega, t \in [0,\infty)\}$; $\mathcal{T}_\infty \in \Re^{m_{\dim}+1}$; $x \in (x_1, \ldots, x_{m_{\dim}})$ is the point of Euclidean space $\Re^{m_{\dim}}$ ($m_{\dim} = 1,2,3$); and t denotes time.

Nonstationary thermal and thermoelastic processes in the bodies are analysed in references [Parkus (1959), Carslaw, Jaeger (1959), Melan, Parkus (1959), Boley, Weiner (1960), Nowacki (1962), Kovalenko (1975), Awrejcewicz, Krysko (2003)]. Various models of the bodies are applied for computations, and they can be found, for instance, in the monographs [Nowacki (1970), Nowacki, Olesiak (1991), Pyryev (1999), Jaeger (2005), Awrejcewicz et al. (2004)], where the models of thermoelastic bodies are also studied.

Let us assume that the loads (mechanical and thermal) generate small strains and stresses, and that for the system under analysis the mechanical and thermal interaction of the bodies in contact occurs. The system of the equations of thermoelasticity, expressed with the use of the vector of displacement $\mathbf{u}^l = (u_1^l, u_2^l, u_3^l)$ $(l = 1, 2)$ and temperature gain $T_l(x, t)$ $(l = 1, 2)$ measured in relation to the surrounding temperature, can be presented in the form (1.1), where $N = 4$, $U^l = \|u_1^l, u_2^l, u_3^l, T_l\|^T$. The following are the components of the matrix differential operator (1.2):

$$\Im_{nm}^l(\partial/\partial x, \partial/\partial t) = \mu^l \delta_{nm}(\nabla^2 - c_{l2}^{-2}\partial^2/\partial t^2) + (\lambda^l + \mu^l)\partial^2/\partial x_n \partial x_m, \quad (1.3)$$

$$\Im_{n4}^l(\partial/\partial x, \partial/\partial t) = -\gamma_l \partial/\partial x_n, \quad (1.4)$$

$$\Im_{4m}^l(\partial/\partial x, \partial/\partial t) = -\eta_l \partial^2/\partial t \partial x_m, \qquad n, m = 1, 2, 3, \quad (1.5)$$

$$\Im_{44}^l(\partial/\partial x, \partial/\partial t) = \nabla^2 - a_l^{-1}\partial/\partial t, \qquad l = 1, 2, \quad (1.6)$$

where

$$\gamma_l = \alpha_l(3\lambda^l + 2\mu^l), \qquad c_{l2} = \sqrt{\mu^l/\rho^l}, \qquad l = 1, 2,$$

$$\delta_{nm} = 1, \qquad n = m; \qquad \delta_{nm} = 0, \qquad n \neq m.$$

In the above $\nabla^2 = \partial^2/\partial x_1^2 + \partial^2/\partial x_2^2 + \partial^2/\partial x_3^2$ is a Laplace operator, λ^l, μ^l are Lamé coefficients, ρ^l is the density of the body material, α_l is a thermal expansion coefficient, a_l is thermal diffusivity, and η_l is a thermodynamic constant. The system of differential Equations (1.1)–(1.6) is coupled by the element (1.5) and therefore these equations are called coupled equations of thermoelasticity. The theoretical studies [Nowacki (1962), Pyryev (1999)] and experiments show that the coupling usually has a small effect on the system behaviour. If one neglects the element $\eta_l \partial^2/\partial t \partial x_m$ in (1.5), Equations (1.1)–(1.6) create the system of differential equations of Lamé–Neumann quasistatic uncoupled thermoelasticity.

Position of a point of the body l $(l = 1, 2)$, occupying space Ω^l and bounded by the surface Γ^l in its unloaded state, is characterized by vector $\mathbf{r}_1^l = x_j^l \mathbf{i}_j = x_1^l \mathbf{i}_1 + x_2^l \mathbf{i}_2 + x_3^l \mathbf{i}_3$. For the power load of the system, bodies can (the main vector of external powers and its moment are not equal to zero) exhibit a motion.

As mentioned before, the mechanical $\mathbf{u}^{\Gamma l}$, $\mathbf{t}^{\Gamma l}$, and thermal loads $T^{\Gamma l}$, $q^{\Gamma l}$ $(l = 1, 2)$ influence the bodies in the initial moment. We assume that the main vector and the main moment of external loads $\mathbf{b}^l$, $\mathbf{t}^{\Gamma l}$ are different from zero,

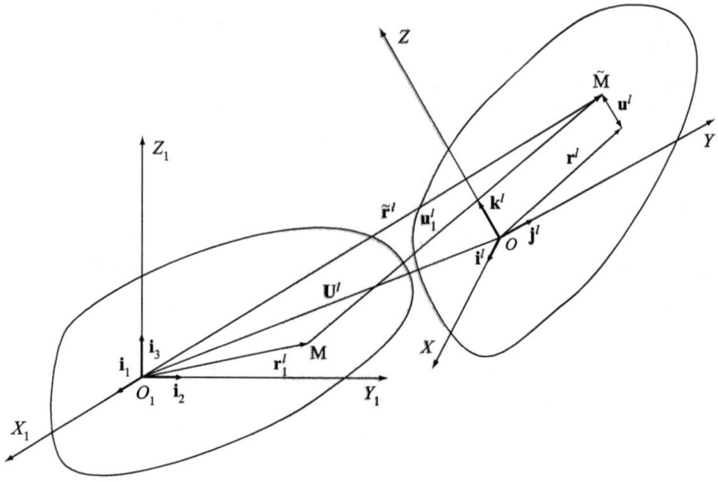

Fig. 1.2: Distribution of general displacement $\mathbf{u}_1^l$ $(l = 1, 2)$.

which causes the motion of bodies. At some point in time $t > 0$ the bodies will have a certain new configuration in space, close to the configuration of that of perfectly rigid bodies, respectively, and their geometry and density correspond to the characteristics of unloaded elastic bodies. The point $\mathbf{r}_1^l$ takes the place of $\tilde{\mathbf{r}}^l$: $\tilde{\mathbf{r}}^l = \mathbf{r}_1^l + \mathbf{u}_1^l$, where $\mathbf{u}_1^l(x_1^l, x_2^l, x_3^l, t)$ is the vector of displacement of the point whose coordinates are $u_{11}^l, u_{12}^l, u_{13}^l$ (Fig. 1.2).

We assume that unit elongation (and shortening) and angular strains are smaller than one. The shift of the bodies and their displacements related to their rotation (we treat them as rigid solids) are not small. In what follows we apply a Green strain tensor $\hat{e}^l = e_{ij}^l \mathbf{i}_i \otimes \mathbf{i}_j$ of the form

$$e_{ij}^l = \frac{1}{2}\left(\frac{\partial u_{1i}^l}{\partial x_j} + \frac{\partial u_{1j}^l}{\partial x_i} + \frac{\partial u_{1m}^l}{\partial x_j}\frac{\partial u_{1m}^l}{\partial x_i}\right), \qquad l = 1, 2, \qquad i, j = 1, 2, 3 \quad (1.7)$$

and nonlinear equations of the motion of elastic bodies according to Lagrange [Kappus (1939), Novogilov (1958), Ulitko (1990)]

$$\mathrm{div}\hat{P}^l + \mathbf{b}^l = \rho^l \frac{\partial^2 \mathbf{u}_1^l}{\partial t^2}, \qquad \left(\frac{\partial P_{ij}^l}{\partial x_i} + b_j^l = \rho^l \frac{\partial^2 u_{1j}^l}{\partial t^2}\right), \tag{1.8}$$

$$P_{ij}^l = \left(\delta_{ij} + \frac{\partial u_{1i}^l}{\partial x_k}\right)\sigma_{kj}^{*l}, \tag{1.9}$$

where $\hat{P}^l = P_{ij}^l \mathbf{i}_i \otimes \mathbf{i}_j$ is the Piola pseudostress tensor; $\hat{\sigma}^{*l} = \sigma_{ij}^{*l} \mathbf{i}_i \otimes \mathbf{i}_j$ is the Kirchhoff pseudostress tensor, defined via stress tensor $\hat{\sigma}^l = \sigma_{ij}^l \mathbf{i}_i \otimes \mathbf{i}_j$ in the Lagrange variables $\sigma_{ij}^{*l} = \sqrt{1/(1+2e_{jj}^l)}(da_i^{*l}/da_i^l)\sigma_{ij}^l$. Observe that da_i^l, da_i^{*l}

denote surfaces of a body deformed element situated vertically to the axis x_i^l, whereas $\mathbf{b}^l$ is the vector of the volume forces expressed in Lagrange's variables. Note that for small deformations the mentioned tensors can be treated as approximately equalones.

Kinematic Equations (1.7) and dynamic Equations (1.8) are supplemented by the physical Duhamel–Neumann equations of the form

$$\hat{\sigma}^l = 2\mu^l\hat{e}^l + (\lambda^l\theta^l - \gamma_l T_l)\hat{I}, \qquad (\sigma_{ij}^l = (\lambda^l\theta^l - \gamma_l T_l)\delta_{ij} + 2\mu^l e_{ij}^l), \qquad (1.10)$$

where $\theta^l = e_{11}^l + e_{22}^l + e_{33}^l$, $(l = 1, 2)$.

The mechanical boundary conditions of the uncoupled bodies (without a contact) can be formulated in the following form,

$$\mathbf{N}^l \cdot \hat{P}^l = \mathbf{t}^{\Gamma l}, \qquad (N_i^l P_{ij}^l = t_j^{\Gamma l}), \qquad x \in \Gamma_\sigma^l, \qquad l = 1, 2, \qquad (1.11)$$

where N_i^l are directional cosines of an undeformed surface Γ_σ^l ($\mathbf{N}^l = N_j^l \mathbf{i}_j$), and $t_j^{\Gamma l}$ are the initial loading components regarding the undeformed surface ($\mathbf{t}^{\Gamma l} = t_j^{\Gamma l}\mathbf{i}_j$).

It should be emphasised that displacements $\mathbf{u}_1^l$ are not small. What is more, the nonlinear problem formulated like this includes the problem related to the dynamics of thermoelastic bodies. There is no need to add the equations of motion and to formulate boundary conditions by introducing the additional equations. For instance, the theorems of the motion of a centroid and of the shift of angular momentum result [Ulitko (1990)] directly from the motion equations and boundary conditions.

By integrating the equations of motion along the initial volume and by changing triple integrals into surface integrals (Gauss–Ostrogradski theorem) and taking into account boundary conditions (1.11), we obtain [Ulitko (1990)] the following vector equations,

$$\iiint\limits_{\Omega^l} \rho^l \frac{\partial^2 \mathbf{u}_1^l}{\partial t^2} dv = \oint\!\!\!\oint\limits_{\Gamma_\sigma^l} \mathbf{t}^{\Gamma l} dS + \oint\!\!\!\oint\limits_{\Gamma_c^l} \mathbf{p}^{cl} dS + \iiint\limits_{\Omega^l} \mathbf{b}^l dv. \qquad (1.12)$$

The position of the centres of elastic and homogeneous masses is determined on the basis of equation

$$\mathbf{U}^l = \frac{1}{\Omega^l} \iiint\limits_{\Omega^l} \tilde{\mathbf{r}}^l dv = \frac{1}{\Omega^l} \iiint\limits_{\Omega^l} (\mathbf{r}_1^l + \mathbf{u}_1^l) dv. \qquad (1.13)$$

The eventual equation describing the motion of centroids will have the following form.

$$M^l \frac{d^2 \mathbf{U}^l}{dt^2} = \mathbf{R}^l, \qquad \mathbf{R}^l = \oint\!\!\!\oint\limits_{\Gamma_\sigma^l} \mathbf{t}^{\Gamma l} dS + \oint\!\!\!\oint\limits_{\Gamma_c^l} \mathbf{p}^{cl} dS + \iiint\limits_{\Omega^l} \mathbf{b}^l dv, \qquad (1.14)$$

where $M^l = \rho^l \Omega^l$, ($\Omega^l = \text{const}$).

The equation governing angular momentum variation possesses the following form.

$$\rho^l \frac{d}{dt} \iiint\limits_{\Omega^l} \left(\tilde{\mathbf{r}}^l \times \frac{\partial \mathbf{u}_1^l}{\partial t} \right) dv$$
$$= \oiint\limits_{\Gamma_\sigma^l} (\tilde{\mathbf{r}}^l \times \mathbf{t}^{\Gamma l}) dS + \oiint\limits_{\Gamma_c^l} (\tilde{\mathbf{r}}^l \times \mathbf{p}^{cl}) dS + \iiint\limits_{\Omega^l} (\tilde{\mathbf{r}}^l \times \mathbf{b}^l) dv. \tag{1.15}$$

In accordance with the study [Ulitko (1990)] we assume that the motion of an elastic body $\mathbf{u}_1^l$ consists of (as for the perfectly rigid body) the motion of the centroid and the rotation around it and of the relative motion. After the introduction of the system of movable coordinates X, Y, Z (Fig. 1.2), next to the initial immovable system of coordinates, the general vector of displacement has the following form,

$$\mathbf{u}_1^l = \mathbf{U}^l + \mathbf{r}^l + \mathbf{u}^l - \mathbf{r}_1^l, \tag{1.16}$$

where $\mathbf{U}^l = O_1O$ is the displacement of centroids ($l = 1, 2$); $\mathbf{r}^l$ is the vector of the point in the system of movable coordinates that had the same coordinates in the initial immovable system ($\mathbf{r}^l = x_j^l \mathbf{e}_j^l$, $\mathbf{e}_1^l = \mathbf{i}^l$, $\mathbf{e}_2^l = \mathbf{j}^l$, $\mathbf{e}_3^l = \mathbf{k}^l$); and $\mathbf{u}^l$ is the vector of displacement that takes the deviation of the points of elastic bodies from the perfectly rigid bodies into account.

Orientation of the movable coordinates X, Y, Z regarding the fixed ones X_1, Y_1, Z_1 is defined by the following equations,

$$\mathbf{i}^l = \mathbf{i}_1 \cos \alpha_1^l + \mathbf{i}_2 \cos \beta_1^l + \mathbf{i}_3 \cos \gamma_1^l,$$
$$\mathbf{j}^l = \mathbf{i}_1 \cos \alpha_2^l + \mathbf{i}_2 \cos \beta_2^l + \mathbf{i}_3 \cos \gamma_2^l,$$
$$\mathbf{k}^l = \mathbf{i}_1 \cos \alpha_3^l + \mathbf{i}_2 \cos \beta_3^l + \mathbf{i}_3 \cos \gamma_3^l.$$

Taking into account vector (1.16) and the Duhamel–Neuman relations, the equations of motion have the following form [Ulitko (1990)],

$$2\frac{1-\nu_l}{1-2\nu_l} \text{grad div} \mathbf{u}^l - \text{rot rot} \mathbf{u}^l + \frac{1}{\mu^l} \mathbf{b}^l = \frac{\gamma_l}{\mu^l} \text{grad} T_l + \frac{\rho^l}{\mu^l} \frac{\partial^2 \mathbf{u}_1^l}{\partial t^2}, \tag{1.17}$$

where the differentiation in the left-hand side of the equation of the vector of relative displacement $\mathbf{u}^l = u_1^l \mathbf{i}^l + u_2^l \mathbf{j}^l + u_3^l \mathbf{k}^l$ takes place in the system of movable coordinates $(\mathbf{i}^l, \mathbf{j}^l, \mathbf{k}^l)$. The right-hand side of this equation includes the total displacement expressed by the vector of general displacement $\mathbf{u}_1^l$. While introducing the vector $\mathbf{u}^l$ to the right-hand side of the equation we assume that the mass moments of inertia for elastic bodies are equal to the mass moments of inertia of perfectly rigid bodies. In addition, we assume that in initial state the fixed Cartesian coordinates coincide with the central axes of an ellipsoid of inertia. Eventually, we obtain:

$$2\frac{1-\nu_l}{1-2\nu_l}\text{grad div}\mathbf{u}^l - \text{rot rot}\mathbf{u}^l + \frac{1}{\mu^l}\left(\mathbf{b}^l - \gamma_l \text{grad}T_l - \frac{\rho^l}{M_l}\mathbf{R}^l\right)$$

$$= \frac{\rho^l}{\mu^l}[\ddot{\mathbf{u}}^l + (\dot{\boldsymbol{\omega}}^l \times \mathbf{r}^l) + \boldsymbol{\omega}^l \times (\boldsymbol{\omega}^l \times \mathbf{r}^l) + 2(\boldsymbol{\omega}^l \times \dot{\mathbf{u}}^l)], \qquad (1.18)$$

$$\nabla^2 T_l = \frac{1}{a_l}\left(\frac{\partial T_l}{\partial t} + \dot{\mathbf{U}}^l \text{grad}T_l\right), \qquad l = 1, 2, \qquad (1.19)$$

where $\boldsymbol{\omega}^l = \tilde{p}^l \mathbf{i}^l + \tilde{q}^l \mathbf{j}^l + \tilde{r}^l \mathbf{k}^l$ is the vector of instantaneous angular velocity described by the following Euler equations,

$$(A^l \dot{\tilde{p}}^l + (C^l - B^l)\tilde{q}^l \tilde{r}^l, B^l \dot{\tilde{q}}^l + (A^l - C^l)\tilde{r}^l \tilde{p}^l, C^l \dot{\tilde{r}}^l + (B^l - A^l)\tilde{p}^l \tilde{q}^l)$$

$$+ 2\iiint_{\Omega^l} \rho^l (\mathbf{r}^l \times (\boldsymbol{\omega}^l \times \dot{\mathbf{u}}^l))dv + \iiint_{\Omega^l} \rho^l (\mathbf{r}^l \times \ddot{\mathbf{u}}^l)dv$$

$$= \oiint_{\Gamma_\sigma^l} (\mathbf{r}^l \times \mathbf{t}^{\Gamma l})dS + \oiint_{\Gamma_c^l} (\mathbf{r}^l \times \mathbf{p}^{cl})dS + \iiint_{\Omega^l} (\mathbf{r}^l \times \mathbf{b}^l)dv, \qquad (1.20)$$

whereas A^l, B^l, C^l are mass moments of inertia of the bodies ($l = 1, 2$)

$$A^l = \iiint_{\Omega^l} \rho^l ((x_2^l)^2 + (x_3^l)^2)dv,$$

$$B^l = \iiint_{\Omega^l} \rho^l ((x_3^l)^2 + (x_1^l)^2)dv,$$

$$C^l = \iiint_{\Omega^l} \rho^l ((x_1^l)^2 + (x_2^l)^2)dv.$$

It is clear that Equations (1.20) are related to Equations (1.18) by the moments of Coriolis forces and inertial forces of relative motion.

Analysis of a stress–strain body state and its 3D motion is reduced to integration of Equation (1.18) and Euler's Equations (1.20) including the following relations,

$$\tilde{p}^l = \dot{\mathbf{j}}^l \cdot \mathbf{k}^l = -\dot{\mathbf{k}}^l \cdot \mathbf{j}^l$$

$$= -\dot{\alpha}_2^l \sin\alpha_2^l \cos\alpha_3^l - \dot{\beta}_2^l \sin\beta_2^l \cos\beta_3^l - \dot{\gamma}_2^l \sin\gamma_2^l \cos\gamma_3^l$$

$$= +\dot{\alpha}_3^l \sin\alpha_3^l \cos\alpha_2^l + \dot{\beta}_3^l \sin\beta_3^l \cos\beta_2^l + \dot{\gamma}_3^l \sin\gamma_3^l \cos\gamma_2^l,$$

$$\tilde{q}^l = \dot{\mathbf{k}}^l \cdot \mathbf{i}^l = -\dot{\mathbf{i}}^l \cdot \mathbf{k}^l$$

$$= -\dot{\alpha}_3^l \sin\alpha_3^l \cos\alpha_1^l - \dot{\beta}_3^l \sin\beta_3^l \cos\beta_1^l - \dot{\gamma}_3^l \sin\gamma_3^l \cos\gamma_1^l$$

$$= +\dot{\alpha}_1^l \sin\alpha_1^l \cos\alpha_3^l + \dot{\beta}_1^l \sin\beta_1^l \cos\beta_3^l + \dot{\gamma}_1^l \sin\gamma_1^l \cos\gamma_3^l,$$

$$\tilde{r}^l = \dot{\mathbf{i}}^l \cdot \mathbf{j}^l = -\dot{\mathbf{j}}^l \cdot \mathbf{i}^l$$

$$= -\dot{\alpha}_1^l \sin\alpha_1^l \cos\alpha_2^l - \dot{\beta}_1^l \sin\beta_1^l \cos\beta_2^l - \dot{\gamma}_1^l \sin\gamma_1^l \cos\gamma_2^l$$

$$= +\dot{\alpha}_2^l \sin\alpha_2^l \cos\alpha_1^l + \dot{\beta}_2^l \sin\beta_2^l \cos\beta_1^l + \dot{\gamma}_2^l \sin\gamma_2^l \cos\gamma_1^l.$$

In order to determine stresses in a body (movable) coordinates, the following Duhamel–Neumann equations are used

$$\hat{\sigma}^l = 2\mu^l \hat{e}^l + (\lambda^l \theta^l - \gamma_l T_l)\hat{I}, \qquad (\sigma_{ij}^l = (\lambda^l \theta^l - \gamma_l T_l)\delta_{ij} + 2\mu^l e_{ij}^l), \tag{1.21}$$

where $\theta^l = e_{11}^l + e_{22}^l + e_{33}^l$, $(l = 1, 2)$; $\hat{\sigma}^l = \sigma_{jk}^l \mathbf{e}_j^l \otimes \mathbf{e}_k^l$, $\mathbf{e}_1^l = \mathbf{i}^l$, $\mathbf{e}_2^l = \mathbf{j}^l$, $\mathbf{e}_3^l = \mathbf{k}^l$. The symmetrical tensor ($\hat{e}^l = \hat{e}^{lT}$, $e_{jk}^l = e_{kj}^l$) of the strain $\hat{e}^l = e_{jk}^l \mathbf{e}_j^l \otimes \mathbf{e}_k^l$, $(l = 1, 2)$ in the body coordinates has the following components

$$e_{jk}^l = \frac{1}{2}\left(\frac{\partial u_j^l}{\partial x_k} + \frac{\partial u_k^l}{\partial x_j}\right), \qquad l = 1, 2, \qquad i, j = 1, 2, 3. \tag{1.22}$$

Taking mechanical $\mathbf{u}^{\Gamma l}$, $\mathbf{t}^{\Gamma l}$, $\mathbf{b}^l$ and thermal loads $T^{\Gamma l}$, $q^{\Gamma l}$, Q^l $(l = 1, 2)$ changing very (sufficiently) slowly in time (time of the change of the load is longer than time b_l/c_{12}, where b_l denotes the size of the body l) allows us to use the quasi-static equations of thermoelasticity (the inertial elements in the equations of motion or the operator $\mu^l c_{l2}^{-2} \partial^2/\partial t^2$ in (1.3) can be neglected). The disturbing pulse for which a quasi-static solution can be applied has been determined in the work [Awrejcewicz, Pyryev (2003b)] on the basis of a classical Lamb problem [Awrejcewicz, Pyryev (2003a)].

It follows that the theoretical analysis of contact dynamics can be fully implemented in some simple cases. It takes place when displacement of the contacting bodies can be reduced either to rotation around an axis (Chapter 2) or to the motion along one axis (Chapters 3–5) and assuming that the studied system characteristics do not depend on the mentioned axes. For the material systems analyzed in this study, the mathematical models with one and two degrees-of-freedom are constructed. Equations (1.14) and (1.20) are expressed in a dimensionless form

$$\ddot{\varphi} + 2h\dot{\varphi} + \Phi(\varphi) = \varepsilon[\zeta\cos(\omega_0\tau) + F(v_r)p(\theta, u^w, \tau)], \tag{1.23}$$

$$\ddot{\phi} = a_M[M(\tau) - F(v_r)p(\theta, u^w, \tau)], \tag{1.24}$$

$$\Phi(\varphi) = -a\varphi + b\varphi^3, \qquad v_r = \dot{\phi} - \dot{\varphi} - \zeta_k \sin(\omega_0\tau), \qquad a \in \Re^1, \tag{1.25}$$

where ϕ, φ are dimensionless displacements of each body; $M(\tau)$ is a mechanical load that drives one body; ζ and ζ_k are the amplitudes of both mechanical and kinematic perturbations of the other body; $F(v_r)$ is a coefficient of kinetic friction; v_r is the velocity of slipping of the bodies in contact; $\Phi(\varphi)$ is a nonlinear conservative force; and $p(\theta, u^w, \tau)$ is contact pressure that depends on contact temperature, wear, and time. In the case $\dot{\phi} = \text{const} = V$ we deal with the motion of one body with the given velocity (Equation (1.23)). In this case, the analysed systems are self-excited ones, which means that in some areas of the plane $(\varphi, \dot{\varphi})$ the following condition will be satisfied: $\varepsilon F(v_r)pv_r - 2hv_r^2 < 0$.

The numerical analysis of the problem has been carried out using the Runge–Kutta method for Equations (1.23) and (1.24) and the quadrature method for integral equations regarding contact pressure computations as well as the use of iterative methods. The numerical computations support in full the conclusions obtained on the basis of our theoretical study.

For each of the analysed bodies (during the analysis of classic problems), first we solve the equations of thermal conduction, which allows us to determine the range of temperature. Next, differential equations are solved for the known load of body forces, $\gamma_l \partial/\partial x_n$ being its elements. Furthermore, it has to be emphasised that during the analysis of contact problems, contact conditions must be modelled.

1.1.2 Boundary and contact conditions

The stress vector $\mathbf{p}^{Nl}$ operating on the surface element $d\Gamma^l$ with a normal $\mathbf{N}^l(N_1^l, N_2^l, N_3^l)$ is determined in the following way: $\mathbf{p}^{Nl} = \mathbf{N}^l \cdot \hat{\sigma}^l$ $(p_j^{Nl} \equiv N_i^l \sigma_{ij}^l)$ (Fig. 1.1), assuming that $\mathbf{N}_c \approx \mathbf{N}_c^1 \approx -\mathbf{N}_c^2$ leads to a relation $\mathbf{N}_c^1 \cdot \hat{\sigma}^1 = -\mathbf{N}_c^2 \cdot \hat{\sigma}^2 = -\mathbf{p}^c$. Furthermore, we also introduce the vector of a relative displacement of bodies in the form of $\mathbf{w} = \mathbf{u}^2 - \mathbf{u}^1$.

Let us resolve the stress factor $\mathbf{p}^c$ into the component $-p\mathbf{N}_c$ in a normal direction and the tangent component $\mathbf{p}_\tau$ in a plane $d\Gamma_c$ (Fig. 1.3a): $p = -\mathbf{p}^c \cdot \mathbf{N}_c$, $\mathbf{p}_\tau = \mathbf{p}^c + p\mathbf{N}_c = (\hat{I} - \mathbf{N}_c \otimes \mathbf{N}_c)\mathbf{p}^c$, $p_\tau = \sqrt{(\mathbf{p}^c)^2 - (p)^2}$ (p, normal pressure; p_τ, tangent stress). Let us resolve the vector of displacement $\mathbf{u}^l$ into the component $u_N^l \mathbf{N}_c$ in a normal direction and the tangent component $\mathbf{u}_\tau^l$ in a plane $d\Gamma_c$: $u_N^l = \mathbf{u}^l \cdot \mathbf{N}_c$, $\mathbf{u}_\tau^l = \mathbf{u}^l - u_N^l \mathbf{N}_c = (\hat{I} - \mathbf{N}_c \otimes \mathbf{N}_c)\mathbf{u}^l$. Furthermore, let us resolve the vector of relative displacement $\mathbf{w}$ into the component $w_N \mathbf{N}_c$ in a normal direction and a tangent component $\mathbf{w}_\tau$ (tangent slip) in a plane $d\Gamma_c$: $w_N = \mathbf{w} \cdot \mathbf{N}_c$, $\mathbf{w}_\tau = \mathbf{w} - w_N \mathbf{N}_c = (\hat{I} - \mathbf{N}_c \otimes \mathbf{N}_c)\mathbf{w}$.

Classical mechanical boundary conditions have the following form.

$$\mathbf{u}^l = \mathbf{u}^{\Gamma l}, \qquad (u_j^l = u_j^{\Gamma l}), \qquad x \in \Gamma_u^l, \tag{1.26}$$

$$\mathbf{p}^{Nl} \equiv \mathbf{N}^l \cdot \hat{\sigma}^l = \mathbf{t}^{\Gamma l}, \quad (p_j^{Nl} \equiv N_i^l \sigma_{ij} = t_j^{\Gamma l}), \quad x \in \Gamma_\sigma^l, \quad l = 1, 2, \tag{1.27}$$

whereas *classical temperature boundary conditions* are governed by the following equations.

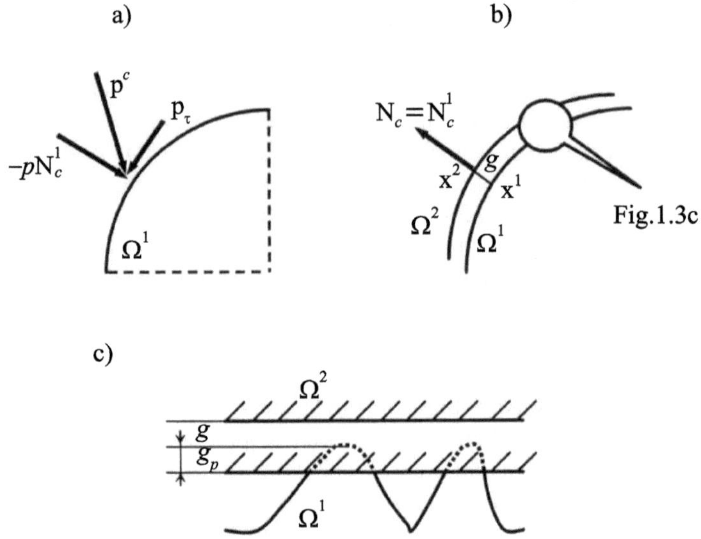

Fig. 1.3: The distribution of a stress vector $\mathbf{p}^c$ into components (a), distance definition g (b), and the definition of proximity of the bodies in contact g_p (c).

$$T_l = T^{\Gamma l}, \qquad x \in \Gamma_T^l, \tag{1.28}$$

$$\lambda_l \partial T_l / \partial N^l = q^{\Gamma l}, \qquad x \in \Gamma_q, \tag{1.29}$$

$$\lambda_l \partial T_l / \partial N^l = \alpha^{lT}(T_{sm} - T_l), \qquad x \in \Gamma_{Tq}, \qquad l = 1, 2, \tag{1.30}$$

where T_{sm} is the temperature of the medium contacting the body with the surface Γ_{Tq}^l; u_j^l are the components of the vector of displacement; $\mathbf{u}^{\Gamma l}$ is a given displacement for Γ_u^l; N_j^l are the components of a unit vector to $\partial\Omega^l$; and $t_j^{\Gamma l}$ are the components of the external load given for Γ_σ^l.

Let us introduce the notion of a *perfect contact* of two bodies 1 and 2 by assuming that walls of the bodies in contact have the same temperatures and densities of the heat flow that crosses those walls in a given point of their contact, and that the equality of displacements and stresses is binding. It is expressed by the following relations

$$T_1 = T_2, \qquad x \in \Gamma_c, \qquad q^1 - q^2 = 0 \quad \text{or} \quad \lambda_1 \frac{\partial T_1}{\partial N_c} = \lambda_2 \frac{\partial T_2}{\partial N_c}, \qquad x \in \Gamma_c, \tag{1.31}$$

$$w_N = 0, \qquad \mathbf{w}_\tau = 0, \qquad x \in \Gamma_c, \tag{1.32}$$

$$p_N^1 = p_N^2, \qquad \mathbf{p}_\tau^1 = \mathbf{p}_\tau^2, \qquad x \in \Gamma_c, \tag{1.33}$$

where $q^l = \lambda_l \partial T_l / \partial N_c$.

In fact, a set of phenomena called the external friction occurs between the bodies in contact.

Friction processes occurring in the joints of machine elements have great effect on the operating costs and durability of machines. Although a lot of effort has already been put into the research of friction (see, for instance, the review [Oden, Martins (1985), Ibrahim (1994)] and a monograph [Kragelsky, Shchedrov (1956)]), and a lot of practical knowledge on this topic has been gained, a homogeneous and general theory of friction so far has not been formulated. We lack proper physical models sufficiently explaining complex processes of friction, and we do not have mathematical relations useful in solving complex problems of modern engineering in machine construction and operation.

Another separate and equally important phenomenon is the interaction of the frictional surfaces with their surroundings. For high velocities, oxidation of the subsurface layer takes place as a result of diffusion of atmospheric oxygen to metals, and then the compounds of oxygen and metal are formed in this layer. The intensity of the ionic diffusion of the molecules of the environment to the frictional body depends mainly on the temperature and strains. Various processes of mechanic thermodiffusion were described in the monographs [Nowacki, Olesiak (1991)] and [Pyryev (1999)]. The axiosymmetrical problems for the cylinder with its general mechanical, temperature, and diffusional load were studied in the work [Olesiak, Pyryev (1997)], and, with the processes of ion exchange taken into account, in the work [Mokrik, Pyryev (1993)].

The surfaces of real bodies always have the microgeometrical condition of irregularity (roughness, corrugation, inexact shape), which results in the geometrical surface of contact of such bodies decomposed into many small areas of contact and clearances. The notions of geometrical, real, and elementary contact surface are introduced (see, for instance [Solski, Ziemba (1965), Chichinadze et al. (1979), Lawrowski (1993)]). The systems with frictional centres with the corrugated contact surface were analysed in the monographs [Shtaerman (1949), Jonson (1985)] and in the works [Váradi et al. (2000), Pauk, Woźniak (1999)].

In order to analyse the characteristics of real surfaces of the bodies in contact, we apply a phenomenological approach that assumes the introduction of the so-called 'third body' [Aleksandrov, Annakulova (1992), Ganghoffer, Schultz (1995), Zmitrowicz (2001), Dragon-Louiset (2001)]. The macroscopic properties of this body include all possible microscopic phenomena that occur during the contact of a surface with microirregularities. The 'third body' is treated as a whole, without analysing its internal structure. Therefore, the mathematical description is simplified and the status of the system can be determined with relatively few parameters.

Friction, as a resisting force, has a great influence on the dynamics of feed motion and on the precision of the positioning of many operating units responsible for the accuracy of shape, size, and the quality of obtained surfaces.

Model of a friction force allows us to determine this force during the macroscopic slip of one of the analysed surfaces along the other one. In this case $F_{fr} \equiv p_\tau = fp$, where F_{fr}, p denote, respectively, a tangent and a normal

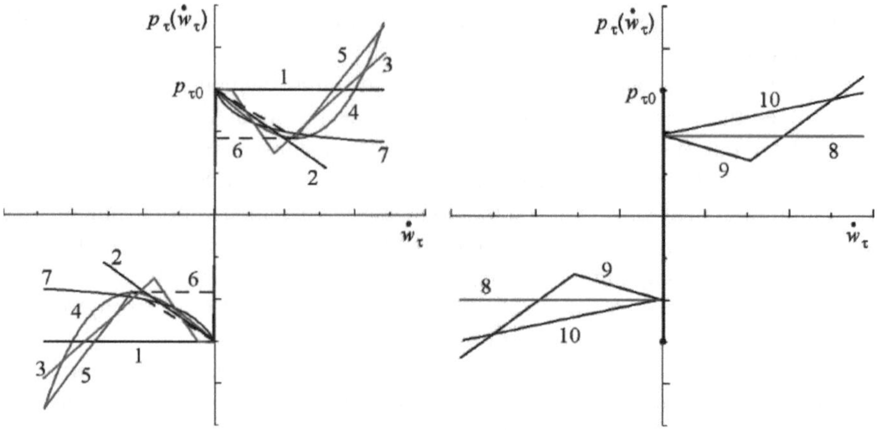

Fig. 1.4: Models of friction forces used in the literature for modelling the stick-slip phenomenon: 1, [Blok (1940)]; 2, [Bell, Burdekin (1969)]; 3, [Chin, Chen (1993)]; 4, [Awrejcewicz, Pyryev (2002)]; 5, [Brockley et al. (1967)]; 6, [Bell, Burdekin (1969)]; 7, [Banerjee (1968)]; 8, [Panovko (1980)]; 9, [Capone et al. (1993)]; 10, [Derjagin et al. (1957)] (numbers relate to the references given in the text).

component of the contact stress, and f is the coefficient of kinetic friction. In a general case [e.g., Chichinadze et al. (1979), Kragelsky, Gitis (1987), Lawrowski (1993)] f depends on the contact pressure and the slip velocity (but it does not depend on the velocity when the frictional surfaces are already sufficiently ground in) and on the temperature of a contact surface; that is, it is the function:

$$f = f(p, \dot{w}_\tau, T). \tag{1.34}$$

In the monographs [Blau (1996), Kragelsky et al. (1982), Kragelsky, Gitis (1987), Chichinadze et al. (1979)] and in the works [Martins et al. (1990)], many phenomena of dry friction are studied, and the computation methods are presented for the estimation of such friction. The results of experimental research of the relation between the friction coefficient and the slip velocity are also presented there. It was discovered that this relation can have various patterns, that is, with its minima and maxima, and with a monotonic drop of value. Also, the patterns with a constant value of a frictional coefficient for a changing slip velocity are provided (Fig. 1.4).

In the monograph [Blau (1996)] a multidisciplinary approach to static and kinetic friction, both with and without lubrication, and reviews of the conventional and novel methods used to measure friction are addressed. The elementary problems found in the mechanics of sliding objects and machine components, and the effects of contact pressure, sliding speed, surface roughness, humidity, and temperature on friction are discussed.

The survey [Awrejcewicz, Olejnik (2005)] is devoted to a significant role of various dry friction laws in engineering sciences. Both advantages and disadvantages of a frictional process are illustrated and discussed, but excluding the nature of friction. It is shown how the classic friction laws and modern friction theories nowadays exist in pure and applied sciences.

In the works [Blok (1940), Brockley et al. (1967), Halling (1975), Cockerham, Cole (1976), You, Hsia (1995)] the linear models of friction were used (curves 1 in Fig. 1.4); in the works [Banerjee (1968), Kauderer (1958)] the parabolic model was applied (curve 7 in Fig. 1.4).

The pattern of friction behaviour in the case of oiled surfaces in the function of slipping velocity and in stationary conditions is best described by the friction characteristic called the Stribeck curve. For small slipping velocities, the friction force is determined mainly by mechanical, structural, and physical–chemical properties of the material of frictional surfaces (dry friction). For larger slipping velocities, grease microclines occur, and the resistance of combined hybrid friction decreases. For the further increase of slipping velocity, the total separation of rubbing surfaces takes place. Only the liquid friction remains, and its value increases together with the increase of velocity. In this study, we consider only dry or semi-dry friction. The Stribeck curve (curve 4 in Fig. 1.4) has already been used in the works [Awrejcewicz, Pyryev (2002) (2003c) (2004a), Pyryev (2004)] to which the authors contributed. The simplified Stribeck curves (curves 3, 9 in Fig. 1.4) are used in the works [Capone et al. (1993), Chin, Chen (1993)]. The model of friction force of the curve 6 presented in Fig. 1.2 is analysed in the work [Van De Velde, De Beats (1998), Cockerham, Symmons (1976), Bo, Pavelesku (1976)]. In the study [Andrzejewski, Awrejcewicz (2005)], the “Magnum” model (based on the phenomenon of friction between the road surface and a tyre) is applied for the simulation of the dynamics of wheeled vehicles. This model embraces interactions between the forces of adhesive, Coloumb, and viscotic friction (adhesive force), and the velocity of a wheel slip. They are presented graphically as a so-called Stribeck curve (curve 4 in Fig. 1.4). For some road surfaces friction modelling with the use of curve 7 (Fig. 1.4) is suggested.

The above-mentioned relation, in accordance with [Moore (1975), Kragelsky, Shchedrov (1956)], has the following form,

$$f = \operatorname{sgn}(\dot{w}_\tau)((a + b|\dot{w}_\tau|)\exp(-c|\dot{w}_\tau|) + d). \tag{1.35}$$

The relation approximating curve 7, according to [Kauderer (1958)], has the form of

$$f = \operatorname{sgn}(\dot{w}_\tau)\left(\frac{f_s - f_{\min}}{1 + c|\dot{w}_\tau|} + f_{\min}\right), \tag{1.36}$$

where $a, b, c, d, f_s, f_{\min}$ are the coefficients with constant values.

In order to emphasise the nonreflectivity of friction forces [e.g., Stefański et al. (2003)], the model of dry friction force that depends on acceleration $\ddot{w}_\tau$ is studied. The relation of the function $f = f(T)$ to the contact temperature is used in the works [Chichinadze et al. (1979), Chichinadze (1995)]. The kinetic friction coefficient can also be presented [Sadowski (1999)] as the sum of coefficients of temperature friction and mechanical friction.

Model of contact rigidity. This generally characterizes the contact of two rubbing surfaces (interference, negative allowance, tension, clamp, and normal contact strains) g_p of the bodies in contact (Fig. 1.3c) in relation to the contact load and contact pressure, contact temperature, and time; that is,

$$g_p = \varepsilon_N(p_\tau, p, T, t). \tag{1.37}$$

For instance, in the work [Shtaerman (1949)], a Winkler model is applied $g_p = k_N p$, where a constant k_N is a parameter determined experimentally. In the work [Pyryev, Mandzyk (1996)], during the analysis of the problem of a frictional contact of two moving thermoelastic cylinders, this model is applied in order to describe their contact with their containment. Sometimes, the models $g_p = k_N p^\alpha$ [Kragelsky et al. (1982), Hess, Soom (1991)] are also used. In the work [Zajtsev, Shchavelin (1989)], in order to take roughness into account, contact rigidity is modelled by the equation $g_p = \alpha(p_\tau + \beta)^{-1} - \gamma$. In the study [Martins et al. (1990)] $p = c_n g_p^{m_n} + b_n g_p^{l_n} \dot{g}_p$ is used, where c_n, b_n, m_n, l_n, α, β, γ are constant parameters dependent on the type of materials and on processing of rubbing surfaces. It can be assumed on the basis of the works [Kudinov (1967)], and also on the basis of the studies of many other researchers [e.g., Hebda, Wachal (1980), Martins et al. (1990)], that in order to understand and describe the behaviour of combined friction, the model of a rubbing pair should be applied, with at least two degrees-of-freedom, and with contact strains in a normal direction in relation to the friction surface taken into consideration. In the case of the systems of feed motion of machine tools, the mathematical model of friction is applicable for determining the friction forces while positioning and starting, and for the study of the conditions of the loss of motion stability and the occurrence of a stick-slip phenomenon.

Model of initial displacement. Initial displacements take place before the process of slipping with the increase of the contact load from zero to the static friction force. If a contact force p_τ is smaller than the developed static friction force $p_{\tau s} = f_s p$ between the bodies in contact under the contact load, the slip will not occur. However, certain slight displacement of the bodies in the direction of the load operation will take place, and the system will take a different balance position. This phenomenon is a subject of research [Kragelsky et al. (1982)]. This work does not take this phenomenon into consideration. Generally, initial displacements depend on the contact load, contact pressure, contact temperature, and time; that is,

$$\varepsilon_\tau = \varepsilon_\tau(p_\tau, p, T, t). \tag{1.38}$$

Model of wear. Friction occurs on the nuclear level, but its effects are observed on the macroscopic one, as the changes of surface and the occurrence of separate particles called the products of wear. The wear process is the process of a gradual change in the body dimensions caused by the process of friction and manifested by the separation of the material from the rubbing surface. Wear is the result of the wear process. While studying the process of attrition of machinery elements on the basis of the changes in the values of attrition occurring in time, the notions of the measure of attrition and the intensity of this process are required. The change of weight, volume, or linear dimensions of the body is usually taken as the measure of attrition (after some period of time of friction). In this work, the thickness U^w of the rubbed layer of material is the measure of attrition.

In the research of tribological wear of machinery, total wear U^w is the basis for establishing the intensity of attrition $\dot{U}^w$ as the quantity that better characterizes the process of occurrence of the products of wear.

The basic methods of research and problems of the theory of wear are described in the works [Solski, Ziemba (1965), Chichinadze et al. (1979), (1995), Lawrowski (1993), Sadowski (1999)]. Friction and wear cause huge energy and material losses. Nevertheless, the general relation between the value of friction force and wear has not been determined thus far.

Empirical models leading to the fuller understanding of the processes are presented in many of the works mentioned above; for example,

$$U^w = g_w(p, \dot{w}_\tau, T, H, t), \tag{1.39}$$

where H is the hardness of the softer body. According to Archard [Archard (1953)], the model of wear has the following form,

$$\dot{U}^w = K^w p |\dot{w}_\tau|. \tag{1.40}$$

The models of wear (1.40) are applied in several works [e.g., Strömberg et al. (1996)]. The model (1.40) is a special case of the model of wear

$$\dot{U}^w = K^w p^\alpha |\dot{w}_\tau|^\beta. \tag{1.41}$$

For the changing external load of a tribomechanical system delay may occur [Hebda, Chichinadze (1989)]. In works [Hebda, Chichinadze (1989), Yevtushenko, Pyryev (1998), (1999b)], models that account for the processes of inheriting and ageing are applied:

$$U^w(t) = K^w |\dot{w}_\tau| \int_0^t K_1(\tau) K_2(t-\tau) p(\tau) d\tau, \tag{1.42}$$

where $K_1(t) = 1 + c\exp(-\gamma t)$, $K_2(t-\tau) = 1 - \exp(-\gamma'(t-\tau))$ are the nuclei of inheriting and memory. The case of the following model of wear ($K_1(t) = 1$) is analysed in the work [Kuzmenko (1981)] and in the study [Yevtushenko, Pyryev (1999a)], where for the model of contact of the thermoelastic layer with a thermally insulated plate, the obtained solution allowed for the heat generation and wear (1.42). The evolution of contact pressure, contact temperature, and wear is presented there. Also, the conditions of the areas of applicability of the analysed model are determined as well as conditions of the occurrence of the so-called frictional thermoelastic instability (FTEI).

Using the models analysed above, we denote the following general contact conditions,

(i) Kinematic contact conditions on the surface:

$$g = u_N^2 - u_N^1 + g_0 + U^w > 0, \qquad \mathbf{p}^{N1} = 0, \qquad \mathbf{p}^{N2} = 0, \qquad x \in \Gamma_s, \tag{1.43}$$

$$g_p = \varepsilon_N(p_\tau, p, T, t), \qquad p \geq 0, \qquad x \in \Gamma_{sl} \cup \Gamma_{st}, \tag{1.44}$$

(ii) Forcing contact conditions on the surface:

$$|p_\tau| < f_s(p, T, t)p, \qquad \dot{w}_\tau = \dot{u}_\tau^2 - \dot{u}_\tau^1 = \dot{\varepsilon}_\tau(p_\tau, p, T, t), \qquad x \in \Gamma_{st}, \tag{1.45}$$

$$|p_\tau| = f(p, \dot{w}_\tau, T)p, \qquad \dot{w}_\tau = -\dot{\lambda} p_\tau / |p_\tau|, \qquad x \in \Gamma_{sl}, \tag{1.46}$$

(iii) Thermal contact conditions:

$$q^1 + q^2 = (T_2 - T_1)/R(p), \qquad q^1 - q^2 = (1 - \eta) f(p, \dot{w}_\tau, T)\dot{w}_\tau p,$$
$$x \in \Gamma_{sl}, \tag{1.47}$$

$$q^1 + q^2 = (T_2 - T_1)/R(p), \qquad q^1 - q^2 = 0, \qquad x \in \Gamma_{st}, \tag{1.48}$$

$$q^1 + q^2 = (T_2 - T_1)/R(u_N^2 - u_N^1 + g_0 + U^w), \qquad q^1 - q^2 = 0, \qquad x \in \Gamma_s, \tag{1.49}$$

where $T = (T_1 + T_2)/2$ denotes contact temperature; $g = u_N^2 - u_N^1 + g_0$ is a clearance (a gap, e.g., an air-gap), if $g > 0$; g_p is the initial interference (Fig. 1.3c) when $g = 0$; g_0 is the initial clearance, when $g_0 > 0$ and it is the initial interference when $g_0 < 0$; $R(\cdot)$ is heat resistance that depends on the contact pressure ($x \in \Gamma_{st}$) or on the clearance ($x \in \Gamma_s$); u_N^l, u_τ^l denote the normal and the tangent component of the displacements of the point Γ, respectively; w_τ is a tangent slip; $\varepsilon_N(p_\tau, p, T, t)$ is contact rigidity; $f(p, \dot{w}_\tau, T)$ is a kinetic coefficient of friction that generally depends on the contact pressure, relative velocity, and contact temperature. Moreover, we assume that $f(p, -\dot{w}_\tau, T) = -f(p, \dot{w}_\tau, T)$ and then the density of generated heat is described by the inequality $f(p, \dot{w}_\tau, T)\dot{w}_\tau > 0$ for the arbitrary slipping direction. The density of the frictional energy stream is $q(t) = f(p, \dot{w}_\tau, T)\dot{w}_\tau p$. Mathematical generalization of the function g is given in the works [Wriggers (1995), Agelet de Saracibar et al. (1999)].

Let us derive the second Equation (1.47). We use the equations of energy balance formulated in relation to the area of the elementary infinitesimal surface of macroscopic contact $d\Gamma$ of solid bodies and infinitesimal time dt of the following form,

$$q + q_0 = q_1 - q_2 + q^w, \tag{1.50}$$

where $q = p_\tau \dot{w}_\tau$ is the density of friction power $[W \cdot m^{-2}]$; $q_l = \lambda_l \partial T_l / \partial N_c$; $\mathbf{q} = -\lambda_l \mathrm{grad} T_l$ is the density of the heat stream $(l = 1, 2)$; $q^w = \eta p_\tau \dot{w}_\tau$ is the density of mechanical dissipation; η is equal to the relation q/q^w; q_0 is the density of the power of the elastic energy gain of the system (for the mechanical and thermal loads changing sufficiently slowly we assume [Aleksandrov, Annakulova (1990)] that $q_0 = 0$). Equation (1.47) is now easily obtained. Let us notice that the parameter η for the boundary condition has a different sign than in the case of the conditions obtained in the works [Strömberg et al. (1996)] (the case of $q_1 = 0$) and it has the same sign as in the works [Aleksandrov, Annakulova (1992), Sadowski (1999)].

1.1.3 Dynamics of contacting bodies

The obtained dimensionless Equations (1.23), (1.24) are nonlinear models that can be encountered in virtually all fields of science, for example, in technology [Kononienko (1964), Moon (1987), Schuster (1995)], biology [Riznichenko (2002)], sociology, and economy [Kapica et al. (2003)], or in the problems of environmental protection of humans (the effect of vibrations and noise on a human being) [Engel (1993)].

For the purpose of the analysis of systems with one or more degrees-of-freedom, asymptotical methods can be applied (method of a small parameter, the Krylov–Bogolubov–Mitropolsky method, equivalent linearization, and others), that are discussed in monographs [Andronov et al. (1966), Osiński (1979), Nayfeh (1981), Awrejcewicz (1989), (1996)]. It turns out, however, that those methods cannot be applied directly in solving the obtained Equations (1.23), (1.24).

As has already been mentioned, the material systems of bodies in contact can show self-excited vibrations. This type of vibration is discussed in many monographs devoted to nonlinear vibrations [e.g., Schmidt, Tondl (1986), Kragelsky, Gitis (1987), Kurnik (1997), Giergiel (1990), Nayfeh, Balachandran (1995), Awrejcewicz (1996)] and in many other articles [e.g., Oden, Martins (1985), Awrejcewicz, Pyryev (2005), (2007)]. Self-excited vibrations belong to nonextinguishing vibrations that are supported by external energy sources in a nonlinear conservative system.

Self-excited stick-slip vibrations occur in many mass-elastic systems with slip friction, both in technology and in everyday life. In the case of machines, self-excited vibrations usually have detrimental effects, as they can lead to the damage of the vibrating object. Here, the following phenomena can be enumerated: the vibrations resulting from the air flow around vibrating strands, rods, and layers (e.g., the wing flutter in a plane or the vibrations of conductors). Self-excited vibrations caused by the frictional contact of the bodies moving in relation to each other (e.g., in a slip bearing with no lubrication) occur frequently, as well as self-excited relaxational vibrations.

The basic methods and problems of the analysis of self-excited vibrations are presented in the monographs [Osiński (1979), Andronov et al. (1966), Kragelsky, Gitis (1987), Awrejcewicz (1996), Awrejcewicz et al. (1998), and others]. The study [Tondl (1970), (1978)] shows that the introduction of additional dry friction to the system produces a stable balance point. The larger is the area of attraction of this solution the greater is the input of dry friction, up to the moment when self-excited vibrations extinguish.

For relative motion of rubbing bodies, the real velocity of slipping can change in steps within the scope of small velocities. This type of stroke–slip motion is called stick-slip. In the case of the tribomechanical systems characterized by this type of motion, mechanical relaxational vibrations that can disturb machine work are observed. For instance, in the system cutter–holder–slide, or in the system: processed element–machine, those vibrations may significantly worsen the quality of the processed surfaces. Such vibrations also disturb the work of a driver while engaging the clutch on a car, as well as the work of slip units in measuring instruments, the vibrations of turbine blades, and the vibrations of suspension bridges. Depending on the characteristics of tribomechanical systems and on operating conditions, relaxational vibrations may have various features, namely they may occur as sudden leaps with the frequency of 8–10 Hz (e.g., in a car clutch) up to 4000–5000 Hz (e.g., in a braking system of a vehicle).

The models discussed can be useful in the analysis of the stick-slip phenomenon that occurs in the power–feed systems of machine tools [Marchelek (1991)].

Stepwise movement is observed with friction in the case of relative slip and relative rest, and it occurs as a result of self-excited vibrations with the reduction of the friction coefficient and the increase of slip velocity. The main source of self-excited vibrations is the positive difference between the rest forces $p_{\tau 0}$ (static friction) and slip (kinetic friction), and the value of friction force decreases when relative velocity increases (see Fig. 1.4).

Most of the theoretical works concerning this field deal with pure self-excited vibrations caused by friction [Awrejcewicz et al. (1998), Awrejcewicz (1996), Kragelsky, Gitis (1987), Martins et al. (1990), Ibrahim (1994), Andronov et al. (1981)].

From the many studies are devoted to stepwise movement, let us enumerate here only the surveys [Kragelsky, Shchedrov (1956), Ibrahim (1966)].

For small slip velocities, nonuniformity of motion is observed, namely periodically changeable motion and rest (interrupts).

Motion of a system with two degrees-of-freedom that is in contact with a movable plane is analysed in the work [Richard, Detournay (2000)]. The mass of a system may vibrate on the vertical spring fixed in the centre of the body that slips horizontally. The friction coefficient is considered a constant quantity. In this case, stick-slip motion may occur.

Unstable feed motion of working units (stepwise movement occurring with low rate of feed and fluctuations of slip velocity) is very troublesome for machine tool constructors, producers, and users. It is still much of a problem, because the requirements related to the precision of machines and their work are very strict.

Self-excited vibrations of an oscillator caused by dry friction are analysed in many works [e.g., Panovko (1980), Balandin (1993)]. In the work [Balandin (1993)], periodic self-excited vibrations of a body moving in the area of the clearance in fitting with perfectly rigid walls are studied (curve 2 in Fig. 1.4 denotes the model of friction force). It is stated in this work that there are four types of periodic motion possible: with one stroke without tacking during a period of vibrations; with one stroke and tacking during a period; with two strokes without tacking during a period; and with two strokes and tacking during a period.

The obtained Equations (1.23), (1.24) describe the models of material systems in which the external mechanical and kinematic excitations are presented as a periodic function whose amplitudes ζ and ζ_k are on the right-hand side of differential equations (case $\ddot{\phi} = 0$). There is, however, no influence of the system on the source of vibrations. Such a model is called a system with a perfect source of energy (a perfect system). When there is an interaction in the system between the source of energy and the vibrating object, it is called *a system with nonperfect source of energy* [Kononienko (1964)]. If we assume for the analysed models that one of the bodies in contact is a source of vibrations for the other one, then the obtained system of Equations (1.23), (1.24) is the system with nonperfect sources of energy. Similar equations for vibrations of a self-excited system with dry friction are studied in the works mentioned above, with various characteristics of the source of the system energy assumed (various $M(\dot{\phi})$). Analytical research was conducted with the use of the method based on the asymptotical Krylov–Mitropolski–Bogolubov method [Bogolubov, Mitropolski (1961)]. The amplitude, phase, and frequency of vibrations are determined in a stationary state and a non-stationary one at the moment of resonance. The influence of self-excited vibrations generated by dry friction and having a nonperfect source of energy is presented in the work [Giergiel (1990)]. The model was comprised of a body placed on a belt driven by an engine with nonperfect characteristics. Self-excited vibration occurs only for certain velocity values of a belt. The application of various analytical methods in the studies on the systems with nonperfect sources of energy is also presented in the work [Nayfeh, Mook (1979)].

The vibrations of a plate that rotates as a result of a transverse friction force (curve 4 in Fig. 1.4) are analysed in the work [Nakai (1998)]. In works [Hess, Soom (1991)], damped vibrations of a body with one degree-of-freedom on a rigid plane (condition (1.44)) with harmonic perturbations are studied.

The model of an oscillator with friction (curve 1 in Fig. 1.4) for harmonic perturbation is considered in the work [Feeny, Moon (1993)].

In the studies by Warmiński [e.g., Warmiński et al. (2000)] regular and chaotic vibrations of parametric – self-excited systems with perfect and nonperfect sources of energy are described.

In the case of machine tools, self-excited vibration makes the work less precise and deteriorates the quality of the obtained surfaces. Therefore, this kind of vibration has been the object of interest for many researchers [Kauderer (1958), Tondl (1978), Kragelsky, Gitis (1987), Grudziński, Wedman (1998), Grudziński et al. (1995), Van De Velde, De Beats (1998), Rozman et al. (1996)].

The stability of an elastic system with a frictional contact is expressed through the function of potential energy and dissipation [Mróz (2000)]. Static and dynamic forms of the loss of stability are also determined.

In the analysed models (Equations (1.23), (1.24) for $a > 0$) chaos may also occur. Nowadays, numerous studies are devoted to the problem of chaos in deterministic systems. For instance, one can find in the works [Awrejcewicz (1988), (1996), Awrejcewicz, Mrozowski (1989)] the examples of a self-excited system of van der Pol–Duffing type, excited inertially, where static load is taken into account.

Let us remember that an irregular and unpredictable evolution of many nonlinear systems is called '*chaos*'. Despite this irregularity, chaotic dynamic systems satisfy deterministic equations that are very sensitive to initial conditions. For studying chaotic systems, various analytical and numerical methods can be used [e.g., Awrejcewicz (1988), (1989), (1991), (1996), Awrejcewicz, Holicke (2007), Ott (1993), Baker, Gollub (1996), Nusse, Yorke (1994)]. The reasons for the occurrence of chaotic vibrations are studied in the work [Awrejcewicz (1996)], where an extensive bibliography of this field is also provided.

In the system described by Equations (1.23), (1.24) ($\dot{\phi} = \text{const}$), chaotic motion may occur [Awrejcewicz, Pyryev (2003c)], and the set of parameters for which it occurs can be determined with the use of the Melnikov method [Melnikov (1963), Guckenheimer, Holmes (1983), Awrejcewicz, Pyryev (2006a), (2006c)]. The basic idea of the Melnikov method is the application of an integrable undisturbed solution ($\varepsilon = 0$, $h = 0$) of the system of two equations for the solution of a disturbed system of linear equations. In the works mentioned above, the Melnikov function is built for $p = 1$ for mechanical perturbations with the amplitude ζ and for kinematical ones with the amplitude ζ_k.

In the work [Litak et al. (1999)], the influence of self-excited and externally excited vibrations is studied with the example of the Froude pendulum excited externally. Critical values of amplitude and frequency of excitation for which the system passes to chaotic motion are determined numerically and on the basis of Melnikov analytical criterion.

In the work [Awrejcewicz et al. (2002)], the analysis of vibrations of a spinning triple physical pendulum, whose motion may be regular, chaotic, or hyperchaotic, is presented. In the work [Pomeau, Manneville (1980)] three types of "intermittent" transition of the system into chaos are determined. The first type is associated with the saddle–node bifurcation, the second one related to the subcritical Hopf bifurcation, and the third type connected with doubling of a period.

In the work [Wagg (2003)], the dynamics of a system with one degree-of-freedom described by Equation (1.23) is considered, for $\Phi(x)$. Chaotic motion is analysed, in the case when there is a nonlinear assymetrical conservative force.

A mechanical model of a rolling disc is presented in the framework of nonsmooth dynamics and convex analysis [Le Saux et al. (2005)].

Examples of the analysis of regular and chaotic vibrations of self-excited systems externally excited are also presented in the works [Guckenheimer, Holmes (1983), Awrejcewicz, Mrozowski (1989), Pyryev (2004)]. The studies [Minorski

(1962), Tondl (1978), Szabelski (1984), (1991), Dao, Dinh (1999)] are devoted to the analysis of the vibrations of parametric–self-excited systems.

In most of the theoretical works devoted to this issue, periodic, chaotic, or self-excited vibrations caused by friction are the subject of study. In this work, in addition, tribological phenomena are also studied and modelled.

1.1.4 Contact thermoelasticity

The contact problems have been mainly introduced by the pioneering works [Hertz (1882, 1895)]. In his works, Hertz analyses the problem of pressing two even elastic surfaces against each other. For the assumption that the main radii of the bodies' curvature are large in comparison to the characteristic linear size of a contact pad, the problem is reduced to the one that describes the contact of two elastic half-spaces, which enabled expressing analytically the contact pressure p and the boundary of a contact surface $\partial\Gamma_c$, that separates various types of boundary conditions (1.43), (1.44). More complex problems, including the change of boundary conditions, require solving the integral Fredholm equations describing contact pressure. In contrast to the mentioned problems, this study considers the models including boundary conditions that do not change along the surface. We do not obtain a Fredholm integral equation describing contact pressure because the models are one-dimensional. The model is developed through the analysis of nonstationary problems and with the variety of boundary conditions taken into account.

Contact problems of the theory of elasticity and of thermoelasticity are the subject of analysis in several monographs [e.g., Belajev (1945), Mindlin (1949), Shtaerman (1949), Timoshenko, Goodier (1951), Sneddon (1966), de Pater, Kalker (1975), Galin (1976), (1980), Johnson (1985), Aleksandrov, Romalis (1986), Goryacheva, Dobychin (1988), Alexandrov, Pozharskii (2001)].

Analysis of the contact problems including heat generation via friction has been carried out in the references [Bowden, Ridler (1935), (1936), Blok (1937), Ling (1959),(1973)].

In most of the contact problems, force is used for clamping the bodies (first type of problems). In the case of one-dimensional problems, this approach results in the fact that constant pressure in the contact pad is known. In this study, the authors consider contact problems when the close proximity of bodies is taken into account (second type of problems). In the case of one-dimensional problems such perturbations make the pressure in a contact pad an unknown value. What is more, in this case the problem of elasticity and thermal conduction are coupled by boundary conditions.

Reducing the issue to a one-dimensional problem enabled avoiding many difficulties, such as those related to the discussion of possible singularities occurring on the boundary of the contact area $\partial\Gamma_c$ [e.g., Galin (1961), Gladwell (1980), Gladwell et al. (1983)] or, for instance, to the paradox of a "cooled sphere" [Barber (1973),

Kulchytsky-Zhyhailo, Olesiak (2000)]. This paradox occurs when heat flows from the body with a higher coefficient of thermal distortion $\delta_l = \alpha_l(1+\nu_l)/\lambda_l$ to the body with a smaller coefficient of thermal distortion. The paradox is to a large extent related to the assumption of the condition of a perfect thermal contact and a singular distribution of a heat flow. The derived conditions that guarantee that the paradox of a 'cooled cylinder' does not occur are given in the works [Kulchytsky-Zhyhailo, Olesiak (2000)].

We believe that all such 'paradoxes' are the effects of imperfect modelling and assumptions resulting from the lack of possibilities of full and adequate modelling of the analysed process in order to obtain any conclusions. Let us remember that even in the problems without the quality change of boundary conditions on the boundary $\partial\Gamma_c$, but with the bump movement of load (e.g., of the vertical one for the horizontal one equal to zero), the singularity of shear stress may occur in the point of the bump movement of load [Timoshenko, Goodier (1951), Grinchenko, Ulitko (1999)]. In this book, attention is focused on the analysis of quasistationary problems, among others, and the thickness of the bodies in contact, their inertia, heat generation, and wear are taken into account.

In the works [Aleksandrov, Annakulova (1990), Zmitrowicz (1987), Moore (1975), Rabinowicz (1965)], the method of construction of general contact conditions is presented, and friction, wear, and heat generated by friction within the framework of thermomechanical theory are taken into account. In the work [Zmitrowicz (2001)], classical formulation of variational contact problems is generalized for the case of wearing solid bodies, with the layer of wear molecules between the bodies in contact taken into account.

Mechanical (1.43)–(1.46) and thermal (1.47)–(1.49) boundary conditions were used by the author in the works [Olesiak, Pyryev (1996a), (1996b), Pyryev, Mokryk (1996), Pyryev (2001), (2002)] during the analysis of a one-dimensional problem of thermoelastic contact of bodies, with the relation between heat resistance and contact pressure or clearance taken into account. It is shown that depending on the parameters, problems may have one stable stationary solution, three stationary solutions of which one is an unstable stationary solution, and one unstable stationary solution. In this last case, contact characteristics are characterized by self-excited vibrations [Pyryev (2001)]. Analogical boundary conditions are studied in the works [Barber, Zhang (1988), Barber, Comninou (1989)].

The analysis of the influence of friction forces that are proportional to Hertz contact pressure (curve 1 in Fig. 1.4) on the stresses in the case of a problem of a sphere slipping along the surface of half space is provided in the work [Hamilton, Goodman (1966)].

The first axial–symmetrical contact problems of thermoelasticity, with heat generation taken into account, were solved in the works [Barber (1975), (1976), Generalov et al. (1976)]. In the work [Barber (1976)], one of the bodies slips along the surface of the other one with constant velocity. In the works [Barber (1975), Generalov et al. (1976)], one of the bodies rotates around the axis of symmetry with constant angular velocity. In both cases, force is used to press the bodies. These problems are analysed in more detail in many different works [e.g., Pauk (1994), Yevtushenko, Kulchytsky-Zhyhailo (1995)]. The studies mentioned above concern the problems with the assumption of constant velocity of slipping. In the case of the second type of contact problems, the assumption of the constant velocity of sliding results in the possibility of occurrence of frictional thermoelastic instability (FTEI) [Burton et al. (1973), Barber (1976), Morov (1985), Pyryev (1994), (2001), Pyryev et al. (1995), Pyryev, Grilitskiy (1996), Pyryev, Mandzyk (1996), Yevtushenko, Pyryev (1997), Pyryev (2004), Ciavarella, Barber (2005)]. In contrast to these works, the author takes into account the inertia of the bodies in contact, which results in the variable value of the velocity of slipping. The velocity of slipping depends on the mechanical load. The system regulates the amount of generated heat on its own, and in this case FTEI does not occur.

The problems of ThermoElastic Instability (TEI) occurring as a result of various thermal distortions $\delta_l = \alpha_l(1+\nu_l)/\lambda_l$ of the bodies in contact and of the relation between thermal resistance and contact pressure (conditions (1.47)–(1.49) for $\dot{w}_\tau = 0$) were considered by J. Barber in his works [e.g., Barber (1999)]. In the studies in which Barber took part, the perturbation method was applied to analyse the conditions of the occurrence of TEI for two thermoelastic half spaces in contact, for the thermoelastic layer and half spaces, for two layers, two half infinite thin-walled cylinders whose endings are in contact, and the contact problem of thermoelasticity concerning a long two-layered cylinder.

The problems of thermoelastic instability generated by the relation between the friction coefficient and the velocity of slipping and contact temperature are studied in the work [Maksimov (1988)].

The problems of instability of a friction contact are considered in the work [e.g., Radi et al. (1999)]. In the works [Michalowski, Mróz (1978), Jarzebowski, Mróz (1994), Mróz, Stupkiewicz (1994), Radi et al. (1999)], the problems related to the friction contact are analysed, without taking tribological processes into account. In the works of Sadowski [e.g., Sadowski (1999)] the thermodynamical nature of friction and of phenomena accompanying it is studied.

Observe that in references [Afferante et al. (2006), Afferrante, Ciavarella (2006), (2007), Yi (2006)] a new form of coupled instability, named Thermo-Elastic Dynamic Instability (TEDI), which can occur by interaction between frictional heating and the natural dynamic modes of sliding bodies, has been introduced. However, in the case of inertial terms neglect (see section 1.1.1), TEDI does not appear.

Modelling of mechanical contact with friction taken into account is conducted with the use of the laws of thermodynamics of irreversible processes (first and second law of thermodynamics, energy balance) and the theory of plasticity in the work [Klarbring (1990)], with heat generation taken into account in the work [Johansson, Klarbring (1993), Laursen (1999)], and with wear taken into account in the work [Strömberg et al. (1996)] for small displacements. The example given in [Strömberg et al. (1996)] concerns the contact of a material point with half-space (heat generation not taken into account). The works mentioned above are based on the thermodynamic method discussed earlier in the works [Onsager (1931), Nowacki (1962), Ziegler (1963)]. Anisotropic models of friction and wear are obtained in the works [Mróz, Stupkiewicz (1994), Strömberg (1999)]. Zmitrowicz [Zmitrowicz (1999)] analyses anisotropic models of friction and heat flows. He also presents the problem of braking of a half-infinite rod on a thermal-insulated half-space.

The work [Bielski, Telega (2001)] describes the models of friction applied in geophysics and proposes a new model of friction. In the work [Matysiak, Yevtushenko (2001)] a review of the problems of heat generation during friction is presented. The discussed problems are divided into stationary, quasistationary, and nonstationary. In the works [Rojek et al. (2001)] a model of adhesion is applied in the description of the bone–implant interphase. The influence of the products of wear on the analysed interphase is studied, and the currently applied models of wear are presented. The work [Stupkiewicz, Mróz (2001)] includes the analysis of the most important effects occurring in processes of deformation of fragile contact layers for cycling loading. New constitutional models for cyclic states are proposed.

The work [Zboiński, Ostachowicz (2001)] presents the summary of the results of the three-dimensional analysis of contact in the points of attachment of moving blades in turbo-machines in the elastic and plastic–elastic range. This work focuses on the formulations of variational contact problems of elasticity and plastic–elasticity, and on their analysis with the use of a finite element method.

Many works [e.g., Willis (1966), Ling, Rice (1966), Conway et al. (1967), Kaczyński, Matysiak (1988), (1993), Matysiak et al. (2000)] are devoted to the problems of the contact of anisotropic or nonhomogenous bodies. In the work [Ling, Rice (1966)], a two-dimensional quasi-established problem of heat conduction concerning heating of half-spaces with a movable flow of heat is analysed, with the relation of the coefficients of heat conduction and specific heat to temperature.

The analysis of the temperature on the surface of frictional contact is presented in the works [Blok (1937), Jaeger (1942), Ling (1959), Kennedy, Ling (1981), Kennedy (1981), Barber, Comninou (1989)]. The temperature of a contact surface of the bodies is important in grinding, mechanical polishing, friction welding, and electric contact, among others.

In the work [Strömberg (1999)], the problem in $\Re^2$ is analysed concerning the contact of a thermoelastic plate ($l = 2$) and a thermally insulated rigid half space ($l = 1$) in a to-and-fro motion. A Coulomb friction model is assumed (curve 1 in Fig. 1.4), as well as an Archard wear model (1.40). The condition of mechanical contact is (1.44) and for the initial clamp ($g_0 < 0$). The dynamics of the bodies in contact is not taken into account there. The problem is solved with the use of a finite element method.

[Barber, Ciavarella (2000), Tichy, Meyer (2000), Klarbring (1986), Telega (1988)] are important reviews in this field.

The instability of frictional contact is analysed in the work [Radi et al. (1999)]. Contact conditions are generalised with the use of the laws of the theory of plasticity in the works [Radi et al. (1999), Haraldsson, Wriggers (2000)].

Numerical modelling (a finite element method) of frictional contact is a subject of many works, such as [Laursen, Simo (1993), Laursen, Oancea (1997), Haraldsson, Wriggers (1995), (2000), Wriggers (2002)], and the works [Agelet de Saracibar, Chiumenti (1999), Gu, Shillor (2001)] take wear into account. The analysis of the works published during the last decade in the journal *Computer Methods in Applied Mechanics and Engineering* shows that applying a finite element method in the analysis of contact problems is very useful, which proves that this issue is still important.

The authors of this study present a theoretical and numerical analysis of the dynamics of the bodies in contact, taking into account the influence of tribological processes. The subject of this work is therefore mathematical modelling of phenomena occurring on the contact surface of the bodies moving next to each other, with frictional heat generation, wear, and the inertia of the bodies taken into account, as well as the solution of certain problems of the mechanics of a solid body that are related to this modelling.

1.1.5 On some equations governing discontinuous systems dynamics

Discontinuous dynamical systems arise due to physical discontinuities such as dry friction, impact, and backlash in mechanical systems or diode elements in electrical circuits. Many publications deal with discontinuous systems [Neimark (1978), Filippov (1988), Ibrahim (1994), Kunze (2000), Awrejcewicz and Lamarque (2002)].

Before proceeding we should clarify what we mean by the term nonsmooth system: (i) systems described by differential equations with a discontinuous right-hand side (also called Filippov systems [Filippov (1988)], or (ii) systems which expose discontinuities in the state, such as impacting systems with velocity reversals [Brogliato (1999)].

Because equations including friction modelling possess discontinuities, we consider the main properties of such a system using a relatively simple Equation (1.23). In what follows we consider the equation governing contact interaction with another body moving with the constant velocity ω_1 of the following form,

$$\ddot{\varphi} = -\Phi(\varphi, \dot{\varphi}, \tau) + F_{fr}(\varphi, \omega_r, \tau), \tag{1.51}$$

where $\omega_r = \omega_1 - \dot{\varphi}(\tau)$ is the relative velocity of the moving bodies, and function $\Phi(\varphi, \dot{\varphi}, \tau)$ describes forces acting on the bodies excluding the friction force $F_{fr}(\varphi, \omega_r, \tau)$.

Owing to Amonton's assumptions the friction force is defined in the following way,

$$F_{fr} = \mathrm{Sgn}(\omega_r) \begin{cases} F(|\omega_r|)p(\tau), & \text{for } \omega_r \neq 0, \\ F_s, & \text{for } \omega_r = 0, \end{cases} \tag{1.52}$$

where $F(\omega_r)$ denotes the nondimensional kinetic friction coefficient (either of a force or a moment of forces), $F_s = f_s p(\tau)$, and f_s denotes the static friction coefficient. Here and in the following we applied the notation $F(0+) = f_s$ (classical Coulomb friction). We use the set-valued sign-function

$$\mathrm{Sgn}(\omega_r) = \begin{cases} 1, & \text{for } \omega_r > 0, \\ [-1,1], & \text{for } \omega_r = 0, \\ -1, & \text{for } \omega_r < 0, \end{cases} \tag{1.53}$$

which is set-valued at $\omega_r = 0$. The maximum static friction force is denoted by F_s. Observe that in the case of zero value velocity of the contacting bodies the friction force may take arbitrary values from the interval $[-F_s, F_s]$. The friction force magnitude depends on the system load, therefore the system dynamics is modelled by the following differential inclusion [Filippov (1988), Leine et al. (2000), Kunze (2000), Van de Wouw, Leine (2004)]

$$\ddot{\varphi} \in -\Phi(\varphi, \dot{\varphi}, \tau) + F_{fr}(\varphi, \omega_r, \tau). \tag{1.54}$$

Owing to Amonton's assumptions the friction force associated with the considered problem is defined in the following way (a signum model with static friction point),

$$F_{fr} = \begin{cases} F(|\omega_r|)\mathrm{sgn}(\omega_r)p(\tau), & \text{for } \omega_r \neq 0, \\ \Phi(\varphi, \omega_1, \tau), & \text{for } \omega_r = 0, \ |\Phi(\varphi, \omega_1, \tau)| \leq F_s, \\ F_s\mathrm{sgn}(\Phi(\varphi, \omega_1, \tau)), & \text{for } \omega_r = 0, \ |\Phi(\varphi, \omega_1, \tau)| > F_s. \end{cases} \tag{1.55}$$

In this case the equation governing dynamics of the body with friction has the following form (see, for instance, [Finigenko (2001)])

$$\ddot{\varphi} = \begin{cases} -\Phi(\varphi, \dot{\varphi}, \tau) + F(|\omega_r|)\mathrm{sgn}(\omega_r)p(\tau), & \text{for } \omega_r \neq 0, \\ 0, & \text{for } \omega_r = 0, \ |\Phi(\varphi, \omega_1, \tau)| \leq F_s, \\ -\Phi(\varphi, \omega_1, \tau) + F_s\mathrm{sgn}(\Phi(\varphi, \omega_1, \tau)), & \text{for } \omega_r = 0, \ |\Phi(\varphi, \omega_1, \tau)| > F_s. \end{cases} \tag{1.56}$$

Note that system dynamics is governed by three different equations. The first and third of them change their form depending on the sign changes of ω_r and $\Phi(\varphi, \omega_1, \tau)$. The mentioned properties are generated by discontinuities of friction force (moment).

The second equation exhibits the case when for certain time instant $\omega_r = 0$ and also for $|\Phi(\varphi, \omega_1, \tau)| \leq F_s$, one gets $\ddot{\varphi}(\tau) = 0$. The relative body motion vanishes and the body remains in relative rest. In the thus-far described case the acting forces cannot overcome the friction force. In the phase plane $(\varphi, \ \dot{\varphi})$ points being in relative rest lie on the straight line $\dot{\varphi} = \omega_1$, which divides the phase plane into two parts. Points being in rest satisfy the condition $|\Phi(\varphi, \omega_1, \tau)| \leq F_s$.

The third equation exhibits the case when for a certain time instant τ_1 both relative velocity $\omega_r = 0$ and $|\Phi(\varphi, \omega_1, \tau)| > F_s$ are satisfied; then $\ddot{\varphi} \neq 0$. The third equation yields $\mathrm{sgn}(\ddot{\varphi}(\tau)) = -\mathrm{sgn}(\Phi(\varphi, \omega_1, \tau))$, and hence for $\tau > \tau_1$ the relative motion exists. The phase curve goes from one to another half-plane of the phase plane depending on $\mathrm{sgn}(\Phi(\varphi, \omega_1, \tau))$. Friction in this case is changed by the amount of $2F_s$.

Signum model. Friction force (1.55) can be also cast in the form

$$F_{fr} = \begin{cases} F(|\omega_r|)\mathrm{sgn}(\omega_r)p(\tau), & \text{for } \omega_r \neq 0, \\ \min(|\Phi(\varphi, \omega_1, \tau)|,\ F_s)\mathrm{sgn}(\Phi(\varphi, \omega_1, \tau)), & \text{for } \omega_r = 0. \end{cases} \tag{1.57}$$

The constitutive relation for F_{fr} is known as the signum model with static friction point. Friction model (1.57) should be understood such that a transition from stick to slip can take place only if $|\Phi(\varphi, \omega_1, \tau)|$ exceeds F_s. The mentioned modelling has been used in references [Leine et al. (2000), Leine, Campen (2002a), (2002b), Batako et al. (2003), Awrejcewicz, Pyryev (2006b)], among others.

The friction force (1.55) can also be described as a sum of static F_{fr}^{st} and kinematic F_{fr}^{d} [Chin, Chen (1993), Tarng, Cheng (1995)] frictions of the form

$$F_{fr}(\varphi, \omega_r, \tau) = F_{fr}^{st}(\varphi, \tau) + F_{fr}^{d}(\omega_r, \tau), \tag{1.58}$$

where the kinetic friction force F_{fr}^{d} has the following form.

$$F_{fr}^{d}(\omega_r, \tau) = F(|\omega_r|)\mathrm{sgn}(\omega_r)p(\tau). \tag{1.59}$$

The static friction force can be cast into the following form,

$$\begin{aligned} F_{fr}^{st}(\varphi, \tau) &= \Phi(\varphi, \omega_1, \tau)H_+(Z)(1 - |\mathrm{sgn}(\omega_r)|) \\ &\quad - F_s\mathrm{sgn}(\Phi(\varphi, \omega_1, \tau))H_+(-Z)\mathrm{sgn}(Z)(1 - |\mathrm{sgn}(\omega_r)|), \end{aligned} \tag{1.60}$$

where

$$Z = F_s - |\Phi(\varphi, \omega_1, \tau)|, \tag{1.61}$$

$$H_+(Z) = \begin{cases} 1, & \text{for } Z \geq 0, \\ 0, & \text{for } Z < 0, \end{cases} \qquad \mathrm{sgn}(Z) = \begin{cases} \{Z/|Z|\}, & \text{for } Z \neq 0, \\ 0, & \text{for } Z = 0. \end{cases} \tag{1.62}$$

The last applied mathematical modelling of friction force (1.60)–(1.62) has been used in reference [Mostaghel (2005)], but the second term in (1.60) has been omitted. In reference [Mostaghel (1999)] the function $H_+(Z)$ has been approximated via function $\mathrm{sgn}(Z)$ through the formula

$$H_+(Z) = 0.5(1 + \mathrm{sgn}(Z))(2 - \mathrm{sgn}(Z)). \tag{1.63}$$

Smoothing method. The friction curve is therefore often approximated by a smooth function. One possible approximation for $\mathrm{sgn}(\omega_r)$ is either

$$\mathrm{sgn}_{\varepsilon_0}(\omega_r) = \begin{cases} \dfrac{\omega_r}{|\omega_r|}, & \text{for } |\omega_r| \geq \varepsilon_0, \\ \left(2 - \dfrac{|\omega_r|}{\varepsilon_0}\right)\dfrac{\omega_r}{\varepsilon_0}, & \text{for } |\omega_r| < \varepsilon_0, \end{cases} \tag{1.64}$$

or

$$\mathrm{sgn}_{\varepsilon_1}(\omega_r) = \frac{2}{\pi}\arctan\left(\frac{\omega_r}{\varepsilon_1}\right), \qquad \varepsilon_m \ll 1, \qquad m = 0, 1. \tag{1.65}$$

The switch model. The relative velocity ω_r will most likely not be exactly zero in digital computation. Instead, an adjoint switch model [Karnopp (1985), Leine et al. (1998), Andreaus, Casini (2002), Leine, Campen (2002a), (2002b)] is studied which is discontinuous but yields a set of ordinary differential equations. The state equation for the switch model reads

$$\ddot{\varphi} = -\Phi(\varphi, \dot{\varphi}, \tau) + F(|\omega_r|)\mathrm{sgn}(\omega_r)p(\tau),$$
$$\text{for} \quad |\omega_r| > \varepsilon_\omega, \quad \text{or} \quad |\Phi(\varphi, \omega_1, \tau)| > F_s, \tag{1.66}$$

$$\frac{d}{d\tau}\left\|\begin{matrix}\varphi \\ \dot{\varphi}\end{matrix}\right\| = \left\|\begin{matrix}\omega_1 \\ \omega_r\omega_d\end{matrix}\right\|,$$
$$\text{for} \quad |\omega_r| < \varepsilon_\omega, \quad \text{and} \quad |\Phi(\varphi, \omega_1, \tau)| < F_s, \tag{1.67}$$

where ω_d is the nondimensional free system frequency. The multiplier ω_d determines how 'fast' the solution is pushed to the centre. The choice of the multiplier ω_d is somewhat arbitrary [Leine et al. (1998)].

A region of near-zero velocity is defined as $\omega_r < \varepsilon_\omega$ ($\varepsilon_\omega \ll 1$). In reference [Awrejcewicz, Pyryev (2002)] $\varepsilon_0 = 10^{-4}$ has been applied, whereas in work [Leine et al. (1998)] the following values have been taken: $\varepsilon_1 = 10^{-6}$, $\varepsilon_\omega = 10^{-6}$.

Continuous friction model. In two of the recent papers [Awrejcewicz et al. (2007), Pyryev at al. (2007)] devoted to mathematical modelling of dry friction the so-called 'continuous friction model' is proposed and studied using a 1-dof model with dry friction. The space (Φ, ω_r) ($\Phi = \Phi(\varphi, \dot{\varphi}, \tau)$) is divided into the following four regions.

$$V_1 : |\omega_r| > \varepsilon_\omega,$$

$$V_2 : [(0 \leq \omega_r \leq \varepsilon_\omega) \cap (\Phi > F_s)] \cup [(-\varepsilon_\omega \leq \omega_r \leq 0) \cap (\Phi < -F_s)],$$

$$V_3 : [(0 < \omega_r \leq \varepsilon_\omega) \cap (\Phi < -F_s)] \cup [(-\varepsilon_\omega \leq \omega_r < 0) \cap (\Phi > F_s)],$$

$$V_4 : (|\omega_r| \leq \varepsilon_\omega) \cap (|\Phi| \leq F_s).$$

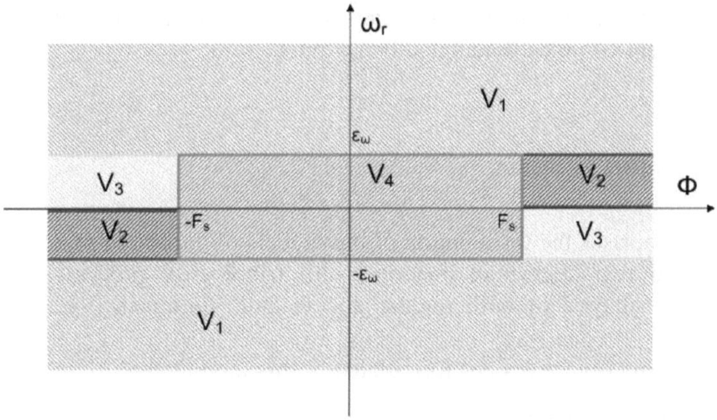

Fig. 1.5: The space (Φ - ω_r) divided into four regions: V_1, V_2, V_3, and V_4.

It is schematically shown in Fig. 1.5. The continuous friction force is defined in the following way.

$$F_{fr} = \begin{cases} F(|\omega_r|)\,\mathrm{sgn}(\omega_r)p(\tau), & \text{for } V_1, \\ F_s\,\mathrm{sgn}(\Phi(\varphi,\dot{\varphi},\tau)), & \text{for } V_2, \\ (2A_3-1)F_s\,\mathrm{sgn}(\omega_r), & \text{for } V_3, \\ A_3(-\Phi(\varphi,\dot{\varphi},\tau)+F_s\,\mathrm{sgn}(\omega_r))+\Phi(\varphi,\dot{\varphi},\tau), & \text{for } V_4, \end{cases} \quad (1.68)$$

where

$$A_3 = \frac{\omega_r^2}{\varepsilon_\omega^2}\left(3 - 2\frac{|\omega_r|}{\varepsilon_\omega}\right) \quad (1.69)$$

is the approximating function.

In a continuous friction model the space (Φ, ω_r) is divided into regions (switch model). The friction force is a continuous function on ω_r (smoothing methods) and for $\omega_r = 0$ the friction force is equal to the forces $\Phi(\varphi, \dot{\varphi}, \tau)$ acting on the bodies excluding friction force $F_{fr}(\varphi, \omega_r, \tau)$ (signum model). In another words, a continuous friction model can be treated as an approximating friction force in switch instant by a smoothing function taking into consideration the forces acting on the bodies excluding friction force.

Some books deal with nonsmooth dynamics but include only mechanical or mathematical aspects (see, e.g., [Brogliato (1999), Kunze (2000), Moreau, Panagiotopoulos (1988)]) and neither the physics of such models nor dynamical behaviour of the models involving chaos, for example. Some of them include thermodynamics for nonsmooth systems (e.g., [Fremond (2002)]). Again the approach can be oriented to modelling of nonsmooth processes mainly with convex or more sophisticated mathematical analysis. The study is devoted to different systems and does not present complex dynamical behaviours.

1.2. Aim and scope

The study comprises the formulation of the models of motion of material systems with one and two degrees-of-freedom, with tribological processes taken into account. The analysis of both regular and chaotic vibrations is also presented in this book.

The following objectives are formulated in this monograph.

- Modelling the dynamics of contact systems with one and two degrees-of-freedom with friction, taking into account the inertia of the bodies, heat generated by friction, and wear, and the analysis of their influence on the solutions of certain problems
- Drawing up the methodology of a nonlinear problem of the motion of thermoelastic bodies in contact
- Examining the phenomena of stick-slip motion in self-excited systems, taking tribological processes into account, and determining the influence of the system parameters on the character of regular vibrations, as well as determining the critical values of parameters, for which a qualitative change in the system's behaviour occurs
- Showing that for external mechanical and harmonic kinematical excitations the system may pass to chaotic motion in certain circumstances
- Determining the conditions of the occurrence of frictional thermoelastic instability
- The analysis of the kinetics of contact characteristics during starting, motion, and braking of the bodies in contact

For the determination of regular vibrations of systems, the analytical method of perturbations is applied, whereas chaotic vibrations are analysed with the use of Melnikov and frequency spectrum methods. The stability of periodic solutions is determined with the use of numerical methods.

The obtained analytical results are presented for the concrete numerical data, and then they are compared with the results of numerical simulation. Numerical methods are the basis for determining the conditions of transition to chaotic vibrations and bifurcational values of parameters for which this transition takes place.

Chapter 2
Thermoelastic Contact of Shaft and Bush in Wear Regime

A classical problem concerning vibration of a friction pair consisting of a rotating shaft and bush fixed to a frame by massless springs (a simple model of a typical braking pad or the so-called Pronny's brake) has been investigated in references [Andronov et al. (1966), Neimark (1978)]. In works of [Pyryev et al. (1995), Pyryev, Grilitskiy (1995)], the so-called thermoelastic contact between a rotating cylinder and a fixed noninertial pad has been studied. Further on, a more complicated axially symmetric problem of both regular and chaotic self-excited vibrations (caused by friction) and wear of the rotating cylinder and bush (fixed to a frame by springs and viscous damping elements) is investigated.

2.1. Analysed system

An elastic and heat-transferring cylinder (shaft) with radius R_1 is inserted into the bush (full bush or braking pad) with initial compression value of $U_0 h_U(t)$ ($h_U(t) \to 1$, $t \to \infty$). The bush located on the shaft has internal radius $R_1 - U_0$ ($U_0/R_1 \ll 1$), whereas the cylinder is compressed on the value of $U_0 h_U(t)$ measured in the radial direction.

The initial compression can be realized applying split-bearing bushing lying on the shaft and initially compressed ([see, e.g. Andronov et al. (1966)]).

A scheme of the analysed system is shown in Fig. 2.1. The bush (pad) is fixed to the frame (housing or base) by springs. It is assumed that in this model the bush is an absolutely rigid body. Furthermore, radial springs are initially compressed and they have stiffness coefficients k_1, whereas tangential springs are characterised by nonlinear stiffness k_2, k_3 of Duffing type per unit bush (pad) length.

In the tangential direction, the bush (pad) is driven by the damping force measured per bush length unit (c denotes the viscous damping coefficient) and force $F_2 = F_* \cos(\omega' t)$ is also measured per bush length unit.

It is assumed that (i) the cylinder (shaft) rotates at an angular velocity $\dot{\varphi}_1(t)$ such that centrifugal forces can be neglected in our system; (ii) the cylinder angular

DOI: 10.1007/978-0-387-09653-7_2,

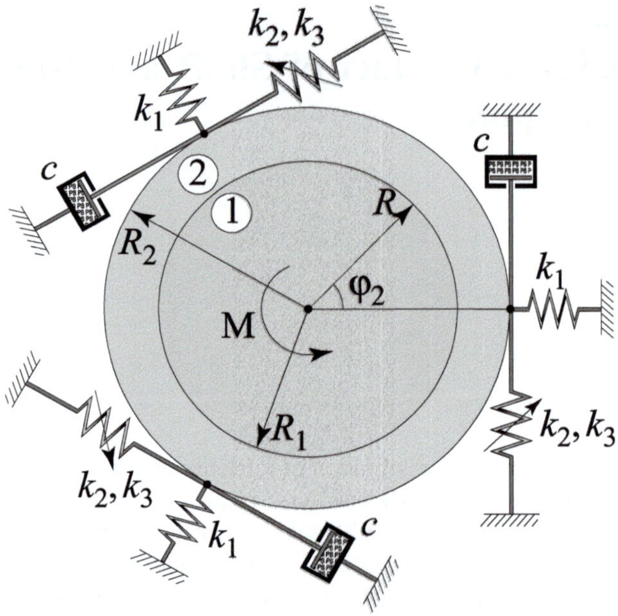

Fig. 2.1: The analysed system.

velocity $\dot{\varphi}_1 = \Omega_* \omega_1(t)$ (Ω_*, constant having units of angular velocity) is governed by the equation $\omega_1 = \omega_* + \zeta_k \sin \omega' t$, where ζ_k is the dimensionless amplitude of kinematic external excitation and it changes according to rotational motion of the cylinder driven by the moment $M = M_0 h_M(t)$ ($h_M(t) \to 1$, $t \to \infty$) measured per cylinder length unit; (iii) between the bush and cylinder dry friction (per length unit) defined by the function $F_{fr}(V_r)$ occurs, where V_r is the relative velocity of the bush and cylinder $V_r = \dot{\varphi}_1 R_1 - \dot{\varphi}_2 R_1$. B_1, B_2 are the inertial mass moments of the cylinder and bush also measured per length unit (cylinder, bush).

It is further assumed that the bush transfers heat ideally, and that at the initial time instant the ambient temperature change is governed by $T_0 h_T(t)$, where T_0 is the constant in temperature units, and $h_T(t)$ stands for a known dimensionless function of time ($h_T(t) \to 1$, $t \to \infty$), and that between the shaft and bush Newton's heat exchange occurs. The shaft starts to expand, and a contact between the shaft and bush is initiated. Another assumption made is that the dry friction occurring between the bush and shaft is defined by the function $F_{fr}(V_r)$, where $V_r = \dot{\varphi}_1 R_1 - \dot{\varphi}_2 R_1$ is the relative velocity of the shaft and bush.

Owing to Amonton's assumption, friction force F_{fr} is the product of normal reaction $N(t)$ and the friction coefficient; that is, $F_{fr} = f(V_r)N(t)$ is the frictional force defining resistance of the two bodies' movement, and $f(V_r)$ denotes the kinetic friction coefficient.

The so-called Stribeck curve [Kragelsky, Gitis (1987), Ibrahim (1994)] shown in Fig. 2.2, has its minimum for $V_r = V_{\min}$ and for $V_r < V_{\min}$ we have ($f'(V_r) < 0$).

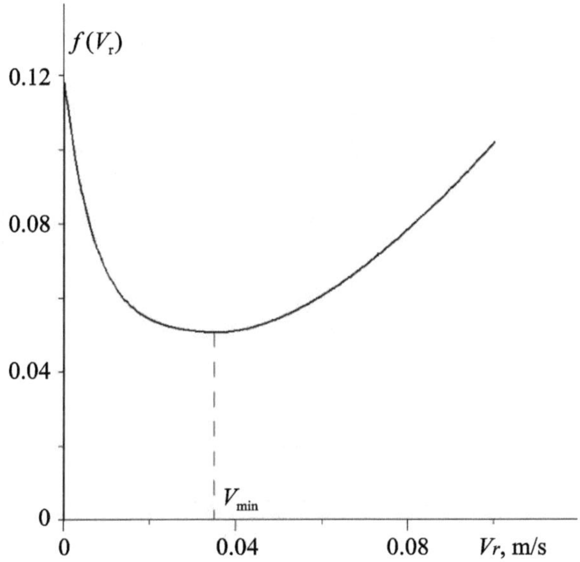

Fig. 2.2: Kinetic friction coefficient versus relative velocity.

The friction force F_{fr} yields heat generated by friction on the contact surface $R = R_1$ and wear U^w.

Usually it is assumed [see, e.g., (1959)] that the frictional work is transformed to heat energy. However, in practice it means the occurrence of two heat streams q_1 and q_2 moving into the insides of the contacting bodies. Let $T_1(r,t)$ denote cylinder temperature and let its initial value be T_{sm}. Furthermore, we assume that the bush transfers heat perfectly, heat transfer occurring between the cylinder and bush is governed by Newton's rule, and that temperature of the surrounding medium changes according to the equation $T_{sm} + T_0 h_T(t)$.

The formulated problem is defined by the equations that govern dynamics of the bodies in the vicinity of the equilibrium configuration, with angular displacements $\varphi_1(t)$, $\varphi_2(t)$ and angular velocities $\dot{\varphi}_2(t)$ for the bush and $\dot{\varphi}_1(t)$ for the shaft, with stresses $\sigma_R(R,t)$ in the shaft, contact pressure $P(t) = N(t)/2\pi R_1 = -\sigma_R(R_1,t)$, temperature $T_1(R,t)$ of the shaft, and finally with displacement $U(R,t)$ in the direction along the R axis.

2.2. Mathematical formulation of the problem

2.2.1 *Equations for rotational movement of absolutely rigid bush*

Let axis Z be the cylinder axis. The equilibrium state of the moments of forces with respect to the shaft axis yields

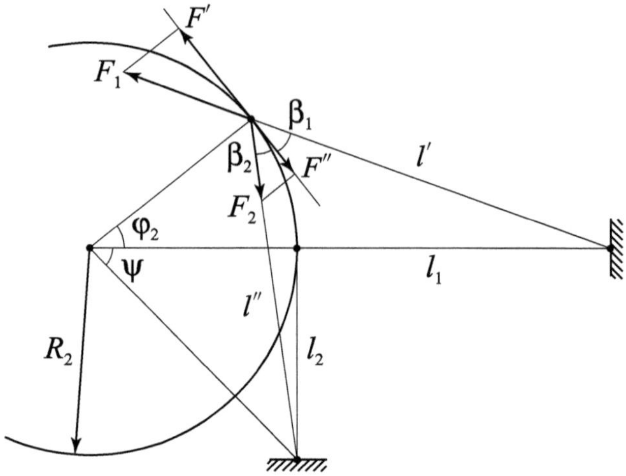

Fig. 2.3: Geometry of spring configuration.

$$B_2\ddot{\varphi}_2 \in M_{fr} - M_e - cR_2^2\dot{\varphi}_2, \tag{2.1}$$

where $M_{fr} = f(R_1(\dot{\varphi}_1 - \dot{\varphi}_2))2\pi R_1^2 P(t)$ denotes the moment of friction force, and $M_e = (F'' - F' - F_2)R_2$ is the moment of elastic and external forces. In addition, the following geometrical relations hold (Fig. 2.3).

$$F'' = (k_2\Delta l_2 + k_3(\Delta l_2)^3)\cos\beta_2, \qquad F' = k_1\Delta l_1\cos\beta_1.$$

According to results shown in the figure, one gets

$$\frac{l''}{\sin(\varphi_2 + \Psi)} = \frac{\sqrt{R_2^2 + l_2^2}}{\sin(\pi/2 - \beta_2)}, \qquad \frac{l'}{\sin\varphi_2} = \frac{l_1 + R_2}{\sin(\pi/2 + \beta_1)},$$

and hence

$$\cos\beta_2 = \sin(\varphi_2 + \Psi)\frac{\sqrt{R_2^2 + l_2^2}}{l''} = \frac{R_2\sin\varphi_2 + l_2\cos\varphi_2}{l''},$$

$$\cos\beta_1 = \frac{l_1 + R_2}{l'}\sin\varphi_2.$$

Furthermore, we obtain:

$$l'' = \sqrt{R_2^2 + (R_2^2 + l_2^2) - 2R_2\sqrt{R_2^2 + l_2^2}\cos(\varphi_2 + \psi)}$$

$$= \sqrt{2R_2^2 + l_2^2 - 2R_2(R_2\cos\varphi_2 - l_2\sin\varphi_2)},$$

$$l' = \sqrt{R_2^2 + (R_2 + l_1)^2 - 2R_2(R_2 + l_1)\cos\varphi_2}.$$

Finally, we have

$$F'' = (k_2\Delta l_2 + k_3(\Delta l_2)^3)(R_2 \sin\varphi_2 + l_2 \cos\varphi_2)/l'', \qquad \Delta l_2 = l'' - l_2,$$

$$l'' = l_2\sqrt{2(R_2/l_2)^2(1 - \cos\varphi_2) + 1 + 2(R_2/l_2)\sin\varphi_2},$$

$$F' = k_1\Delta l_1 \sin\varphi_2(l_1 + R_2)/l', \quad \Delta l_1 = l_0 - l',$$

$$l' = \sqrt{R_2^2 + (R_2 + l_1)^2 - 2R_2(R_2 + l_1)\cos\varphi_2},$$

where l_0 denotes the radial spring length without initial compression; l' denotes the length of the compressed radial spring during bush (pad) rotation with respect to angle φ_2; l_1 denotes the length of the compressed spring, where $\varphi_2 = 0$; $l_0 - l'$ is the compression of radial spring length; l'' is the tangential spring length related to bush (pad) rotation with respect to angle φ_2; l_2 is the tangential spring length, where $\varphi_2 = 0$.

Keeping an accuracy of φ_2^3, one gets:

$$\sin\varphi_2 = \varphi_2 - (1/6)\varphi_2^3 + O(\varphi_2^4), \qquad \cos\varphi_2 = 1 - (1/2)\varphi_2^2 + O(\varphi_2^4),$$

$$l' = l_1 + (1/2)R_2(l_1 + R_2)/l_1\varphi_2^2 + O(\varphi_2^4),$$

$$l'' = l_2 + R_2\varphi_2 - (1/6)R_2\varphi_2^3 + O(\varphi_2^4),$$

$$F' = k_1(l_0/l_1 - 1)(R_2 + l_1)\varphi_2 + (1/6)(R_2 + l_1)(1 - (l_0/l_1)(1 + 3(R_2/l_1)$$
$$+ 3(R_2/l_1)^2))k_1\varphi_2^3 + O(\varphi_2^4),$$

$$F'' - k_2R_2\varphi_2 + (R_2^3k_3 - (2/3)R_2k_2)\varphi_2^3 + O(\varphi_2^4).$$

Let the initial conditions be

$$\varphi_2(0) = \varphi_2^\circ, \qquad \dot{\varphi}_2(0) = \dot{\varphi}_2^\circ, \tag{2.2}$$

and let us introduce the dimensionless parameters

$$\tau = \frac{t}{t_*}, \quad p = \frac{P}{P_*}, \quad \varphi(\tau) = \varphi_2(t_*\tau), \quad \phi(\tau) = \varphi_1(t_*\tau), \quad \varphi^\circ = \varphi_2^\circ, \quad \dot{\varphi}^\circ = \frac{\dot{\varphi}_2^\circ}{\Omega_*},$$

$$\varepsilon = \frac{P_* t_*^2 2\pi R_1^2}{B_2}, h = \frac{cR_2^2}{2B_2t_*}, \zeta = \frac{R_2F_*}{2\pi R_1^2 P_*}, t_D^2 = \frac{B_2}{k_2R_2^2}, \chi = \frac{t_*^2 k_* R_2^2}{B_2} = \frac{t_*^2 k_*}{t_D^2 k_2},$$

$$t_T = R_1^2/a_1, \quad \omega_0 = \omega' t_*, \quad F(\dot{\phi} - \dot{\varphi}) = f(V_*(\dot{\phi} - \dot{\varphi})), \quad V_* = R_1/t_*, \tag{2.3}$$

$$b = \frac{\left(k_3 R_2^4 - \frac{2}{3}k_2 R_2^2 - (l_1 + R_2)R_2\left(1 - \frac{l_0}{l_1}\left(1 + 3\frac{R_2}{l_1} + 3\left(\frac{R_2}{l_1}\right)^2\right)\right)\frac{k_1}{6}\right)t_*^2}{B_2},$$

$$a = \frac{k_1}{k_*}\left(\frac{l_0}{l_1} - 1\right)\left(1 + \frac{l_1}{R_2}\right) - \frac{k_2}{k_*}, \tag{2.4}$$

where τ is the dimensionless time, $t_D = \sqrt{B_2/k_2}/R_2$ is the characteristic system time related to the bush oscillation period $T_p = 2\pi t_D$, P_* is the characteristic contact pressure at shaft temperature T_*, and V_* is the characteristic velocity of the system, $\Omega_* = t_*^{-1}$.

As exhibited by Equation (2.4), parameter a can take arbitrary values ($a \in \Re^1$) depending on the values of spring parameters. For example, for stretched radial springs $l_0/l_1 > 1$ parameter $a > 0$, if $l_0/l_1 - 1 > k_2 k_1^{-1}(1 + l_1/R_2)^{-1}$. In the case of nonstretched radial springs one always has $a < 0$.

Observe that introduction of a spring with stiffness coefficient k_1 to our system allows us to get an unstable system equilibrium which enables the occurrence of chaotic dynamics. In the case of a lack of radial springs ($k_1 = 0$) the nondimensional coefficient standing by φ equals $a = -k_2/k_*$, whereas the coefficient standing by φ^3 is $b = (k_3 R_2^4 - (2/3)k_2 R_2^2)t_*^2/B_2$. Introduction of the nonlinear stiffness coefficient k_3 into our system (the Duffing nonlinearity) makes it possible: (i) to take into account positive and negative values of the parameter b; (ii) to introduce either a stiff ($k_3 > 0$) or weak ($k_3 < 0$) spring characteristic; (iii) to take into account physical and geometrical nonlinearity.

The dimensionless equations (inclusion) governing the system dynamics have the form

$$\begin{aligned}&\ddot{\varphi}(\tau) + 2h\dot{\varphi}(\tau) - a\chi\varphi(\tau) + b\varphi^3(\tau) \\ &\quad \in \varepsilon[\zeta\cos(\omega_0\tau) + F(\dot{\phi} - \dot{\varphi})p(\tau)], \qquad 0 < \tau < \infty,\end{aligned} \tag{2.5}$$

with the initial conditions:

$$\varphi(0) = \varphi^\circ, \qquad \dot{\varphi}(0) = \dot{\varphi}^\circ = \omega^\circ. \tag{2.6}$$

In order to solve the motion equations (inclusion) one needs to know $\dot{\varphi}(\tau)$, $\dot{\phi}(\tau)$, and contact pressure $p(\tau)$. The latter can be obtained from the thermoelasticity equation that includes also tribological processes.

2.2.2 *Thermoelastic shaft*

In the analysed case, the inertial terms in the equation of motion can be omitted and the problem may be considered as a quasi-static one. For axially symmetric stress of the shaft, the equations used belong to the theory of thermal stresses for an isotropic body, formulated by Nowacki [Nowacki (1962)] and applying cylindrical coordinates (R, ϕ, Z):

$$\frac{\partial^2 U(R,t)}{\partial R^2} + \frac{1}{R}\frac{\partial U(R,t)}{\partial R} - \frac{1}{R^2}U(R,t) = \alpha_1 \frac{1+\nu_1}{1-\nu_1}\frac{\partial T_1(R,t)}{\partial R}, \tag{2.7}$$

$$\frac{\partial^2 T_1(R,t)}{\partial R^2} + \frac{1}{R}\frac{\partial T_1(R,t)}{\partial R} = \frac{1}{a_1}\frac{\partial T_1(R,t)}{\partial t}, \quad 0 < R < R_1, \quad 0 < t < t_c, \tag{2.8}$$

with the attached mechanical

$$U(0,t) = 0, \qquad U(R_1,t) = -U_0 h_U(t) + U^w(t), \qquad 0 < t < t_c, \tag{2.9}$$

and thermal boundary conditions

$$\lambda_1 \frac{\partial T_1(R_1,t)}{\partial R} + \alpha_T (T_1(R_1,t) - T_{sm} - T_0 h_T(t)) = (1-\eta) f(V_r) V_r P(t), \tag{2.10}$$

$$R\frac{\partial T_1(R,t)}{\partial R}\bigg|_{R\to 0} = 0, \qquad 0 < t < t_c \tag{2.11}$$

and with the following initial conditions

$$T_1(R,0) = T_{sm}, \qquad 0 < R < R_1. \tag{2.12}$$

Velocity of the bush (pad) wear is proportional to a certain power of friction force. According to Archard's assumption [Archard (1953), Aleksandrov, Annakulova (1990), (1992)] we have

$$\dot{U}^w(t) = K^w |V_r(t)| P(t). \tag{2.13}$$

Radial stress $\sigma_R(R,t)$ in the cylinder may be found with the use of radial displacement $U(R,t)$ and temperature $T_1(R,t)$ by the application of the following formula

$$\sigma_R(R,t) = \frac{E_1}{1-2\nu_1}\left[\frac{1-\nu_1}{1+\nu_1}\frac{\partial U(R,t)}{\partial R} + \frac{\nu_1}{1+\nu_1}\frac{U(R,t)}{R} - \alpha_1 (T_1(R,t) - T_{sm})\right]. \tag{2.14}$$

The following notation is used: $P(t) = N(t)/2\pi R_1 = -\sigma_R(R_1,t)$: contact pressure; $V_r = \dot{\varphi}_1 R_1 - \dot{\varphi}_2 R_1$, E_1: Young's modulus; ν_1: Poisson's ratio; a_1: thermal diffusivity, α_1: thermal expansion coefficient; λ_1: thermal conductivity; K^w: wear constant, t_c: time of contact; $h_T(t)$: ambient temperature.

Integrating Equation (2.7), with (2.9) and (2.14) taken into account, the contact pressure is determined:

$$P(t) = \frac{2E_1\alpha_1}{1-2\nu_1}\frac{1}{R_1^2}\int_0^{R_1}[T_1(\xi,t) - T_{sm}]\xi d\xi$$

$$+ \frac{E_1}{(1-2\nu_1)(1+\nu_1)R_1}[U_0 h_U(t) - U^w(t)]. \tag{2.15}$$

Let us introduce the following dimensionless parameters.

$$\tau_c = \frac{t_c}{t_*},\ r = \frac{R}{R_1},\ u = \frac{U}{U_*},\ u^w = \frac{U^w}{U_*},\ \theta = \frac{T_1 - T_{sm}}{T_*},\ Bi = \frac{\alpha_T R_1}{\lambda_1},$$

$$h_T(\tau) = h_T(t_*\tau), \qquad h_U(\tau) = h_U(t_*\tau),$$

$$k^w = \frac{P_* K^w R_1}{U_*}, \qquad \gamma = \frac{2(1-\eta)E_1\alpha_1 R_1^2}{\lambda_1(1-2\nu_1)t_*}, \qquad \tilde{\omega} = \frac{t_*}{t_T}, \tag{2.16}$$

where

$$T_* = \frac{U_*}{2\alpha_1(1+\nu_1)R_1}, \qquad P_* = \frac{E_1 U_*}{(1+\nu_1)(1-2\nu_1)R_1} = \frac{2\alpha_1 E_1 T_*}{1-2\nu_1}. \tag{2.17}$$

Note, that physically P_* denotes the contact pressure between the cylinder and bush with radius R_1 during heating up of the cylinder to temperature T_*, or it describes the cylinder compression in the radial direction on the amount of U_*.

The thermoelastic problem under consideration takes the following dimensionless form.

$$\frac{\partial^2\theta(r,\tau)}{\partial r^2} + \frac{1}{r}\frac{\partial\theta(r,\tau)}{\partial r} = \frac{1}{\tilde{\omega}}\frac{\partial\theta(r,\tau)}{\partial\tau}, \qquad 0 < \tau < \tau_c, \qquad 0 < r < 1; \tag{2.18}$$

$$\frac{\partial\theta(1,\tau)}{\partial r} + Bi\theta(1,\tau) = \gamma F(\dot{\phi}(\tau) - \dot{\varphi}(\tau))(\dot{\phi}(\tau) - \dot{\varphi}(\tau))p(\tau)$$

$$+Bi\frac{T_0}{T_*}h_T(\tau),\ 0 < \tau < \tau_c; \tag{2.19}$$

$$r\frac{\partial\theta(r,\tau)}{\partial r}\bigg|_{r\to 0} = 0, \qquad 0 < \tau < \tau_c; \tag{2.20}$$

$$\theta(r,0) = 0, \qquad 0 < r < 1. \tag{2.21}$$

$$p(\tau) = \frac{U_0}{U_*}h_U(\tau) - u^w(\tau) + \int_0^1 \theta(\xi,\tau)\xi d\xi, \qquad 0 < \tau < \tau_c; \tag{2.22}$$

$$u^w(\tau) = k^w \int_0^\tau |\dot{\phi}(\tau) - \dot{\varphi}(\tau)| p(\tau) d\tau, \qquad 0 < \tau < \tau_c. \tag{2.23}$$

The cylinder radial displacement is

$$u(r,\tau)/r = \frac{1}{2(1-\nu_1)} \left[\frac{1}{r^2} \int_0^r \theta(\eta,t)\eta d\eta - \int_0^1 \theta(\eta,t)\eta d\eta \right] - \frac{U_0}{U_*} h_U(\tau) + u^w(\tau).$$

Observe, that in order to solve the problem defined by Equations (2.18)–(2.23), one should know time-dependent velocities of the bush and shaft. Note also that problems (2.5), (2.6), and (2.18)–(2.23) are mutually adjoined and require simultaneous solution.

2.2.3 Rotational motion of the shaft

Let axis Z coincide with the shaft axis. The moments of force related to the shaft axis give

$$B_1 \ddot{\varphi}_1 \in M - M_{fr}, \tag{2.24}$$

where $M_{fr} = f(V_r) 2\pi R_1^2 P(t)$ denotes the moment of friction force, M is the moment acting on the shaft, and $\varphi_1(t)$ stands for the angle of the shaft position. In order to solve Equation (2.24), the following initial conditions are attached.

$$\varphi_1(0) = \varphi_1^\circ, \qquad \dot{\varphi}_1(0) = \dot{\varphi}_1^\circ. \tag{2.25}$$

Introducing the dimensionless parameters

$$u_M = \frac{t_*^2 P_* 2\pi R_1^2}{B_1}, \quad m_0 = \frac{M_0}{P_* 2\pi R_1^2}, \quad \phi^\circ = \varphi_1^\circ, \quad \dot{\phi}^\circ = \frac{\dot{\varphi}_1^\circ}{\Omega_*}, \quad \phi = \varphi_1(t_*\tau),$$

$$\dot{\phi} = \frac{d\phi}{d\tau} = \omega_1(\tau), \quad h_M(\tau) = h_M(t_*\tau), \quad F(\dot{\phi} - \dot{\varphi}) = f(V_*(\dot{\phi} - \dot{\varphi})); \tag{2.26}$$

one gets the dimensionless equation governing the shaft dynamics:

$$\ddot{\phi}(\tau) \in a_M \left[m_0 h_M(\tau) - F(\dot{\phi} - \dot{\varphi}) p(\tau) \right], \qquad 0 < \tau < \infty, \tag{2.27}$$

$$\phi(0) = \phi^\circ, \qquad \dot{\phi}(0) = \dot{\phi}^\circ = \omega_1^\circ. \tag{2.28}$$

For computation purposes, the multivalued relation $\mathrm{sgn}(\cdot)$ is approximated by the function $\mathrm{sgn}_{\varepsilon_0}(\cdot)$ defined by (1.64), where the regularisation parameter ε_0 is a 'small' positive real number. The differential inclusions (2.5), (2.27) are then approximated by the system of equations.

It is seen that again contact pressure needs to be defined. It will be found following the solving procedure for the thermoelastic equation that takes into account the tribologic processes. Note that problems (2.5)–(2.6), (2.18)–(2.23), (2.27), and (2.28) are coupled and require a common solution.

2.2.4 Laplace transform

Let us apply the Laplace transform to Equations (2.18)–(2.23).

$$\{\bar{\theta}(r,s), \bar{p}(s), \bar{u}^w(s), \bar{h}_U(s), \bar{q}(s)\}$$
$$= \int_0^\infty \{\theta(r,\tau), p(\tau), u^w(\tau), h_U(\tau), q(\tau)\} e^{-s\tau} d\tau,$$

where s is the transformation parameter. Taking into account the boundary (2.2.2), (2.20) and the initial conditions (2.21) we obtain

$$\bar{\theta}(r,s) = sG_\theta(r,s)\bar{q}(s), \qquad \bar{p}(s) = \frac{U_0}{U_*}\bar{h}_U(s) - \bar{u}^w(s) + sG_p(s)\bar{q}(s),$$

$$G_\theta(r,s) = \frac{I_0(s_\omega r)}{\tilde{\omega} s \Delta_1(s)}, \qquad G_p(s) = \frac{\Delta_2(s)}{\tilde{\omega} s \Delta_1(s)},$$

$$\Delta_2(s) = \frac{I_1(s_\omega)}{s_\omega}, \qquad \Delta_1(s) = s_\omega^2 \Delta_2(s) + BiI_0(s_\omega), \qquad s_\omega = \sqrt{\frac{s}{\tilde{\omega}}}, \quad (2.29)$$

where $I_n(x) = i^{-n} J_n(ix)$ is the modified first-order Bessel function with argument x, and the nonlinear part of boundary problem (2.2.2) has the form

$$q(\tau) = \gamma\tilde{\omega}F(\dot{\phi} - \dot{\varphi})(\dot{\phi} - \dot{\varphi})p(\tau) + Bi\tilde{\omega}\frac{T_0}{T_*}h_T(\tau). \quad (2.30)$$

One of the commonly used methods to find inverse integrals of Laplace transformation applies the following relation,

$$\{G_p(\tau), G_\theta(r,\tau)\} = \frac{1}{2\pi i} \int_{-i\infty + c_L}^{i\infty + c_L} \{G_p(s), G_\theta(r,s)\} e^{s\tau} ds,$$

allowing for computation of the residual sum of function $G_\theta(r,s)e^{s\tau}$, $G_p(s)e^{s\tau}$ in the complex plane to the left of straight line $s = c_L$. Using (2.29), applying the inverse Laplace transform, and using the theorem of convolution [Carslaw, Jaeger (1959)], the following function is found.

$$p(\tau) = \frac{U_0}{U_*} h_U(\tau) - u^w(\tau) + Bi\tilde{\omega}\frac{T_0}{T_*}\int_0^\tau G_p(\tau-\xi)\dot{h}_T(\xi)d\xi$$

$$+ \gamma\tilde{\omega}\int_0^\tau \dot{G}_p(\tau-\xi)F(\dot{\phi}-\dot{\varphi})p(\xi)(\dot{\phi}-\dot{\varphi})d\xi, \tag{2.31}$$

$$\theta(r,\tau) = Bi\tilde{\omega}\frac{T_0}{T_*}\int_0^\tau G_\theta(r,\tau-\xi)\dot{h}_T(\xi)d\xi$$

$$+ \gamma\tilde{\omega}\int_0^\tau \dot{G}_\theta(r,\tau-\xi)F(\dot{\phi}-\dot{\varphi})p(\xi)(\dot{\phi}-\dot{\varphi})d\xi, \tag{2.32}$$

where

$$\{G_p(\tau), G_\theta(1,\tau)\} = \frac{\{0.5, 1\}}{Bi\tilde{\omega}} - \sum_{m=1}^{\infty}\frac{\{2Bi, 2\mu_m^2\}}{\mu_m^2\tilde{\omega}(Bi^2+\mu_m^2)}e^{-\mu_m^2\tilde{\omega}\tau}, \tag{2.33}$$

and μ_m are the roots of the characteristic equation ($m = 1, 2, 3, \ldots$),

$$Bi\, J_0(\mu) - \mu J_1(\mu) = 0. \tag{2.34}$$

Note that the investigated problem has been transformed to the set of nonlinear differential Equations (2.5) and (2.27), integral equation (2.31) describing angular velocities $\dot{\varphi}(\tau)$, $\dot{\phi}(\tau)$, and contact pressure $p(\tau)$. Temperature is defined by (2.32) and (2.23).

Values of parameters t_*, U_*, k_* are defined when considering particular cases of the studied problem.

2.3. Kinematic external shaft excitations

We consider first the case of kinematic shaft excitation; that is, we assume that both rotation velocity $\omega_1(\tau)$ is known and the bush temperature is constant: $h_T(\tau) = 0$, $h_U(\tau) = H(\tau)$; $H(\tau) = 1$, $\tau > 0$; $H(\tau) = 0$, $\tau < 0$. The bush (or the pad) is attached to the housing (a frame or base) through springs with stiffness k_2 ($k_1 = k_3 = h = 0$). Furthermore, it is assumed that $U_* = U_0$, any arbitrary parameter $k_* = k_2$ which gives constant $a = -1$ (see Equation (2.5)). In the considered case, Equations (2.27) and (2.28) are neglected.

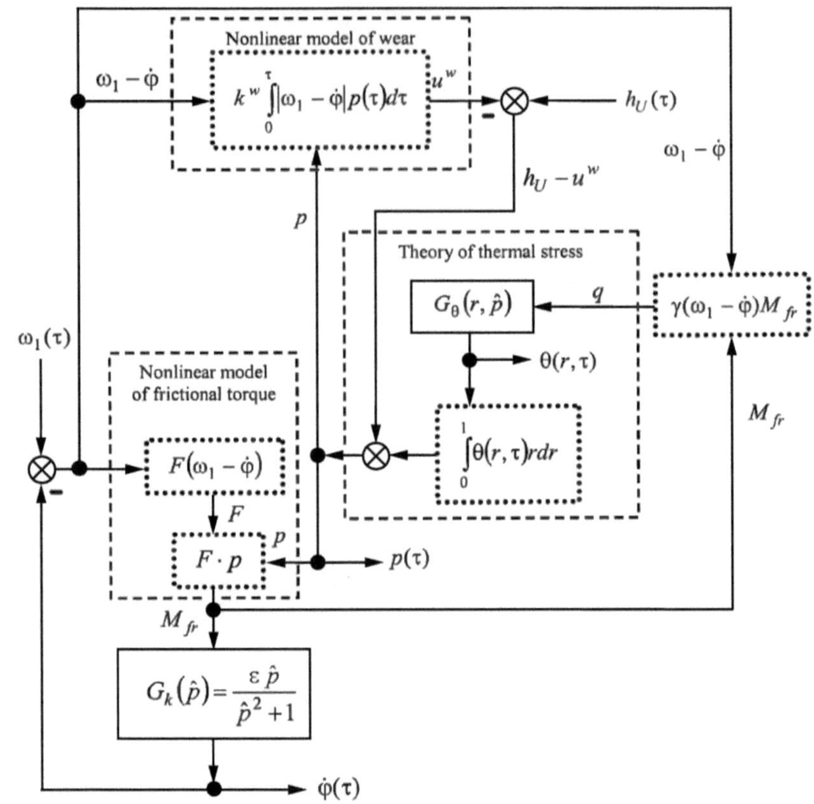

Fig. 2.4: Block diagram of the system.

2.3.1 *Block diagram*

The system shown in Fig. 2.1 is presented as a block diagram in Fig. 2.4 ($\hat{p} \equiv d/d\tau$). One may easily trace various relations and interactions between input and output parameters.

2.3.2 *Stationary process associated with a constant shaft velocity rotation*

We consider the case when the shaft rotates at constant velocity $\omega_1(\tau) = \dot{\phi}(\tau) = \omega_1^0 H(\tau)$. The lack of external excitation $\zeta = \zeta_k = 0$ and $\chi = 1$ is assumed. (The latter relation yields $t_* = \sqrt{B_2/(k_2 R_2^2)}$). In this case t_* has a physical sense. Namely, it is the period of free bush vibrations.

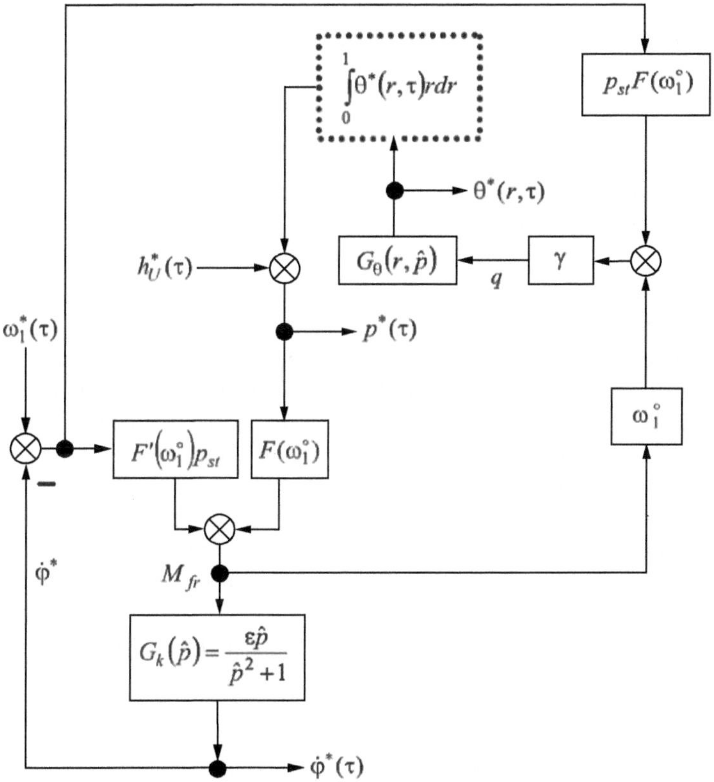

Fig. 2.5: Block diagram of the linearized system.

2.3.2.1 Block diagram of the linearized system

The corresponding linearized system is shown in Fig. 2.5.

Using the block diagram one can easily derive transmittances $G_{\varphi h}(s)$, $G_{\varphi\omega}(s)$, $G_{ph}(s)$, $G_{p\omega}(s)$, $G_{\theta h}(s)$, $G_{\theta\omega}(s)$ coupling the disturbance signals $h_U^*(t)$, $\omega_1^*(t)$ and input signals $\varphi^*(t)$, $p^*(t)$, $\theta^*(t)$.

2.3.2.2 Stationary process without wear

The stationary solution to the considered problem (in Equations (2.5) and (2.18) differential terms related to time or omitted) without wear ($k^w = 0$) is governed by the equation

$$p_{st} = \frac{1}{1-v}, \qquad \theta_{st} = \frac{2v}{1-v}, \qquad \varphi_{st} = \frac{\gamma_1 \omega_1^\circ}{1-v}, \tag{2.35}$$

$$\gamma_1 = \frac{\varepsilon F(\omega_1^\circ)}{\omega_1^\circ}, \qquad v = \frac{\gamma \omega_1^\circ F(\omega_1^\circ)}{2Bi} = \frac{E_1 \alpha_1 R_1^2 \omega_1^\circ F(\omega_1^\circ)}{2Bi\lambda_1(1-2\nu_1)}.$$

The stationary solution (2.35) has a physical meaning ($p_{st} > 0$), when the inequality $v < 1$ holds. Observe that the stationary solution (contrary to the function describing a stationary solution) does not have a singularity. When $v \to 1$, a trajectory (unsteady solution) may tend either to periodic motion or it may increase exponentially.

The parameter v, as shown in the following, will play an essential role while determining Frictional Thermoelastic Instability (FTEI). It is proportional to the Young modulus E_1, shaft extension coefficient α_1, relative velocity $\Omega_* \omega_1^0 R_1$, body dimension R_1, and friction coefficient $F(\omega_1^0)$, but it decreases with an increase of heat transfer coefficient λ_1 and the Biot number Bi.

Furthermore, we focus on the analysis of perturbations of the stationary solution (2.35). Introducing perturbations $h_U(\tau) = 1 + h_U^*(\tau)$, $\omega_1(\tau) = \omega_1^0 + \omega_1^*(\tau)$, $|h_U^*| \ll 1$, $|\omega_1^*| \ll 1$, the solutions are given in the form

$$\varphi(r,\tau) = \varphi_{st} + \varphi^*(\tau), \quad \theta(r,\tau) = \theta_{st}(r) + \theta^*(r,\tau), \quad p(\tau) = p_{st} + p^*(\tau), \tag{2.36}$$

where $|\varphi^*| \ll 1$, $|\theta^*| \ll 1$, $|p^*| \ll 1$.

First, the right-hand sides of Equations (2.5) and the boundary condition (2.2.2) are linearized giving the equations

$$\ddot{\varphi}^*(t) + \varphi^*(t) = \varepsilon F(\omega_1^\circ) p^*(\tau) + \varepsilon F'(\omega_1^\circ) p_{st}(\omega_1^* - \dot{\varphi}^*), \qquad 0 < \tau < \infty, \tag{2.37}$$

$$\varphi^*(0) = 0, \qquad \dot{\varphi}^*(0) = 0; \tag{2.38}$$

$$p^*(\tau) = h_U^*(\tau) + \int_0^1 \theta^*(\xi,\tau)\xi d\xi, \qquad 0 < \tau < \infty; \tag{2.39}$$

$$\frac{\partial^2 \theta^*(r,\tau)}{\partial r^2} + \frac{1}{r}\frac{\partial \theta^*(r,\tau)}{\partial r} = \frac{1}{\tilde{\omega}}\frac{\partial \theta^*(r,\tau)}{\partial \tau}, \qquad 0 < \tau < \infty, \qquad 0 < r < 1; \tag{2.40}$$

$$\frac{\partial \theta^*(1,\tau)}{\partial r} + Bi\theta^*(1,\tau) = \gamma[\omega_1^\circ F(\omega_1^\circ) p^* + p_{st}(\omega_1^* - \dot{\varphi}^*)(F(\omega_1^\circ) + \omega_1^\circ F'(\omega_1^\circ))], \tag{2.41}$$

$$r\left.\frac{\partial \theta^*(r,\tau)}{\partial r}\right|_{r\to 0} = 0, \qquad 0 < \tau < \infty; \qquad \theta^*(r,0) = 0, \qquad 0 < r < 1. \tag{2.42}$$

When the Laplace transform is applied to the linear system (2.37)–(2.42), the following solution in the Laplace transform notation is obtained.

$$\{\bar{\theta}^*(r,s), \bar{p}^*(s), \bar{\varphi}^*(s), \bar{h}_U^*(s), \bar{\omega}_1^*(s)\} = \int_0^\infty \{\theta^*, p^*, \varphi^*, h_U^*, \omega_1^*\} e^{-s\tau} d\tau.$$

This gives the solution in the Laplace transform domain

$$\bar{\varphi}^*(s) = G_{\varphi h}(s)\bar{h}_U^*(s) + G_{\varphi\omega}(s)\bar{\omega}_1^*(s), \tag{2.43}$$

$$\bar{p}^*(s) = G_{ph}(s)\bar{h}_U^*(s) + G_{p\omega}(s)\bar{\omega}_1^*(s), \tag{2.44}$$

$$\bar{\theta}^*(r,s) = G_{\theta h}(r,s)\bar{h}_U^*(s) + G_{\theta\omega}(s)\bar{\omega}_1^*(s), \tag{2.45}$$

where

$$\{G_{\varphi h}(s), G_{\varphi\omega}(s), G_{ph}(s), G_{p\omega}(s), G_{\theta h}(s), G_{\theta\omega}(s)\} = \frac{L(s)}{\Delta^*(s)}, \tag{2.46}$$

$$\begin{aligned} L(s) = \{&F(\omega_1^\circ)\Delta_1(s), p_{st}[\gamma F^2(\omega_1^\circ)\Delta_2(s) + F'(\omega_1^\circ)\Delta_1(s)], \Omega_2(s)\Delta_1(s), \\ &\gamma\varepsilon(s^2+1)\Delta_2(s)(F(\omega_1^\circ) + \omega_1^\circ F'(\omega_1^\circ))p_{st}, I_0(s_\omega)\gamma\omega_1^\circ F(\omega_1^\circ)\Omega_1(s), \\ &\gamma\varepsilon I_0(s_\omega)(s^2+1)p_{st}(F(\omega_1^\circ) + \omega_1^\circ F'(\omega_1^\circ))\}. \end{aligned}$$

The characteristic equation of the linearized system is

$$\Delta^*(s) = 0, \qquad \Delta^*(s) = \Delta_1(s)\Omega_2(s) - 2Bi\ v\Delta_2(s)\Omega_1(s) \tag{2.47}$$

$$\Omega_1(s) = s^2 - \frac{\gamma_1}{1-v}s + 1, \qquad \Omega_2(s) = s^2 + \frac{\gamma_2}{1-v}s + 1, \qquad \gamma_2 = \varepsilon F'(\omega_1^\circ).$$

Observe that the roots s_m ($\mathrm{Re}s_1 > \mathrm{Re}s_2 > \cdots > \mathrm{Re}s_m > \cdots$, $m = 1, 2, 3, \ldots$) of the characteristic equation (2.47) lie either on the left-hand plane ($\mathrm{Re}s < 0$ and a stationary solution is stable) or on the right-hand plane ($\mathrm{Re}s > 0$ and a stationary solution is unstable) of the complex variable s. The parameters separating two half-planes are further referred to as critical ones. Let us analyse the stable stationary solution. In this case, the characteristic function has the form

$$\Delta^*(s) = \sum_{m=0}^{\infty} \left(\frac{s}{\tilde{\omega}}\right)^m d_m, \tag{2.48}$$

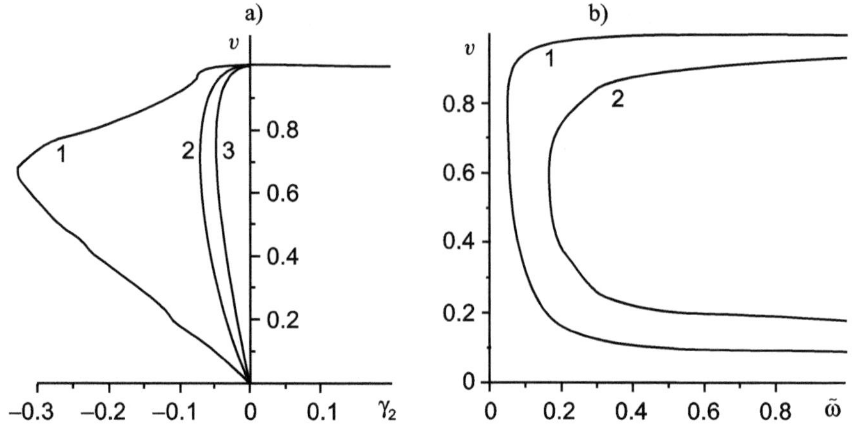

Fig. 2.6: Critical values of v versus γ_2 (a) and $\tilde{\omega}$ (b). (a) Curve 1 – $\tilde{\omega} = 0.05$, curve 2 – $\tilde{\omega} = 0.1$, curve 3 – $\tilde{\omega} = 1$. (b) Curve 1 – $\gamma_2 = -0.05$, curve 2 – $\gamma_2 = -0.08$.

$$d_m = (1 - \delta_{m1})\tilde{\omega}^2(d_{m-2}^{(1)} - 2Bi\ v d_{m-2}^{(2)}) + \frac{\tilde{\omega}}{1 - v}(\gamma_2 d_{m-1}^{(1)} + 2Bi\ v\gamma_1 d_{m-1}^{(2)}) + d_m^{(1)} - 2Bi\ v d_m^{(2)},$$

$$m = 1, 2, \ldots; \qquad \delta_{mn} = 1,\ m = n; \qquad \delta_{mn} = 0,\ m \neq n;$$

$$d_0 = Bi(1 - v), \qquad d_m^{(1)} = \frac{Bi + 2m}{2^{2m}(m!)^2}, \qquad d_m^{(2)} = \frac{1}{2^{2m+1}m!(1+m)!}.$$

Equation (2.47) has been studied numerically. Figure 2.6a shows the dependence of v on γ_2 for given values of $\tilde{\omega} = 0.05; 0.1; 1$ (curves 1–3). Figure 2.6b illustrates the dependence v on $\tilde{\omega}$ for fixed values of $\gamma_2 = -0.05; -0.08$ (curves 1–2). In both cases, $Bi = 10$ and $\gamma_1 = 0.586$. The parameters located inside the mentioned curves are associated with a stable stationary solution. Observe that decreasing γ_2, a stable zone is narrowed owing to v, whereas it is extended owing to an increase of the critical value of $\tilde{\omega}$.

When $v = 0$ and neglecting heat expansion of the cylinder, we obtain a model of self-oscillations [Andronov et al. (1966)], with the characteristic equation of the linearized problem $\Omega_2(s) = 0$. In this case, a steady-state solution is stable, when $\gamma_2 > 0$. When $\tilde{\omega} = 0$ (the bush is immovable) we obtain a model [Pyryev et al. (1995)] with the characteristic equation $\Delta_1(s) - 2Bi\ v\Delta_2(s) = 0$. In this case, when $v > 1$, a steady solution is unstable (root $s_1 > 0$). Frictional thermal instability [Pyryev, Grilitskiy (1995)] or thermal explosion [Aleksandrov, Annakulova (1990)] takes place.

The analysis of particular cases of the considered model shows that the steady-state solution is stable when $\gamma_2 > 0$. This corresponds to $V_r > V_{\min}$ (see Fig. 2.6). On the other hand, a steady solution is stable when $v < 1$. The specific parameters of the system, the view of Stribeck's curve, and analysis of the characteristic equation roots (2.47) can help definitely to answer the question of stability of the steady-state solution. If the steady-state solution (2.35) is unstable, then leaving a transient process it can reach either a stable limit cycle similar to frictional self-oscillations or it can increase with time (it depends on other nonlinear terms).

2.3.2.3 Analysis of steady-state solution in the presence of wear

The steady-state solution of the problem in the case of the presence of wear ($k^w \neq 0$) is found as (in Equations (2.5) and (2.18) the terms with time derivatives and $\dot{u}^w(\tau) = 0$ are neglected)

$$p_{st} = 0, \qquad \theta_{st} = 0, \qquad \varphi_{st} = 0, \qquad u^w_{st} = 1. \tag{2.49}$$

Performing a procedure similar to the previous case, the characteristic equation of the linearized problem is obtained in the vicinity of the steady solution (2.49):

$$\Delta^*(s) = 0, \qquad \Delta^*(s) = (s + k^w \omega_1^\circ)\Delta_1(s) - 2Bi\ vs\Delta_2(s). \tag{2.50}$$

To analyse the regions of parameters in which steady-state solution (2.49) is stable, a characteristic function is sought in the form

$$\Delta^*(s) = \sum_{m=0}^{\infty} \left(\frac{s}{\tilde{\omega}}\right)^m b_m, \tag{2.51}$$

$$b_0 = kd_0^{(1)}, \qquad k = k^w \omega_1^\circ,$$

$$b_m = kd_m^{(1)} + \tilde{\omega}(d_{m-1}^{(1)} - 2Bi\ vd_{m-1}^{(2)}), \qquad m = 1, 2, \ldots. \tag{2.52}$$

First, a root of the characteristic equation for small wear $k \ll 1$ can be presented in the form

$$s_1 = -\frac{k}{1 - v}. \tag{2.53}$$

As can be seen from (2.53), for low wear when $v < 1$, the steady solution (2.49) is stable and in the opposite case, unstable. In the conditions of wear presence contact time t_c of the system is limited. The material of the bush will wear with time. That is why we should accurately state the conditions of stability. Parameters of the problem under which roots s_1, s_2 of the characteristic equation have positive real part and are complex conjugate, are related to intensive wear. Under these parameters, the oscillation amplitude increases according to exponential law but the system contact time is limited. The system will leave the contact faster than the contact characteristics will reach critical values (initial assumptions will lose sense). Parameters of the problem under which the roots of the characteristic equation s_1, s_2 have only positive real part are referred to as the critical parameters.

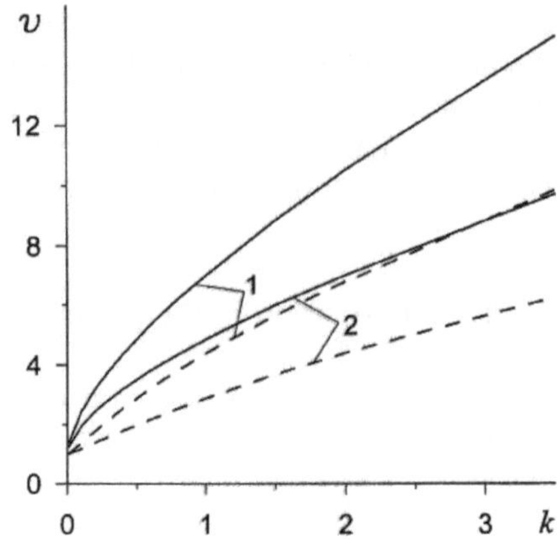

Fig. 2.7: Critical values v versus k for different $\tilde{\omega}$ ($k^w \neq 0$). Curves 1: $\tilde{\omega} = 0.05$, 2: $\tilde{\omega} = 0.1$. Regions under solid curves are stable. Parameter regions between solid and dashed curves correspond to intensive wear.

Under these parameters there will be no oscillations but contact characteristics will rise according to the exponential law. The system will be overheated. The rate of heat expansion is bigger than the wear rate. Figure 2.7 shows the dependence of critical value v (solid curves) on parameter k for different values of $\tilde{\omega} = 0.05, 0.1$ (curves 1 and 2, respectively).

Taking into account three terms in the decomposition (2.51) the formula for critical values is derived (roots coincide on real axis $\mathrm{Re}s_1 = \mathrm{Re}s_2 > 0$, $\mathrm{Im}s_1 = \mathrm{Im}s_2 = 0$),

$$v_{\mathrm{cr}} \approx 1 + \frac{k + \sqrt{2Bi\, k\tilde{\omega}(4 + Bi)}}{2Bi\, \tilde{\omega}}. \tag{2.54}$$

Formula (2.54) yields better results for small values of the wear parameter k. For the parameters that are situated under the solid curve $v < v_{cr}$, the system is stable. Moreover, between solid and dashed curves $v_0 < v < v_{cr}$, the time when the system is in contact is limited and intensive wear takes place. The formula for the dashed curve is also found (roots pass through an imaginary axis $\mathrm{Re}s_1 = \mathrm{Re}s_2 = 0$, $\mathrm{Im}s_1 = -\mathrm{Im}s_2 \neq 0$),

$$v_0 = 1 + \frac{k(2 + Bi)}{4Bi}. \tag{2.55}$$

Again formula (2.55) yields better results for small values of the wear parameter k.

2.3.3 *Numerical analysis of the transient solution*

A numerical analysis of the problem is performed using the Runge–Kutta method by taking into account the following asymptotes

$$\dot{G}_\theta(1,\tau) \approx \sqrt{1/\pi\tau\tilde{\omega}}, \qquad \dot{G}_p(\tau) \approx 1, \qquad \tau \to 0. \tag{2.56}$$

Dependence of the kinematic friction coefficient on relative velocity (see Fig. 2.2) dependence is approximated by the formula

$$f(V_r) = \mathrm{Sgn}(V_r) \begin{cases} f_{\min} + (f_s - f_{\min}) \exp(-b_1|V_r|), & \text{for } |V_r| < V_{\min}, \\ f_s, & \text{for } V_r = 0, \\ f_{\min} + (f_s - f_{\min}) \exp(-b_1|V_{\min}|) \\ \quad + \frac{b_2 b_3 (|V_r| - V_{\min})^2}{1 + b_2(|V_r| - V_{\min})}, & \text{for } |V_r| > V_{\min}, \end{cases} \tag{2.57}$$

where $f_s = 0.12$, $f_{\min} = 0.05$, $b_1 = 140\,\mathrm{sm}^{-1}$, $b_2 = 10\,\mathrm{sm}^{-1}$, $b_3 = 2\,\mathrm{sm}^{-1}$, $V_{\min} = 0.035\,\mathrm{mc}^{-1}$. Function $\mathrm{sgn}(x)$ is approximated by [Martins et al. (1990)]

$$\mathrm{sgn}_{\varepsilon_0}(x) = \begin{cases} 1, & \text{for } x > \varepsilon_0, \\ \left(2 - \frac{|x|}{\varepsilon_0}\right) \frac{x}{\varepsilon_0}, & \text{for } |x| < \varepsilon_0, \\ -1, & \text{for } x < -\varepsilon_0, \end{cases} \tag{2.58}$$

where the regularization parameter $\varepsilon_0 = 0.0001$.

The results of calculations for different values of the parameter $\gamma = 0, 200, 400$ for the steel cylinder ($\alpha_1 = 14 \times 10^{-6\circ}\mathrm{C}^{-1}$, $\lambda_1 = 21\,\mathrm{W\,m}^{-1\circ}\mathrm{C}^{-1}$, $\nu_1 = 0.3$, $a_1 = 5.9 \times 10^{-6}\,\mathrm{m}^2\,\mathrm{s}^{-1}$, $E_1 = 19 \times 10^{10}\,\mathrm{Pa}$) and $R_1 = 0.03\,\mathrm{m}$, $\Omega_*\omega_1 = 1\,\mathrm{rad\,s}^{-1}$, $\varepsilon = 10$, $\tilde{\omega} = 0.1$, $Bi = 10$, $\varphi^\circ = 0$, $\omega^\circ = 0$ are shown in Figs. 2.8 to 2.12. Solid curves correspond to the case of wear absence $k^w = 0$, dashed – to the case of wear with the dimensionless wear coefficient $k^w = 0.1$. In this case $t_* = 0.25\,\mathrm{s}$, $P_* = 1.22 \times 10^4\,\mathrm{Pa}$, $\gamma_1 = 0.51$, $\gamma_2 = -196$.

In the case of heat expansion absence time evolutions of the dimensionless speed $\dot{\varphi}(\tau)$ (curve 1), dimensionless displacement $\varphi(\tau)$ (curve 2) of the bush, and dimensionless friction force $\varepsilon F(\omega_1^\circ - \dot{\varphi})p(\tau)$ (dashed curve 3) for $\gamma = 0$, ($v = 0$, $s_1 = 0.0051$ the steady-state solution is unstable) are shown in Fig. 2.8.

Figure 2.9a illustrates the behaviour of the phase trajectory in the phase plane without taking into account heat expansion of the cylinder $\gamma = 0$ and wear of the bush $k^w = 0$ (solid curve) and with wear (dashed curve). In the conditions of wear absence, a transient solution leads with time to a stable limit cycle with the period 6.608. The stick-slip phenomenon takes place.

Let us point out that (as can be seen in this figure) the friction force has a jump at the instant of time when the bush speed reaches the speed of the cylinder $V_r = 0$. At that moment the rest friction force changes the sign. Then the dimensionless friction force becomes equal to dimensionless displacement and this lasts to the moment when it reaches a maximal value of the static friction force. From that instant of time the friction force decreases. After that, it reaches a local minimum when the bush acceleration is equal to zero. Then, the friction force increases till it

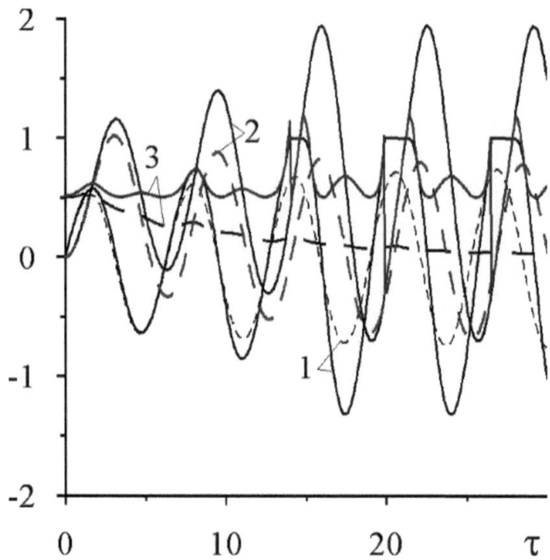

Fig. 2.8: Dependence of the bush movement dimensionless velocity $\dot{\varphi}$ (curve 1), dimensionless displacements φ (curves 2), and dynamic friction force $\varepsilon F(\omega_1^\circ - \dot{\varphi})$ (curve 3) versus dimensionless time τ in conditions of heat expansion absence $\gamma = 0$. Solid curves: $k^w = 0$, dashed curves: $k^w = 0.1$.

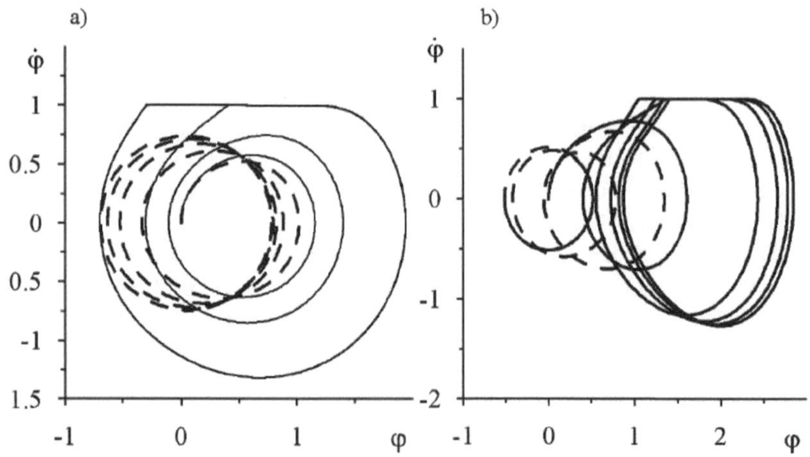

Fig. 2.9: Phase trajectory of bush movement in conditions of heat expansion absence $\gamma = 0$ (a) and $\gamma = 200$ (b). Solid curves: $k^w = 0$, dashed curves: $k^w = 0.1$.

reaches cylinder velocity $V_r = 0$. The process continues cyclically. In the presence of wear the friction force will tend to zero with time and the bush will perform its own oscillations with the period of 2π.

Time evolutions of the dimensionless speed $\dot{\varphi}$ (curve 1), dimensionless displacement $\varphi(\tau)$ (curve 2) of the bush, and dimensionless dynamic friction force

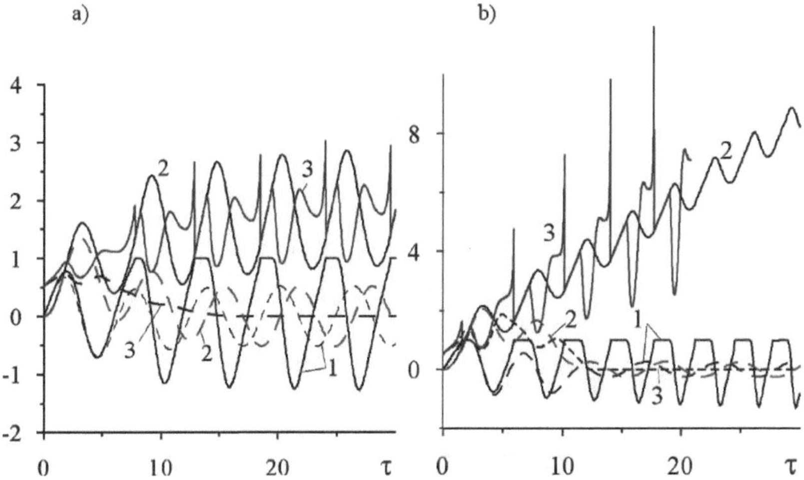

Fig. 2.10: Dependence of bush dimensionless velocity $\dot{\varphi}$ (curve 1), dimensionless displacements φ (curve 2), and dynamic friction force $\varepsilon F(\omega_1^\circ - \dot{\varphi})$ (curve 3) on dimensionless time τ, for $\gamma = 200$ (a) and $\gamma = 400$ (b). Solid curves: $k^w = 0$, dashed curves: $k^w = 0.1$.

$\varepsilon F(\omega_1^\circ - \dot{\varphi})p(\tau)$ (dashed curve 3) are shown in Fig. 2.10a for $\gamma = 200$ (for $v = 0.51$, $s_1 = 0.0012$ the steady-state solution is unstable). Figure 2.9b illustrates the behaviour of the phase trajectory in the phase plane without taking into account heat expansion of the cylinder $\gamma = 200$ and $k^w = 0$ (solid curve) and with wear (dashed line). As can be seen, in the absence of bush wear, the contact characteristics tend with time to the limit cycle with the period equal to 5.5.

Figure 2.10b illustrates time changes of the dimensionless speed $\dot{\varphi}$ (curve 1), dimensionless displacement $\varphi(\tau)$ (curve 2) of the bush, and dimensionless dynamic friction force $\varepsilon F(\omega_1^\circ - \dot{\varphi})p(\tau)$ (dashed curve 3) for $\gamma = 400$ ($v = 1.02$, $s_1 = 2 \times 10^{-6}$). Figure 2.11 illustrates the behaviour of the phase trajectory on the phase plane without taking into account the heat expansion of the cylinder $\gamma = 400$ and $k^w = 0$ (solid curve) and with wear (dashed curve). In the case of wear absence thermoelastic instability takes place. System characteristics do not tend to stable limit cycles with time but exponentially increase. In that case, the cylinder does not have enough time to become cool. Wear presence (dashed curves) leads to loss of the thermoelastic instability.

Time-changing laws for the contact pressure, temperature, and wear are shown in Figs. 2.12 by solid (dashed) curves for the case of wear absence (presence), respectively. Curve 1 corresponds to the case of absence of heat expansion $\gamma = 0$, curve 2 to $\gamma = 200$, curve 3 to $\gamma = 400$. In the last case, contact characteristics increase and the cylinder does not have time for cooling down.

Although parameter $\tilde{\omega}$ decreases and $v < 1$, the time necessary for the leading of contact characteristics on the limit cycle increases. The wear presence leads to decreasing of contact characteristic values and to limiting of system contact time (curves 2 and 3 in Fig. 2.12). With an increase of the parameter γ the system

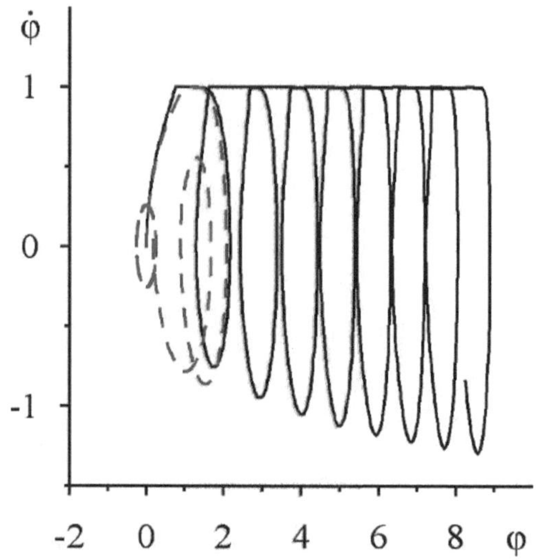

Fig. 2.11: Phase trajectory of bush movement in conditions for $\gamma = 400$. Solid curves: $k^w = 0$, dashed curves: $k^w = 0.1$.

contact time decreases and, correspondingly, the bush wear increases and its value becomes bigger than the value of initial deformation caused by an initial stress state of the cylinder.

A new model of thermoelastic contact is considered, in which changes of the bush movement velocity, contact pressure, friction force, contact temperature, and wear are mutually connected, are considered. Conditions in which frictional self-oscillations arise are investigated. In the wear absence these conditions being at the same time the conditions of steady-state solution stability are also the conditions of the linearized problem. The characteristic equation roots transition from the left-hand side ($\mathrm{Re}s_1 < 0$) to the right-hand side ($\mathrm{Re}s_1 > 0$) of the Laplace transform as regards the parameter complex plane associated with the Hopf bifurcation. In conditions of wear presence stick-slip movements disappear with time. The system contact time is limited. The upper bound of the solution stability conditions can be the conditions of coinciding of first roots of characteristic equation $\mathrm{Im}s_1 = \mathrm{Im}s_2 = 0$, $\mathrm{Re}s_1 = \mathrm{Re}s_2 > 0$ on the right-hand side of the complex plane. Taking into account heat expansion of the cylinder and wear of the bush extends regions for the parameters in which the steady-state solution is stable; that is, heat expansion and wear play the role of stabilising factors.

The possibility of occurrence of self-excited vibrations for the considered system with a lack of tribological processes ($k_1 = 0$, $k_3 = 0$, $\gamma = 0$) has been pointed out by [Andronov et al. (1966), Neimark (1978)].

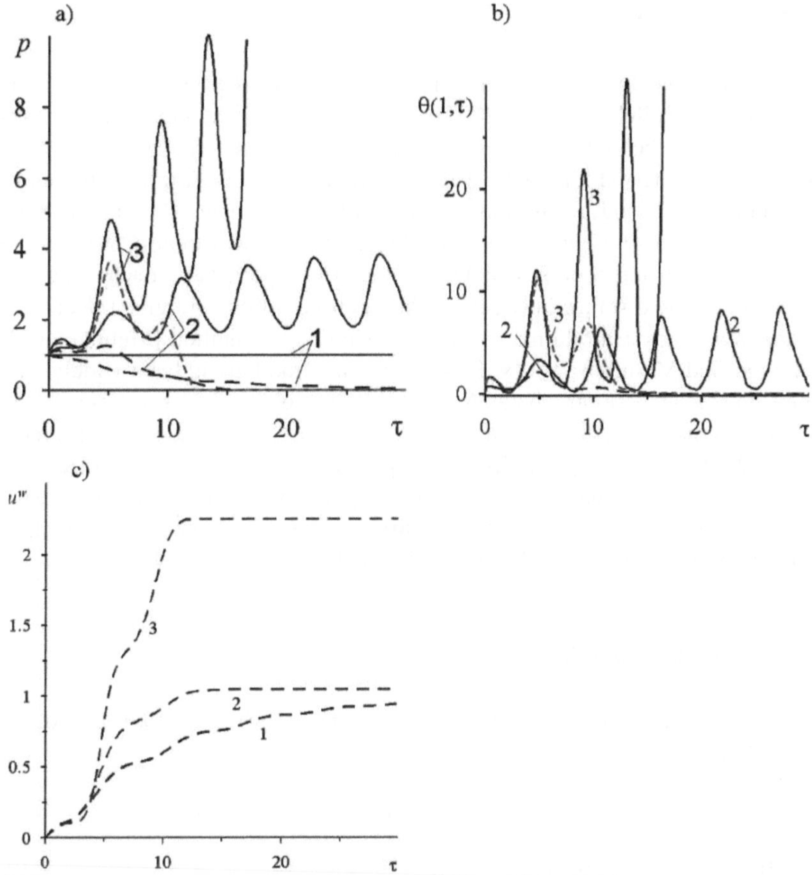

Fig. 2.12: Behaviour of dimensionless contact pressure p (a), dimensionless contact temperature $\theta(1,\tau)$ (b), and time evolution of dimensionless wear u^w (c) versus dimensionless time τ for different values of γ. Curve 1: $\gamma = 0$, 2: $\gamma = 200$, 3: $\gamma = 400$. Solid curves: $k^w = 0$; dashed curves: $k^w = 0.1$.

2.3.4 Chaotic motion of the shaft/bush with kinematic external excitations

In this section, vibrations of a rigid body (bush, pad) attached on the rotating shaft at nonconstant velocity and initially without taking account of tribological processes are considered ($\gamma = 0$, $k^w = 0$). It is assumed that an arbitrary parameter $k_* = k_1(l_0/l_1 - 1)(1 + l_1/R_2) - k_2$, and $\chi = 1$, which yields $a = 1$, $t_* = \sqrt{B_2/(k_* R_2^2)}$. In addition, we assume that $h_U(\tau) = H(\tau)$, $U_* = U_0$, $h_T(\tau) = 0$. In this case, the dimensionless contact pressure $p(\tau) = 1$. In order to predict the bush chaotic dynamics, Melnikov's technique is applied. The analytical form of Melnikov's function is formulated, and then a numerical analysis is performed.

2.3.4.1 Introduction

Because nonlinear dynamic systems may exhibit either regular or chaotic motions [Awrejcewicz, Holicke (1999),(2007)], one of the recent challenging tasks is focused on their behaviour control. Such control can be realized, for instance, by applying an external kinematic excitation (perturbation). Although the majority of the methods aiming at analysis and control of Nonlinear Dynamical Systems (NDS) are realized via numerical algorithms, an analytical treatment seems to be the most powerful and economical. One of the often-applied approaches that allows us to calculate the distance between the homoclinic orbits, and give the conditions of chaos in a nearly Hamiltonian system, is Melnikov's method [Melnikov (1963)] or the modified Melnikov theory. Despite this method having been widely applied in the analysis of smooth dynamic systems, its successful application to analyse simple dynamic systems with friction has been illustrated only quite recently [Grudziński, Wedman (1998)]. The latter approach has even been extended to study more complicated regular and chaotic stick-slip dynamics of a rotating shaft with a rigid bush in wear and heat transfer conditions. Numerical experiments confirmed well the analytical prediction based on the analysis of Melnikov's function.

2.3.4.2 Melnikov function

For the considered case, differential equation (inclusion) (2.5) takes the following nondimensional form.

$$\ddot{\varphi}(\tau) + 2h\dot{\varphi}(\tau) - \varphi(\tau) + b\varphi^3(\tau) \in \varepsilon F(\omega_1 - \dot{\varphi}), \qquad 0 < \tau < \infty, \tag{2.59}$$

$$\varphi(0) = \varphi^\circ, \qquad \dot{\varphi}(0) = \dot{\varphi}^\circ = \omega^\circ. \tag{2.60}$$

Dependence of friction on relative velocity is approximated by the function

$$F(y) = f_s \mathrm{Sgn}(y) - \alpha y + \beta y^3, \qquad \text{where } \mathrm{Sgn}(y) = \begin{cases} \{y/|y|\} & \text{for } y \neq 0, \\ [-1, 1] & \text{for } y = 0. \end{cases} \tag{2.61}$$

The dimensionless angular bush velocity is governed by the equation

$$\omega_1 = \omega_* + \zeta_k \sin \omega_0 \tau, \tag{2.62}$$

where ζ_k is the dimensionless amplitude of the kinematic external excitation.

In order to apply the combined Melnikov's and numerical methods, a perturbation of the Hamiltonian system, where the inclusion $\mathrm{Sgn}(y)$ occurs, has been approximated by continuous perturbation with the application of a small parameter. The multivalued relation $\mathrm{Sgn}(y)$ is approximated by the function $\mathrm{sgn}_{\varepsilon_0}(y)$ defined by (2.58), where the regularization parameter ε_0 is a 'small' positive real number. The differential inclusion (2.59) is then approximated by the equation.

In the so-called first improvement of Melnikov's function $M(\tau_0)$ (see the expression near ε) for $0 < \varepsilon \ll 1$ of the expression representing a distance between stable and unstable manifolds of the critical saddle point, a transition of the parameter ε_0 to zero ($\varepsilon_0 \to 0$) can be realized. In order to be sure of neglecting the so-called second improvement of Melnikov's function near ε^2 (the function under the integral includes the differential of the approximated perturbation), the following condition should be satisfied, $\varepsilon/\varepsilon_0 \ll 1$. Then if the mentioned condition is satisfied only the first improvement of Melnikov's function can be applied to estimate the distance between stable and unstable manifolds of the critical point.

Introducing the new variables

$$x = \varphi, \qquad y = \dot{\varphi}, \tag{2.63}$$

the equations of motion can be transformed into the form,

$$\begin{aligned} \dot{x} &= p_0(x, y) + \varepsilon p_1(x, y, \omega_0\tau, \varepsilon), \\ \dot{y} &= q_0(x, y) + \varepsilon q_1(x, y, \omega_0\tau, \varepsilon), \end{aligned} \tag{2.64}$$

where

$$\begin{aligned} &q_0(x, y) = x - bx^3, \qquad q_1(x, y, \omega_0\tau, \varepsilon) = F(\omega_* + \zeta_k \sin\omega_0\tau - y) - h_1 y, \\ &p_0(x, y) = y, \qquad p_1(x, y, \omega_0\tau, \varepsilon) = 0, \qquad h_1 = 2h/\varepsilon. \end{aligned} \tag{2.65}$$

For a sufficiently small parameter $0 < \varepsilon \ll 1$,

$$x_0(\tau) = \sqrt{\frac{2}{b}}\frac{1}{\cosh(\tau)}, \qquad y_0(\tau) = -\sqrt{\frac{2}{b}}\frac{\sinh(\tau)}{\cosh^2(\tau)}, \tag{2.66}$$

Melnikov's function is defined by the formula [Awrejcewicz (1996)]

$$M(\tau_0) = \int_{-\infty}^{+\infty} (q_0 p_1 - q_1 p_0)\big|_{\substack{x=x_0(\tau-\tau_0)\\ y=y_0(\tau-\tau_0)}} d\tau = -\int_{-\infty}^{+\infty} q_1 p_0\big|_{\substack{x=x_0(\tau-\tau_0)\\ y=y_0(\tau-\tau_0)}} d\tau, \tag{2.67}$$

where $x_0(\tau)$, $y_0(\tau)$ is the solution of a nondisturbed system of equations ($\varepsilon = 0$), which corresponds to the homoclinic orbits, and τ_0 is the parameter that characterizes positions of the point moving in this orbit. In accordance with Melnikov's theory, if the function $M(\tau_0)$ has simple zeros, then for a sufficiently small parameter ε, the motion governed by system (2.64) can be chaotic.

Introducing the change of variable $\tau - \tau_0 = t$, Melnikov's function is

$$M(\tau_0) = -\int_{-\infty}^{+\infty} y_0(t)[f_s \mathrm{Sgn}(\omega_r) - \alpha\omega_r + \beta\omega_r^3]dt, \tag{2.68}$$

where the dimensionless relative velocity has the form

$$\omega_r(t) = \omega_* + \zeta_k \sin(\omega_0(t + \tau_0)) - y_0(t). \tag{2.69}$$

The substitution of (2.66) and (2.69) into (2.68) gives Melnikov's function of the form

$$M(\tau_0) = I(\tau_0) + J(\tau_0), \tag{2.70}$$

where

$$\begin{aligned}
J(\tau_0) &= 2C + 2\zeta_k\sqrt{A^2 + B^2}\sin(\omega_0\tau_0 + \varphi) \\
&\quad + 6\beta\zeta_k^2(I_{220}\cos^2\omega_0\tau_0 + I_{202}\sin^2\omega_0\tau_0 - 2\omega_* I_{111}\sin\omega_0\tau_0\cos\omega_0\tau_0) \\
&\quad + 2\beta\zeta_k^3(-I_{130}\cos^3\omega_0\tau_0 - 3I_{112}\sin^2\omega_0\tau_0\cos\omega_0\tau_0), \\
A &= (\alpha - 3\beta\omega_*^2)I_{110} - 3\beta I_{310}, \quad B = 6\beta\omega_* I_{201}, \\
C &= \beta I_{400} - (\alpha - h_1 - 3\beta\omega_*^2)I_{200}, \quad \varphi = \arctan(A/B), \\
I_{njk} &= \int_0^\infty [y_0(t)]^n[\sin(\omega_0 t)]^j[\cos(\omega_0 t)]^k dt.
\end{aligned} \tag{2.71}$$

After integration of Equation (2.71) we obtain

$$I_{200} = \frac{2}{3b}, \; I_{400} = \frac{8}{35b^2}, \; I_{201} = \frac{\pi\omega_0(2 - \omega_0^2)}{6b\sinh(\pi\omega_0/2)}, \; I_{110} = -\frac{\pi\omega_0}{\sqrt{2b}\cosh(\pi\omega_0/2)},$$

$$\begin{aligned}
I_{310} &= \frac{\omega_0(11 + 10\omega_0^2 - \omega_0^4)}{120b\sqrt{2b}}\left\{\psi\left(\frac{1 - i\omega_0}{4}\right) - \psi\left(\frac{3 - i\omega_0}{4}\right)\right. \\
&\quad \left. + \psi\left(\frac{1 + i\omega_0}{4}\right) - \psi\left(\frac{3 + i\omega_0}{4}\right)\right\}, \\
I_{130} &= -\frac{3\pi\omega_0}{8\sqrt{2b}}\left\{\cot\left(\frac{\pi(1 - i\omega_0)}{4}\right) + \cot\left(\frac{3\pi(1 - i\omega_0)}{4}\right)\right. \\
&\quad \left. - \cot\left(\frac{\pi(3 - i\omega_0)}{4}\right) - \cot\left(\frac{\pi(1 - 3i\omega_0)}{4}\right)\right\},
\end{aligned}$$

$$I_{112} = \frac{\pi\omega_0\cosh(\pi\omega_0/2)}{\sqrt{2b}(1 - 2\cosh(\pi\omega_0))}, \qquad I_{111} = -\frac{\pi\omega_0}{\sqrt{2b}\cosh(\pi\omega_0)},$$

$$I_{220} = \frac{\pi\omega_0(2\omega_0^2 - 1) + \sinh(\pi\omega_0)}{3b\sinh(\pi\omega_0)},$$

$$I_{202} = \frac{\pi\omega_0(1-2\omega_0^2)+\sinh(\pi\omega_0)}{3b\sinh(\pi\omega_0)}, \qquad \psi(z) = \frac{\Gamma'(z)}{\Gamma(z)},$$

where $\Psi(z)$ denotes the derivative of the natural logarithm of the function $\Gamma(z)$.

In (2.70) the term $I(\tau_0)$ is defined by the formula

$$I(\tau_0) = -f_s \int\limits_{-\infty}^{+\infty} y_0(t)\mathrm{Sgn}(\omega_r)dt = 2f_s\sqrt{\frac{2}{b}}\sum_m \frac{\mathrm{sgn}(\omega_r'(t_m))}{\cosh t_m}, \tag{2.72}$$

where t_m are the roots of the equation

$$\omega_r(t_m) = \omega_* + \zeta_k \sin(\omega_0(t_m+\tau_0)) - y_0(t_m) = 0, \tag{2.73}$$

whereas $\omega_r'(t) = \zeta_k\omega_0\cos(\omega_0(t+\tau_0)) - x_0(t) + bx_0^3(t)$.

If Melnikov's function (2.70) changes its sign, then one may expect chaos. Observe that in our case, Melnikov's function has a rather complex structure and its direct theoretical analysis is not easy. Therefore, in this section our considerations are limited to the analysis of the function $J(\tau_0)$ for large values of b and small values of ζ_k ($h_1 = 0$) of the form

$$\begin{aligned} J(\tau_0) = &\left(-\frac{4}{3}(\alpha - 3\beta\omega_*^2) + \frac{16}{35}\frac{\beta}{b} + O\left(\frac{1}{b^2}\right)\right)\frac{1}{b} \\ &+ \left[-\frac{\sqrt{2}\pi\omega_0(\alpha-3\beta\omega_*^2)}{\cosh(\pi\omega_0/2)}\cos(\omega_0\tau_0)\right. \\ &\left. - \frac{2\pi\omega_0\beta\omega_*(\omega_0^2-2)}{\sqrt{b}\sinh(\pi\omega_0/2)}\sin(\omega_0\tau_0) + O\left(\frac{1}{b}\right)\right]\frac{1}{\sqrt{b}}\zeta_k + O(\zeta_k^2). \end{aligned} \tag{2.74}$$

Equation (2.74) provides an observation that the function $J(\tau_0)$ changes its sign (for large b and small ζ_k) when the following inequality holds

$$\zeta_{ch}^{(1)} < \zeta_k, \quad \text{where } \zeta_{ch}^{(1)} = \frac{4\cosh(\pi\omega_0/2)}{3\sqrt{2b}\pi\omega_0}. \tag{2.75}$$

Note that for large values of b and small values of ζ_k we have $I(\tau_0) \equiv 0$, and the remaining terms of $J(\tau_0)$ in (2.70) are small. The value of the mentioned terms changes its sign when (2.75) holds. For a certain value of ζ_k, afterwards referred to as ζ_{nz}, the value of integral $I(\tau_0)$ is not equal to zero, and starts to play a dominant role in $M(\tau_0)$. The function $M(\tau_0)$ begins to change sign when $\zeta_k \geq \zeta_{nz}$ and when $I(\tau_0)$ is not equal to zero (the function $\omega_r(t)$ starts to change its sign). One may find the corresponding estimated value of the parameter ζ_k with the use of the following formula,

$$\zeta_{ch}^{(2)} < \zeta_k, \quad \text{where } \zeta_{ch}^{(2)} \approx \zeta_{nz} = \omega_* - 1/\sqrt{2b}. \tag{2.76}$$

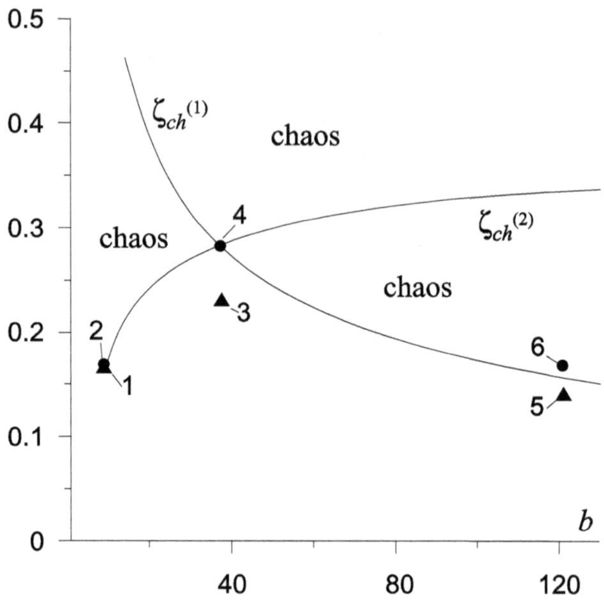

Fig. 2.13: Chaotic threshold in the (ζ_k, b) plane ($\varepsilon = 0.1$, $\omega_0 = 2$, $\omega_* = 0.4$, $f_s = 0.3$, $\alpha = 0.3$, $\beta = 0.3$).

Using both formulas (2.75) and (2.76), the functions $\zeta_{ch}^{(i)}$, $i = 1, 2$ versus the parameter b are shown in Fig. 2.13. One may conclude that the region corresponding to chaos occurs above the curves.

2.3.4.3 Numerical results

In order to verify the analytical conditions in which chaotic oscillations occur, some numerical tests have been carried out. The following initial condition $x(0) = 0$, $y(0) = 0$ and the following parameters $\varepsilon = 0.1$, $\omega_0 = 2$, $\omega_* = 0.4$, $h_1 = 0$, $f_s = 0.3$, $\alpha = 0.3$, $\beta = 0.3$ are fixed.

Arbitrary values of *b*. Formula (2.70) holds for any value of b. The numerically obtained Melnikov's function $M(\tau)$ is shown in Fig. 2.14 for $b = 1$.

Because Melnikov's function can change its sign (it has simple zeros) for $\xi_{ch} \approx 3.6$, chaos may occur according to Melnikov's theory. To confirm the analytical prediction, the bifurcational diagram $x(\zeta_k)$ has been constructed (projection of the Poincaré section into the x-axis for $\zeta_k \in (0, 12)$ and $\zeta_k \in (3.5, 4.0)$). The obtained results are shown in Fig. 2.15.

Observe that for $\zeta_k < \omega_*$ ($\omega_* = 0.4$) the periodic motion occurs with free frequency of the system, which undergoes changes as ζ_k increases. However, this motion vanishes when $\zeta_k \to \omega_*$. With the increase of the parameter ζ_k, period doubling occurs for $\xi_{ch} \approx 3.6$, and a bifurcation cascade leading to chaos follows.

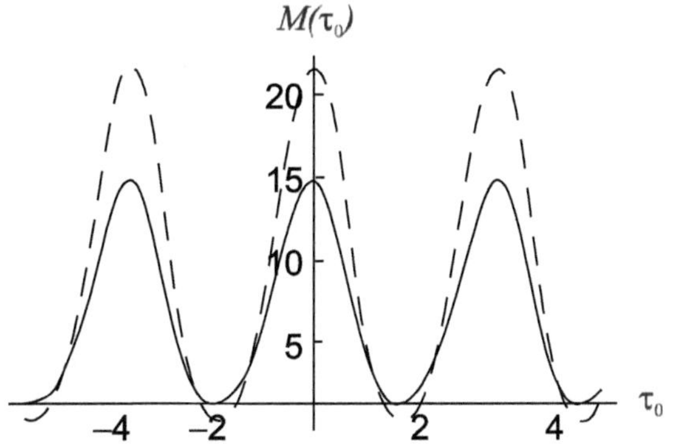

Fig. 2.14: Melnikov's function $M(\tau_0)$ versus parameter τ_0 for $\zeta_k = 3.2$ (solid curves) and for $\zeta_k = 3.81$ (dashed curves), $b = 1$.

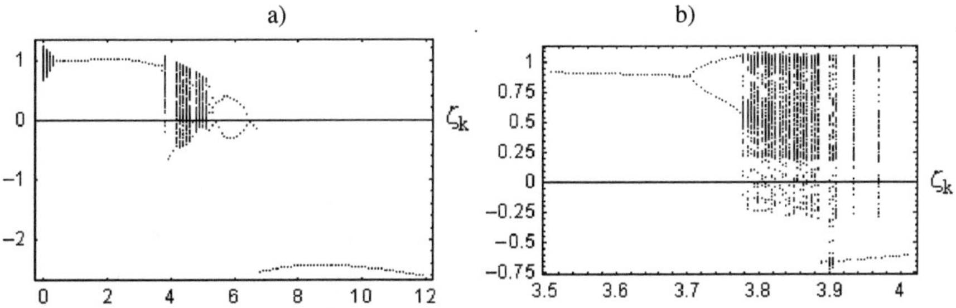

Fig. 2.15: Bifurcation diagrams using ζ_k as a control parameter; $h_1 = 0, \quad \gamma = 0, \quad k^w = 0:$ (a) $\zeta_k \in (0, 12)$; (b) $\zeta_k \in (3.5, 4.0)$.

In addition, for the same parameters the Fast Fourier Transform (FFT) of the process $x(\tau)$ for large values of $\tau \in (\tau_1, \tau_N)$ is reported. The computational results of the obtained power spectrum are shown in Fig. 2.16. The following relations are applied to estimate power spectra.

$$L(\omega) = 20 \log |X(\omega)|, \qquad X(\omega) = X(\Delta_\omega m) = X_m,$$

$$\Delta_\omega = 2\pi/(\Delta_\tau N), \qquad m = 1, 2, 3, \ldots, N,$$

$$X_m = \sum_{n=1}^{N} (x_n - x_0) e^{-2\pi i (n-1)(m-1)N},$$

$$x_n = x(n\Delta_\tau), \qquad n = 1, 2, 3, \ldots, N, \qquad x_0 = \frac{1}{N} \sum_{n=1}^{N} x_n.$$

The following values are fixed during computations: $N = 4000$ and $\Delta_\tau = 0.02$.

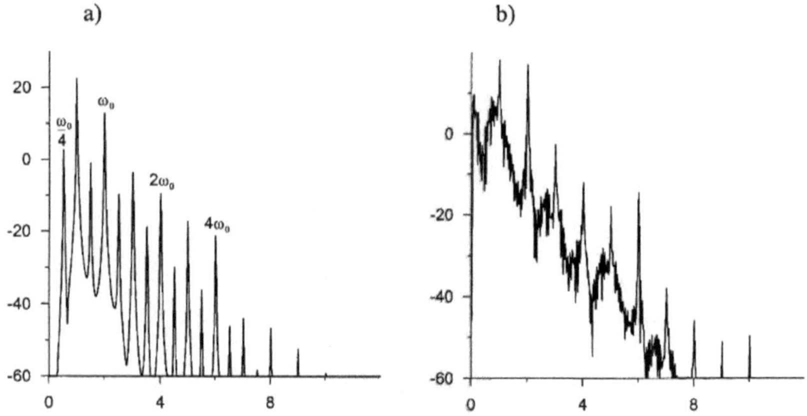

Fig. 2.16: Power spectrum: (a) $\zeta_k = 3.2$; (b) $\zeta_k = 3.81$.

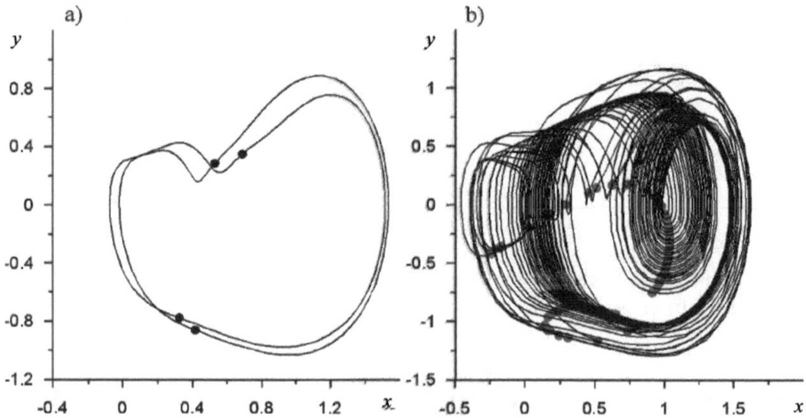

Fig. 2.17: Phase planes and Poincaré sections: (a) $\zeta_k = 3.2$; (b) $\zeta_k = 3.81$.

Observe that for $\zeta_k = 3.2$ the periodic-four motion occurs, whereas for $\zeta_k = 3.81$ the motion is chaotic. The phase planes and Poincaré sections for $\zeta_k = 3.2$ (Fig. 2.17a) and for $\zeta_k = 3.81$ (Fig. 2.18b) are shown. The points of Poincaré maps are obtained with respect to the period of kinematic excitation $2\pi/\omega_0$.

Large values of *b*. In general, for large values of the parameter b the values of ζ_k responsible for chaos are smaller in comparison with the parameter ω_*. As the numerical analysis shows, for $\zeta_k \in [0, \omega_*)$ the system motion is periodic with the free system frequency ω_d. For example, for fixed value of $b = 1$, the following values of ω_d are found; $\zeta_k = 0$ gives $\omega_d \approx 0.902$; $\zeta_k = 0.1$ gives $\omega_d \approx 0.805$; $\zeta_k = 0.2$ gives $\omega_d \approx 0.675$; $\zeta_k = 0.3$ gives $\omega_d = 1.43175$; $\zeta_k = 0.4$ gives $\omega_d = \omega_0 = 2$.

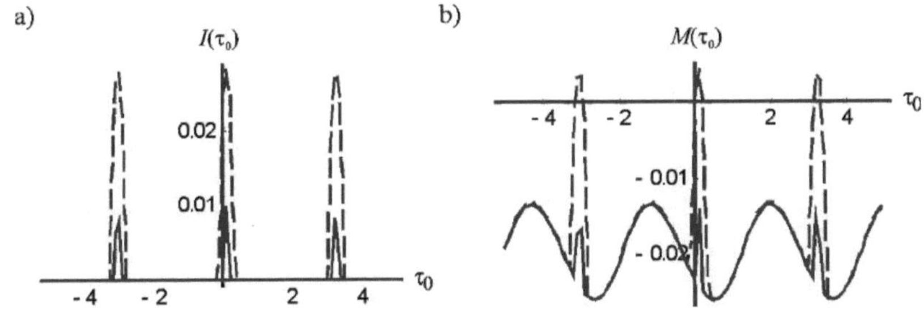

Fig. 2.18: Function $I(\tau_0)$ (a) and Melnikov's function $M(\tau_0)$ (b) versus the parameters τ_0 for $\zeta_k = 0.165$ (solid curves) and $\zeta_k = 0.17$ (dashed curves); $b = 9$.

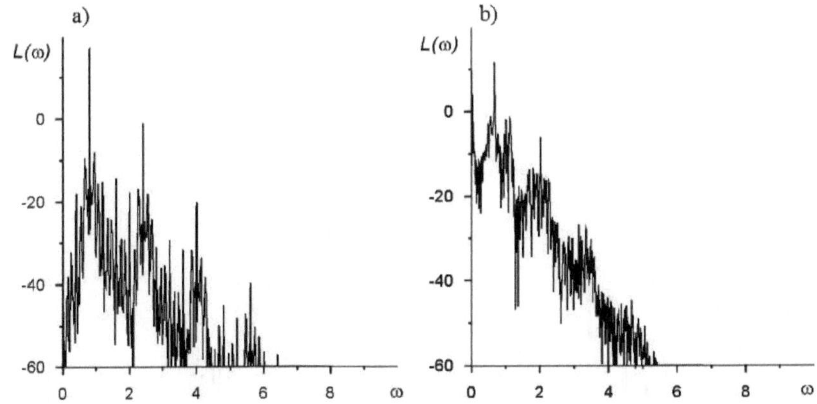

Fig. 2.19: Power spectrum for $b = 9$: (a) $\zeta_k = 0.1$; (b) $\zeta_k = 0.2$.

Numerical analysis of Melnikov's function is carried out for the parameters associated with points 1–6 lying in the chaotic area (points 2, 4, 6) and out of chaos (points 1, 3, 5); see Figs. 2.18, 2.20 and 2.22. For $\zeta_k < \zeta_{nz}$, we have $I(\tau_0) = 0$. The obtained values of ζ_k give $\zeta_{nz} = 0.164$ (formula (2.76)) and $\zeta_{ch}^{(1)} = 0.58$ (formula (2.75)). Figure 2.18 shows the dependence of the function $I(\tau_0)$ (a) and Melnikov's function $M(\tau_0)$ (b) on the parameter τ_0 for point 1 ($b = 9$, $\zeta_k = 0.165$; solid curves) and for point 2 ($b = 9$, $\zeta_k = 0.17$; dashed curves) from Fig. 2.13. For the mentioned parameters condition (2.75) is satisfied. Although for $\zeta_k = 0.165$ the function $I(\tau_0)$ has nonzero values, the function $M(\tau_0)$ does not change its sign yet. The numerical analysis shows that the function $M(\tau_0)$ changes its sign for $\zeta_{ch}^{(2)} \approx 0.168$. Hence, both obtained values of ζ_{nz} and $\zeta_{ch}^{(2)}$ are close to each other. The (numerical) analysis confirms the value of (2.76) used to obtain the parameter $\zeta_{ch}^{(2)}$ responsible for the occurrence of chaos according to Melnikov's rule.

Numerical analysis of Equation (2.64) is carried out for the parameters in the vicinity of points 1 and 2 (see Fig. 2.13). Figure 2.19 illustrates the power spectra for the parameter $\zeta_k = 0.1$ (it corresponds to the point situated below point 1

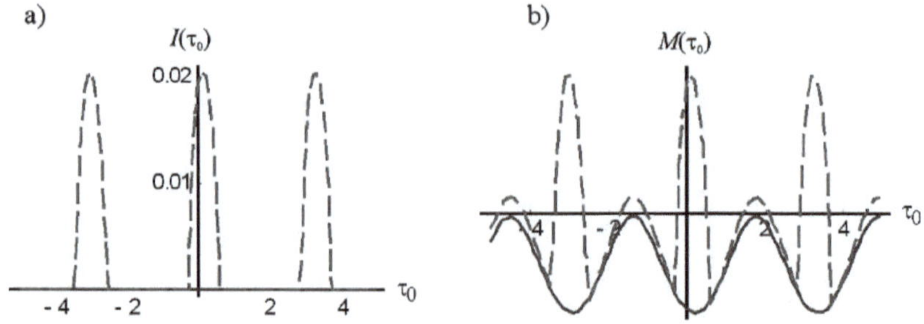

Fig. 2.20: $I(\tau_0)$ (a) and Melnikov's $M(\tau_0)$ (b) functions versus the parameter τ_0 for $\zeta_k = 0.23$ (solid curves) and for $\zeta_k = 0.30$ (dashed curves); $b = 37.41$.

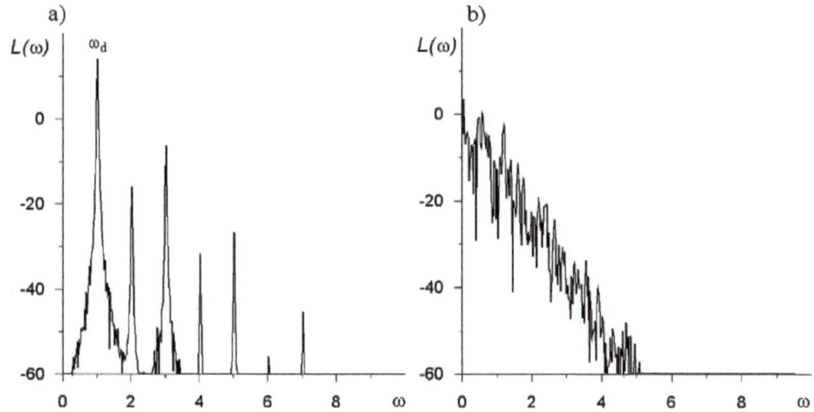

Fig. 2.21: Power spectra for $b = 37.41$: (a) $\zeta_k = 0.23$; (b) $\zeta_k = 0.29$.

and is associated with the quasi-periodic motion of the bush), and for $\zeta_k = 0.2$ (it corresponds to the point lying above point 2 and is associated with chaos).

Note that for $b = 37.41$ we have $\zeta_{ch}^{(1)} = \zeta_{ch}^{(2)}$ (see Fig. 2.13). Figure 2.20 shows $I(\tau_0)$ (a) and Melnikov's (b) $M(\tau_0)$ functions versus parameter τ_0 for the parameters associated with point 3 (see Fig. 2.13) ($b = 37.41$, $\zeta_k = 0.23$; solid curves) and with point 4 ($b = 37.41$, $\zeta_k = 0.30$; dashed curves). Although the function $I(\tau_0)$ is equal to zero, the function $M(\tau_0)$ still does not change its sign ($I(\tau_0) = 0$ for $\zeta_k < \zeta_{ch}^{(2)}$). It begins to change the sign for $\zeta_k = 0.30$. Therefore, the numerical tests confirm the values defined by Equations (2.75) and (2.76) and giving the parameters $\zeta_{ch}^{(1)}$ and $\zeta_{ch}^{(2)}$ responsible for chaos occurrence, according to Melnikov's theory.

The numerical tests of Equations (2.64) are performed for the parameter sets associated with points 3 and 4 (Fig. 2.13). Figure 2.21 presents the power spectra for the parameters $\zeta_k = 0.23$ (point 3 corresponds to the 2π-periodic bush motion) and $\zeta_k = 0.29$ (point 4 corresponds to the chaotic motion). Chaotic behaviour in the vicinity of the homoclinic orbits (2.66) is clearly visible for $\zeta_k = 0.29$.

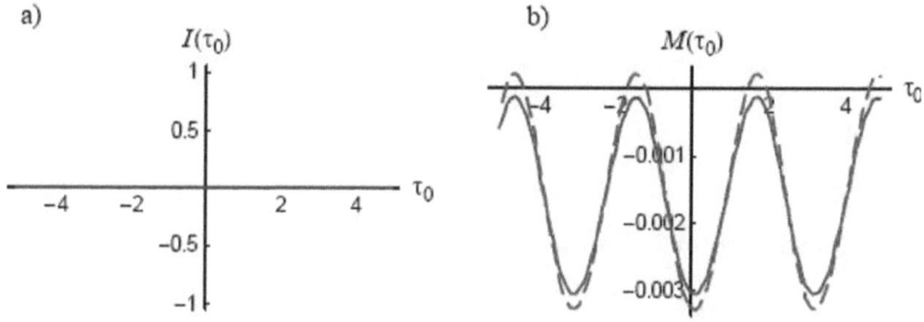

Fig. 2.22: $I(\tau_0)$ (a) and Melnikov's function $M(\tau_0)$ (b) versus the parameter τ_0 for $\zeta_k = 0.14$ (solid curves) and for $\zeta_k = 0.17$ (dashed curves); $b = 121$.

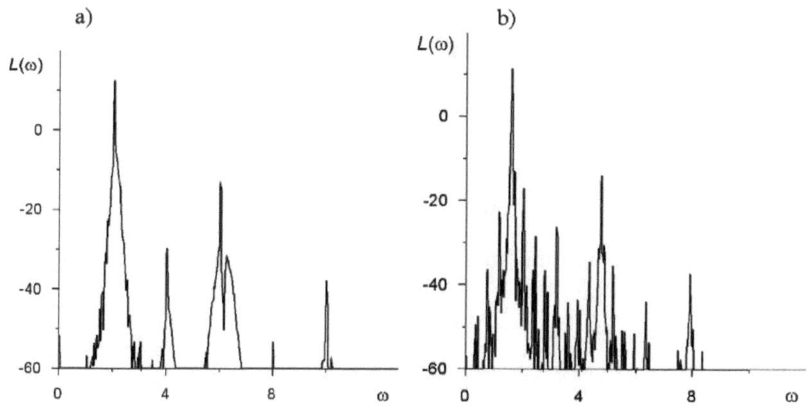

Fig. 2.23: Power spectra for $b = 121$: (a) $\zeta_k = 0.1$; (b) $\zeta_k = 0.17$.

In the next computational step we have taken $b = 121$. Note that $I(\tau_0) = 0$ for $\zeta < \zeta_{ch}^{(2)}$. Formula (2.76) gives $\zeta_{ch}^{(2)} = 0.336$, whereas formula (2.75) provides $\zeta_{ch}^{(1)} = 0.158$. Figure 2.22 presents the function $I(\tau_0)$ (a) and Melnikov's function $M(\tau_0)$ (b) versus τ_0 for the parameters associated with points 5 ($b = 121$, $\zeta_k = 0.14$; solid curves) and 6 ($b = 121$, $\zeta_k = 0.17$; dashed curves). In this case $I(\tau_0) = 0$ for $\zeta_k = 0.14$, but the function $M(\tau_0)$ does not change its sign ($I(\tau_0) = 0$ for $\zeta_k < \zeta_{ch}^{(2)}$) (it starts to change it for $\zeta_k = 0.17$). Again numerical computations confirm the applied value of (2.75) to define the parameter $\zeta_{ch}^{(1)}$ responsible for the prediction of analytical chaos.

Finally, the numerical analysis of Equation (2.64) is carried out for the parameters associated with points 5 and 6 in Fig. 2.13. For the parameter $\zeta_k = 0.1$ (the investigated point lies below the point 5 and corresponds to the periodic motion of the bush) and $\zeta_k = 0.17$ (point 6 corresponds to the narrow region of chaos) the associated power spectra are shown in Fig. 2.23.

It has been shown that for our investigated nonlinear system with friction the associated Melnikov's function can be constructed. As a result, one may control the nonlinear dynamics by analytical prediction of either a regular or chaotic system state. For some parameter sets the complicated analytical structure of Melnikov's functions can be simplified to provide simple analytical conditions for the occurrence of chaos.

2.3.4.4 Calculation of Lyapunov exponents

Vibrations of the bush being in thermoelastic contact with the rotating shaft are governed by the following nondimensional Equation (2.5),

$$\ddot{\varphi}(\tau) + 2h\dot{\varphi}(\tau) - \varphi(\tau) + b\varphi^3(\tau) = \varepsilon F(\omega_1 - \dot{\varphi})p(\tau), \qquad 0 < \tau < \infty, \quad (2.77)$$

with the initial condition $\varphi(0) = \varphi^\circ$, $\dot{\varphi}(0) = \omega^\circ$, where the nondimensional contact pressure is defined through solutions to Equation (2.31),

$$p(\tau) = h_U(\tau) - u^w(\tau) + 2\gamma\tilde{\omega}\int_0^\tau \dot{G}_p(\tau - \xi)F(\omega_1 - \dot{\varphi})p(\xi)(\omega_1 - \dot{\varphi})d\xi. \quad (2.78)$$

The bush wear $u^w(\tau)$ and the shaft temperature $\theta(r, \tau)$ are defined through the Equations (2.23) and (2.32)

$$u^w(\tau) = k^w \int_0^\tau |\omega_1 - \dot{\varphi}(\tau)|p(\tau)d\tau, \qquad 0 < \tau < \tau_c, \quad (2.79)$$

$$\theta(r, \tau) = \gamma\tilde{\omega}\int_0^\tau \dot{G}_\theta(r, \tau - \xi)F(\omega_1 - \dot{\varphi})p(\xi)(\omega_1 - \dot{\varphi})d\xi. \quad (2.80)$$

In Equations (2.77)–(2.80) the following nondimensional quantities are introduced

$$\tau = \frac{t}{t_*}, \quad r = \frac{R}{R_1}, \quad p = \frac{P}{P_*}, \quad \theta = \frac{T_1 - T_{sm}}{T_*}, \quad \varphi(\tau) = \varphi_2(t_*\tau),$$

$$u^w = \frac{U^w}{U_*}, \quad \varepsilon = \frac{P_* t_*^2 2\pi R_1^2}{B_2}, \quad h = \frac{cR_2^2}{2B_2 t_*}, \quad k^w = \frac{P_* K^w R_1}{U_*},$$

$$\gamma = \frac{(1-\eta)E_1\alpha_1 R_1^2}{\lambda_1(1-2\nu_1)t_*}, \quad Bi = \frac{\alpha_T R_1}{\lambda_1}, \quad \tau_c = \frac{t_c}{t_*}, \quad \tilde{\omega} = \frac{t_* a_1}{R_1^2},$$

$$\omega_0 = \omega' t_*, \quad h_U(\tau) = h_U(t_*\tau), \quad F(\omega_1 - \dot{\varphi}) = f(V_*(\omega_1 - \dot{\varphi})),$$

where

$$V_* = \frac{R_1}{t_*}, \qquad t_* = \sqrt{\frac{B_2}{k_* R_2^2}}, \qquad k_* = k_1\left(\frac{l_0}{l_1} - 1\right)\left(1 + \frac{l_1}{R_2}\right) - k_2,$$

$$T_* = \frac{U_*}{\alpha_1(1+\nu_1)R_1}, \qquad P_* = \frac{\alpha_1 E_1 T_*}{(1-2\nu_1)},$$

and l_0 is the nonstretched spring length, l_1 is the length of the compressed spring for $\varphi_2 = 0$, $(k_* > 0)$, E_1 is the elasticity modulus, ν_1 is the Poisson coefficient, α_1 is the coefficient of thermal expansion of the shaft, α_T is the heat transfer coefficient, a_1 is the thermal diffusivity, λ_1 is the heat transfer coefficient, $\varphi_2(t)$ is the angle of bush rotation, K^w is the wear coefficient, η denotes the part of heat energy associated with wear $\eta \in [0,1]$, and t_c is the time of contact $(0 < t < t_c, P(t) > 0)$.

Note that the stated problem is modelled by both the nonlinear differential Equation (2.77) and integral equation (2.78) governing rotational velocity $\dot{\varphi}(\tau)$ and contact pressure $p(\tau)$. Temperature and wear are defined by Equations (2.80) and (2.79), respectively.

A particular case of our problem is further studied ($\gamma = 0$, $k^w = 0$, $p(\tau) \to 1$). The dependence of kinematics friction on relative velocity is approximated by the function $F(y) = f_s \mathrm{sgn}(y) - \alpha y + \beta y^3$. Because the latter is nonsmooth due to the presence of the $\mathrm{sgn}(y)$ function in the kinematic friction, the methods commonly used to compute the exponents require smoothness of the vector fields as a necessary condition. Nonsmooth systems yield only approximations for the Lyapunov exponents, which can be considered valid as long as we do not bother too much with the vicinity of the nonsmoothness points [Awrejcewicz, Lamargue (2003)]. The function $\mathrm{sgn}(y)$ is approximated by the following one (2.58).

Note that while computing Lyapunov exponents, besides the following equations

$$\dot{x} = y, \quad \dot{y} = x - bx^3 + \varepsilon[f_s\,\mathrm{sgn}_{\varepsilon_0}(\mathrm{v}_r) - \alpha \mathrm{v}_r + \beta \mathrm{v}_r^3] - \varepsilon h_1 y, \quad \dot{z} = \omega_0, \tag{2.81}$$

three additional systems of equations $(n = 1,2,3)$ with respect to perturbations are also solved:

$$\begin{aligned} &\dot{\tilde{x}}^{(n)} = \tilde{y}^{(n)}, \\ &\dot{\tilde{y}}^{(n)} = \tilde{x}^{(n)} - 3bx^2\tilde{x}^{(n)} + \varepsilon[f_s\delta_{\varepsilon_0}(\mathrm{v}_r) - \alpha + 3\beta \mathrm{v}_r^2]\tilde{\mathrm{v}}_r^{(n)} - \varepsilon h_1 \tilde{y}^{(n)}, \\ &\dot{\tilde{z}}^{(n)} = 0, \end{aligned} \tag{2.82}$$

where $x = \varphi(\tau)$, $y = \dot{\varphi}(\tau)$, $z = \omega_0\tau$, $\mathrm{v}_r = \omega_k + \zeta_k \sin z - y$, $\tilde{\mathrm{v}}_r^{(n)} = \zeta_k \tilde{z}^{(n)} \cos z - \tilde{y}^{(n)}$, $h_1 = 2h/\varepsilon$, and

$$\delta_{\varepsilon_0}(y) = \begin{cases} 0, & |y| > \varepsilon_0, \\ (2\varepsilon_0)(1 - |y|/\varepsilon_0), & |y| < \varepsilon_0. \end{cases} \tag{2.83}$$

Twelve equations of system (2.81) and (2.82) are solved using the fourth-order Runge–Kutta method and Gram–Schmidt reorthonormalisation procedure.

Let $\tilde{\mathbf{x}}_0^0$, $\tilde{\mathbf{y}}_0^0$, $\tilde{\mathbf{z}}_0^0$ be initial values of perturbation vectors that are orthonormal. After time T, an orbit $\mathbf{x}(\tau)$ reaches the point $\mathbf{x}_1$ with the associated perturbations $\tilde{\mathbf{x}}_1$, $\tilde{\mathbf{y}}_1$, $\tilde{\mathbf{z}}_1$. Then, the so-called Gram–Schmidt reorthonormalization procedure is carried out and the following new initial set of conditions is formulated.

$$\tilde{\mathbf{x}}_1^0 = \frac{\tilde{\mathbf{x}}_1}{||\tilde{\mathbf{x}}_1||}, \tag{2.84}$$

$$\tilde{\mathbf{y}}_1^0 = \frac{\tilde{\mathbf{y}}_1'}{||\tilde{\mathbf{y}}_1'||}, \quad \tilde{\mathbf{y}}_1' = \tilde{\mathbf{y}}_1 - (\tilde{\mathbf{y}}_1, \tilde{\mathbf{x}}_1^0)\tilde{\mathbf{x}}_1^0, \tag{2.85}$$

$$\tilde{\mathbf{z}}_1^0 = \frac{\tilde{\mathbf{z}}_1'}{||\tilde{\mathbf{z}}_1'||}, \quad \tilde{\mathbf{z}}_1' = \tilde{\mathbf{z}}_1 - (\tilde{\mathbf{z}}_1, \tilde{\mathbf{x}}_1^0)\tilde{\mathbf{x}}_1^0 - (\tilde{\mathbf{z}}_1, \tilde{\mathbf{y}}_1^0)\tilde{\mathbf{y}}_1^0. \tag{2.86}$$

Next, after time interval T, a new set of perturbation vectors $\tilde{\mathbf{x}}_2$, $\tilde{\mathbf{y}}_2$, $\tilde{\mathbf{z}}_2$ is defined, which is also reorthonormalized due to the Gram–Schmidt procedure (2.84)–(2.86). This algorithm is repeated M times. Note that $(\tilde{\mathbf{x}}_1^0, \tilde{\mathbf{y}}_1^0) = 0$, $(\tilde{\mathbf{x}}_1^0, \tilde{\mathbf{z}}_1^0) = 0$, $(\tilde{\mathbf{y}}_1^0, \tilde{\mathbf{z}}_1^0) = 0$ and if $\mathbf{x} = (x, y, z)$, $\mathbf{y} = (x_1, y_1, z_1)$ then $||\mathbf{x}|| = \sqrt{x^2 + y^2 + z^2}$, and the scalar product $(\mathbf{x}, \mathbf{y}) = xx_1 + yy_1 + zz_1$. Finally, a spectrum of three Lyapunov exponents is computed via formulas

$$\lambda_1 = \frac{1}{MT}\sum_{i=1}^{M} \ln ||\tilde{\mathbf{x}}_i||, \qquad \lambda_2 = \frac{1}{MT}\sum_{i=1}^{M} \ln ||\tilde{\mathbf{y}}_i'||, \qquad \lambda_3 = \frac{1}{MT}\sum_{i=1}^{M} \ln ||\tilde{\mathbf{z}}_i'||, \tag{2.87}$$

where the occurring vectors are taken before the normalization procedure.

2.3.4.5 Numerical analysis

Our numerical computations are carried out for the particular case ($\gamma = 0$, $k^w = 0$). The following nondimensional parameters are taken: $f_s = \alpha = \beta = 0.3$, $\omega_0 = 2$, $\omega_k = 0.4$, $b = 1$, $\varepsilon = 0.1$. Numerical analysis is carried out for the bifurcation diagram with respect to x versus ζ_k for $\zeta_k \in (0, 12)$ and $\zeta_k \in (3.5, 4.0)$). The obtained results are shown in Figs. 2.15a and 2.15b for $h_1 = 0$, in Fig. 2.24a for $h_1 = 0.5$ and in Fig. 2.24b for $h_1 = 1$. The Lyapunov exponents in time interval $\tau \in (1200, 1514)$ ($\tilde{\mathbf{x}}_0^0 = (1, 0, 0)$, $\tilde{\mathbf{y}}_0^0 = (0, 1, 0)$, $\tilde{\mathbf{z}}_0^0 = (0, 0, 1)$, $T = 0.005$, $M = 80000$, $\varepsilon_0 = 0.01$) are computed due to formulas (2.87) for the same values of the parameters. In Figs. 2.25a, b, and 2.26a, b dependencies of Lyapunov exponents on the control parameter ζ_k are reported. A study of both Lyapunov exponents

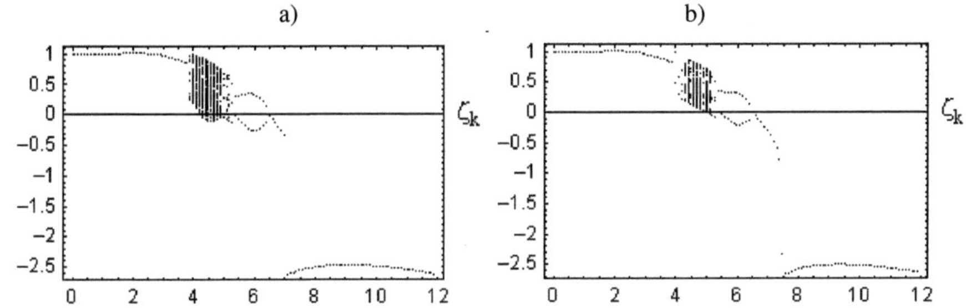

Fig. 2.24: Bifurcation diagrams (a),(b) using ζ_k as control parameter; $\gamma = 0, \quad k^w = 0, \quad b = 1$: (a) $h_1 = 0.5$; (b) $h_1 = 1$.

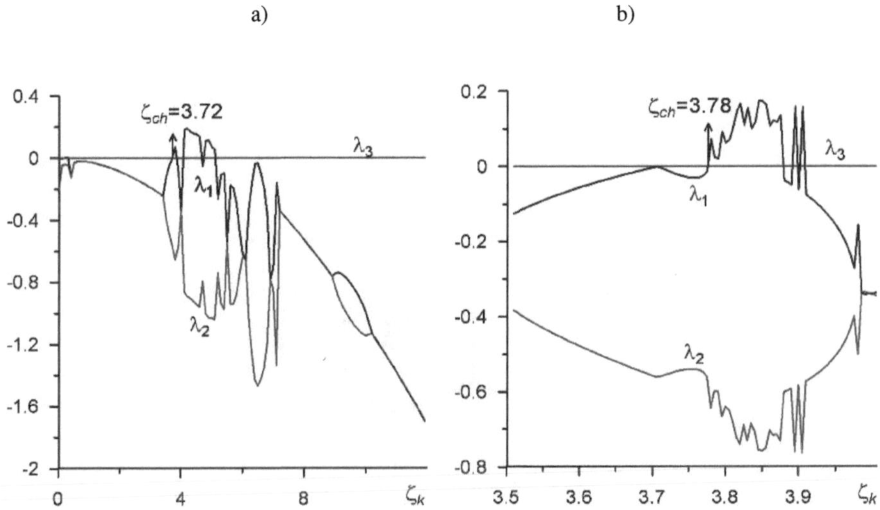

Fig. 2.25: Lyapunov exponents (a), (b) using ζ_k as control parameter; $h_1 = 0, \quad \gamma = 0, \quad k^w = 0, \quad b = 1$: (a) $\zeta_k \in (0, 12)$; (b) $\zeta_k \in (3.5, 4.0)$.

and bifurcation diagrams implies that chaos begins for (i) $\zeta_k = 3.78$, for $h_1 = 0$; (ii) for $\zeta_k = 3.8$, for $h_1 = 0.5$; (iii) for $\zeta_k = 4.25$, for $h_1 = 1$ (note that the largest Lyapunov exponent λ_1 is positive). An increase of the parameter h_1 responsible for damping yields an increase of the amplitude of the bush, where chaos is born.

Note that because our system (2.81) is autonomous, one of the Lyapunov exponents is always zero.

In a general case, numerical analysis is carried out on a steel-made shaft ($\alpha_1 = 14 \cdot 10^{-6}{}^\circ\mathrm{C}^{-1}$, $\lambda_1 = 21\,\mathrm{W}/(\mathrm{m} \cdot {}^\circ\mathrm{C}^{-1})$, $\nu_1 = 0.3$, $a_1 = 5.9 \cdot 10^{-6}\,\mathrm{m}^2/s$, $E_1 = 19 \cdot 10^{10}\,\mathrm{Pa}$). Observe that no accounting of tribological processes ($h_1 = 0.5$, $\zeta_k = 3.9$, $\gamma = 0$, $k^w = 0$) yields chaotic dynamics (Fig. 2.27, curve 2). For $h_1 = 0.5$, $\zeta_k = 3.5$, $\gamma = 0$, $k^w = 0$ regular motion takes place (Fig. 2.27, curve 1). An account of thermal shaft extension ($\gamma = 1.87$) removes the chaotic

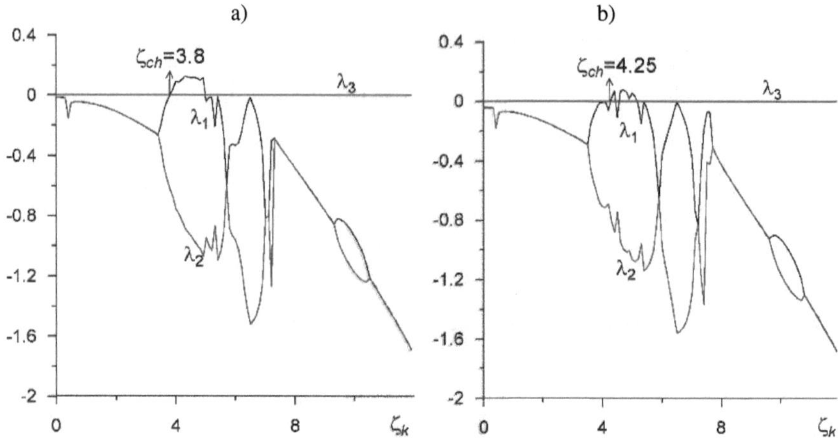

Fig. 2.26: Lyapunov exponents (a), (b) using ζ_k as control parameter; $\gamma = 0$, $k^w = 0$, $b = 1$: (a) $h_1 = 0.5$; (b) $h_1 = 1$.

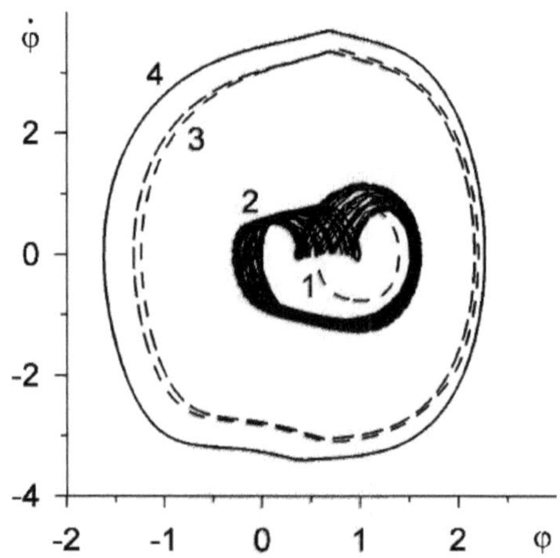

Fig. 2.27: Phase plane of bush motion for $h_1 = 0.5$, $k^w = 0$: curve 1, $\zeta_k = 3.5$, $\gamma = 0$; curve 2, $\zeta_k = 3.9$, $\gamma = 0$; curve 3, $\zeta_k = 3.5$, $\gamma = 1.87$; curve 4, $\zeta_k = 3.9$, $\gamma = 1.87$.

behaviour of our system (Fig. 2.27, curves 3 and 4). For $\zeta_k = 3.5$ a subharmonic motion with frequency $\omega_0/2$ is obtained (Fig. 2.27, curve 3), whereas for $\zeta_k = 3.9$ periodic motion is exhibited (Fig. 2.27, curve 4).

Owing to an account of wear ($k^w = 0.01$) and neglecting shaft thermal extension ($\gamma = 0$), contact pressure tends to zero, whereas cylinder wear approaches $U_*(p(\tau) \rightarrow 0,\ u^w(\tau) \rightarrow 1)$. The nondimensional bush wear is presented in Fig. 2.28, curve 1. In addition, in Fig. 2.28, curves 1 and 2 represent time histories of the nondimensional contact pressure.

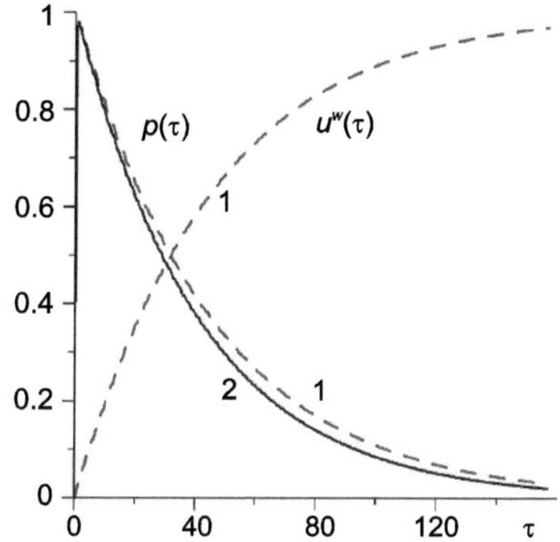

Fig. 2.28: Dimensionless contact pressure $p(\tau)$ and wear $u^w(\tau)$ versus dimensionless time τ: curves 1, $\zeta_k = 3.5$, $\gamma = 0$, $k^w = 0.01$, $h_1 = 0.5$; curves 2, $\zeta_k = 3.9$, $\gamma = 0$, $k^w = 0.01$, $h_1 = 0.5$.

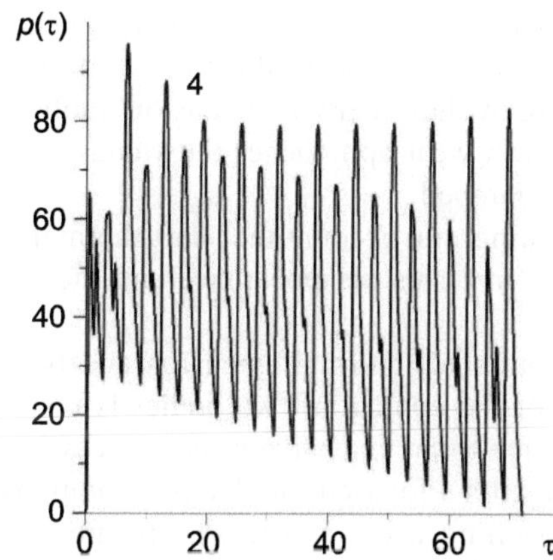

Fig. 2.29: Dimensionless contact pressure $p(\tau)$ versus dimensionless time τ: curve 4, $\zeta_k = 3.9$, $\gamma = 1.87$, $k^w = 0.01$, $h_1 = 0.5$.

A simultaneous account of shaft extension and bush wear yields a finite time of contact between both bodies. For instance, for $h_1 = 0.5$, $\zeta_k = 3.9$, $\gamma = 1.87$, $k^w = 0.01$ contact pressure versus time is exhibited by curve 4 in Fig. 2.29. The nondimensional time contact interval is $\tau_c = 72$. For $\zeta_k = 3.5$ time contact is $\tau_c = 65.8$. In Fig. 2.30, curves 3 and 4 represent the dependence of nondimensional wear on the nondimensional time in a general case. Curve 3 corresponds to $h_1 = 0.5$, $\zeta_k = 3.5$, $\gamma = 1.87$, $k^w = 0.01$, whereas curve 4 is associated with the parameters: $h_1 = 0.5$, $\zeta_k = 3.9$, $\gamma = 1.87$, $k^w = 0.01$. Owing to heat shaft extension, the wear of the bush is increased 30 times (see curves 4 and 2 in Fig. 2.30).

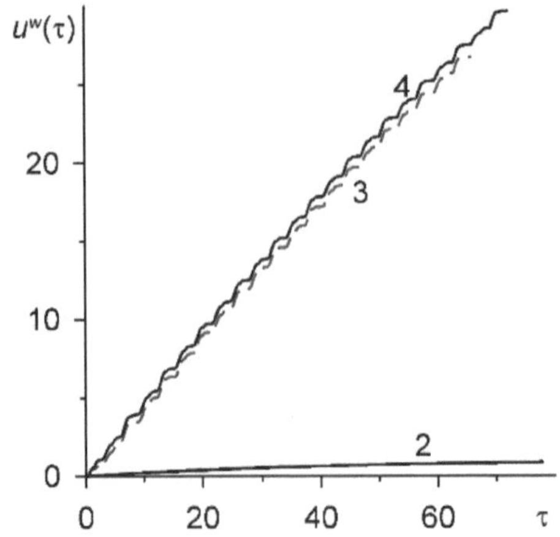

Fig. 2.30: Time history of dimensionless wear $u^w(\tau)$, $h_1 = 0.5$, $k^w = 0.01$; curve 2, $\zeta_k = 3.9$, $\gamma = 0$; curve 3, $\zeta_k = 3.5$, $\gamma = 1.87$; curve 4, $\zeta_k = 3.9$, $\gamma = 1.87$.

The influence of tribological processes on dynamic behaviour of the analysed system in the vicinity of chaos has been illustrated and discussed. An account of bush wear and neglecting of shaft thermal expansion implies that the contact pressure tends to zero, the bush wear approaches the values of the shaft compressing, and bush vibrations are damped.

On the other hand, taking into account the shaft thermal extension and neglecting of bush wear results in chaos disappearance and the occurrence of a regular motion.

In a general case (both shaft thermal extension and bush wear are taken into account), time interval of the contact of two bodies is bounded. With the lack of contact, the bush stops due to an extensive wear process.

Experimental investigations focused on chaotic vibrations in the kinematically excited system and modelling a drive–feed system are carried out in reference [Grudziński, Wedman (1998)].

2.3.5 *Chaotic motion of the bush subject to mechanical external excitations*

In this section the previous model of the bush lying on the shaft rotating at constant velocity is considered (without heat generation; i.e., $\gamma = 0$). As assumed earlier $k_* = k_1(l_0/l_1 - 1)(1 + l_1/R_2) - k_2$, and $\chi = 1$, which defines $a = 1$, $t_* = \sqrt{B_2/(k_* R_2^2)}$. We assume also that $h_U(\tau) = H(\tau)$, $U_* = U_0$, $h_T(\tau) = 0$. According to the latter assumptions, the dimensionless contact pressure is $p(\tau) = 1$.

For this case, Melnikov's function is formulated analytically giving the onset of chaos. Examples of nonlinear system behaviour are illustrated and tested numerically.

2.3.5.1 Melnikov's method

For the considered case differential equations (inclusions) (2.5) take the following form,

$$\ddot{\varphi}(\tau) - \varphi(\tau) + b\varphi^3(\tau) \in -2h\dot{\varphi} + \varepsilon[\zeta\cos(\omega\tau) + F(\omega_1 - \dot{\varphi})p(\tau)], \quad 0 < \tau < \infty, \tag{2.88}$$

$$\varphi(0) = \varphi^\circ, \qquad \dot{\varphi}(0) = \omega^\circ, \tag{2.89}$$

where $p(\tau) = h_U(\tau) - u^w(\tau)$. Friction versus the relative velocity is approximated by function (2.61). Consider first the case without wear ($k^w = 0$) and $h_U(\tau) = H(\tau)$. Equations of motion are governed by (2.64), where

$$q_0(x,y) = x - bx^3, \qquad q_1(x,y,\omega\tau,\varepsilon) = \zeta\cos(\omega\tau) - h_1 y + F(\omega_* - y),$$

$$p_0(x,y) = y, \qquad p_1(x,y,\omega\tau,\varepsilon) = 0, \qquad h_1 = 2h/\varepsilon. \tag{2.90}$$

In this case, Melnikov's function (2.67) takes the form

$$M(\tau_0) = -\int_{-\infty}^{+\infty} y_0(t)[\zeta\cos(\omega(t+\tau_0)) - h_1 y_0(t) + f_s\mathrm{sgn}(\omega_r) - \alpha\omega_r + \beta\omega_r^3]dt, \tag{2.91}$$

and the dimensionless relative velocity is

$$\omega_r(t) = \omega_* - y_0(t). \tag{2.92}$$

Taking into account (2.66), (2.92), and (2.91), the following Melnikov's function is defined,

$$M(\tau_0) = A - \zeta B\sin(\omega\tau_0), \tag{2.93}$$

where

$$A = \frac{16\beta}{35b^2} + h_1\frac{4}{3b} - \frac{4\alpha}{3b} + \frac{8\beta V^2}{b^2}$$
$$+ 2f_s\sqrt{\frac{2}{b}}\left[\sqrt{\frac{1}{2} + \sqrt{\frac{1}{4} - V^2}} - \sqrt{\frac{1}{2} - \sqrt{\frac{1}{4} - V^2}}\right] H\left(\frac{1}{2} - V\right), \tag{2.94}$$

$$H(x) = 1, \qquad x > 0, \qquad H(x) = 0, \qquad x \leq 0,$$

$$B = \frac{2\pi\omega}{\sqrt{2b}\cosh(\pi\omega/2)}, \qquad V = \omega_1\sqrt{\frac{b}{2}}. \tag{2.95}$$

Owing to Melnikov's theory, if the function $M(\tau_0)$ has simple zeros, then for sufficiently small parameter ε, motion governed by (2.64) should be chaotic. It occurs when

$$|A| < \zeta B. \tag{2.96}$$

Owing to wear occurrence $k^w > 0$, after some time, the bush wear amount of U_0 appears (contact pressure $p(\tau) = 0$). Thus, owing to (2.92), the bush starts to vibrate in the chaotic manner when the following condition is satisfied.

$$\frac{h_1}{\zeta} < \frac{3\sqrt{b}}{2\sqrt{2}} \frac{\pi\omega}{\cosh(\pi\omega/2)}. \tag{2.97}$$

Note that the obtained condition is in agreement with that reported by [Guckenheimer, Holmes (1983)]. Furthermore, in the case without damping (see (2.97)), that is, for $h_1 = 0$, a solution to the studied equation ($\zeta > 0$) is always chaotic.

2.3.5.2 Numerical analysis

The condition (2.96) is rewritten in a more suitable form

$$\psi(\alpha, \beta, f_s, b, \omega, \omega_1, h_1) < \zeta, \tag{2.98}$$

where $\psi(\alpha, \beta, f_s, b, \omega, \omega_1, h_1) = |A|/B$.

Figure 2.31 shows the zones of parameters associated with chaos. According to Fig. 2.31a, if for chosen parameters (ω, ω_1) the dimensionless amplitude lies over the shown surface, then these parameters provide a chaotic response.

First, ζ plays the role of a control parameter in the case without account of wear. Numerical analysis is carried out for Equations (2.64) for $\varepsilon = 1$, $k^w = 0$. The bifurcation diagrams are shown in Figs. 2.32a and 2.32b for various values of the parameter b. For $b = 1$, according to (2.98), the critical dimensionless amplitude values $\zeta_{ch} = 0.637$ for a chaotic threshold are estimated.

For $b = 9$ the critical value of the control parameter dramatically decreases and achieves $\zeta_{ch} = 0.084$. The bifurcation diagrams shown in Figs. 2.32 verify the results obtained through Melnikov's theory.

When the contact pressure $p(\tau) = 0$ and there is no damping, then according to (2.97), chaos always exists and the chaotic zone is decreased with an increase of b. Fig. 2.33 confirms the latter remark.

In the case of damping occurrence, the bifurcation diagrams for the control parameter $h_2 = h_1/\zeta$ are also studied. They are given in Fig. 2.34. Critical values

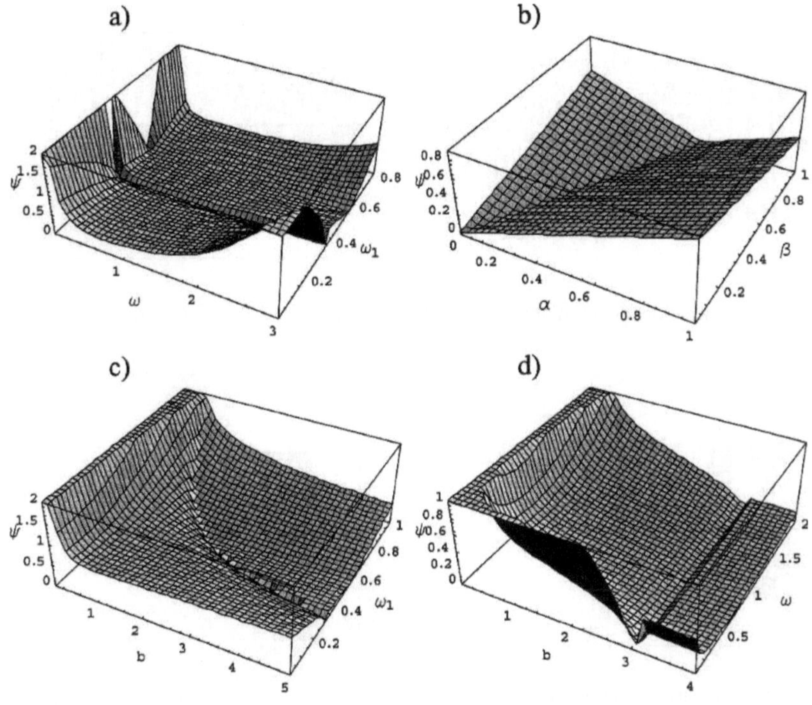

Fig. 2.31: Chaotic threshold in various parameter spaces: (a) (ψ, ω, ω_1) ($\alpha = \beta = f_s = 0.3$, $b = 5$, $h_1 = 0$), (b) (ψ, α, β) ($f_s = 0.3$, $b = 4$, $\omega = 2$, $\omega_1 = 0.4$, $h_1 = 0$), (c) (ψ, b, ω_1) ($\alpha = \beta = f_s = 0.3$, $\omega = 2$, $h_1 = 0$), (d) (ψ, b, ω) ($\alpha = \beta = f_s = 0.3$, $\omega_1 = 0.4$, $h_1 = 0$).

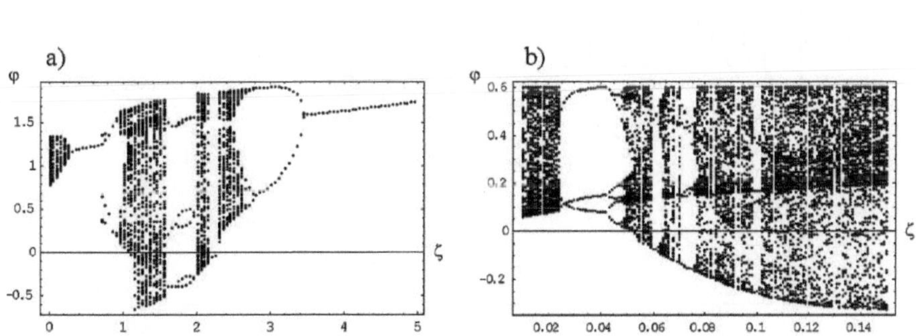

Fig. 2.32: Bifurcation diagram for the variable φ: (a) $b = 1$, $\zeta_{ch} = 0.637$; (b) $b = 9$, $\zeta_{ch} = 0.084$; ($\alpha = \beta = f_s = 0.3$, $\omega = 2$, $\omega_1 = 0.4$).

h_{ch} of the parameter h_2, below which chaotic motion occurs, are computed through (2.97).

Figure 2.35 shows the evolution in time of the dimensionless wear u^w for various parameters (curves 1, 3, 5, 7 (2, 4, 6, 8) correspond to $b = 1$ (9)).

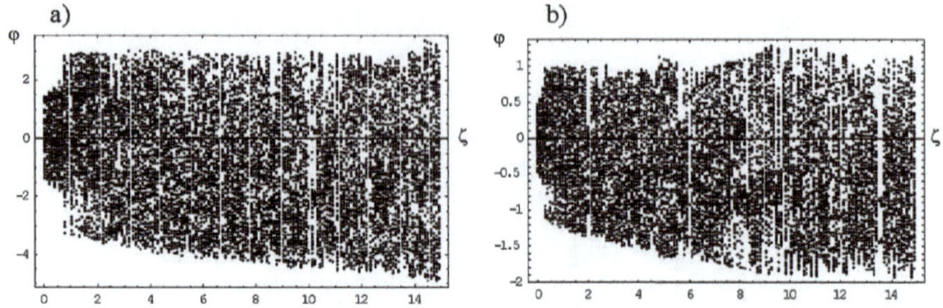

Fig. 2.33: Bifurcation diagram for the variable φ: (a) $b = 1$, $\zeta_{ch} = 0$; (b) $b = 9$, $\zeta_{ch} = 0$; ($\omega = 2$, $k^w \neq 0$).

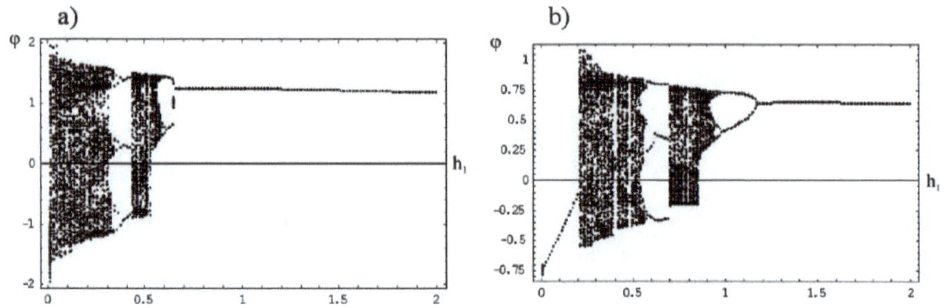

Fig. 2.34: Bifurcation diagram for the variable φ: (a) $b = 1$, $h_{ch} = 0.575$; (b) $b = 4$, $h_{ch} = 1.15$ ($\alpha = \beta = f_s = 0.3$, $\omega = 2$, $\omega_1 = 0.4$).

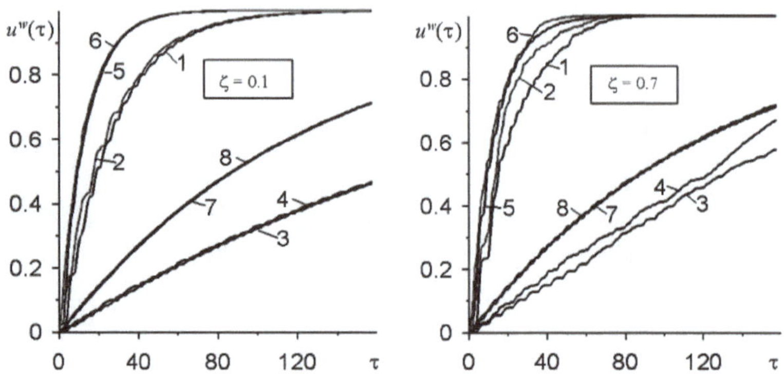

Fig. 2.35: Dimensionless wear versus τ for $k^w = 0.1$, $b = 1$, $\omega_1 = 0.4$ (curves 1); $k^w = 0.1$, $b = 9$, $\omega_1 = 0.4$ (curves 2); $k^w = 0.01$, $b = 1$, $\omega_1 = 0.4$ (curves 3); $k^w = 0.01$, $b = 9$, $\omega_1 = 0.4$ (curves 4); $k^w = 0.1$, $b = 1$, $\omega_1 = 0.8$ (curves 5); $k^w = 0.1$, $b = 9$, $\omega_1 = 0.8$ (curves 6); $k^w = 0.01$, $b = 1$, $\omega_1 = 0.8$ (curves 7); $k^w = 0.01$, $b = 9$, $\omega_1 = 0.8$ (curves 8) ($\alpha = \beta = f_s = 0.3$, $\omega = 2$, $\zeta = 0.7$).

Bush phase trajectory accounted wear is illustrated in Fig. 2.36. In Fig. 2.36c bush dynamics without wear is shown. Figure 2.36b (2.36a) refers to the wear coefficient $k^w = 0.01$ ($k^w = 0.1$).

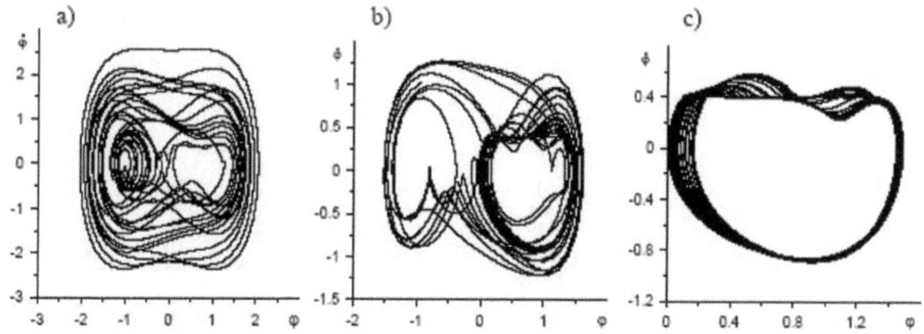

Fig. 2.36: Bush phase trajectory (a) $k^w = 0.1$, $b = 1$, $\omega_1 = 0.4$; (b) $k^w = 0.01$, $b = 1$, $\omega_1 = 0.4$; (c) $k^w = 0$, $b = 1$, $\omega_1 = 0.4$ ($\alpha = \beta = f_s = 0.3$, $\omega = 2$, $\zeta = 0.7$).

From the mentioned figures, one may conclude that wear essentially influences a stick-slip motion.

2.3.6 *Analysis of the bush motion with wear and cylinder kinematic excitations*

In this section, heat extension is not considered; that is, $\gamma = 0$. Additionally, we take $\zeta = 0$, $\chi = 1$ which gives $t_* = \sqrt{B_2/(k_* R_2^2)}$. We also take $h_U(\tau) = H(\tau)$, $U_* = U_0$, $h_T(\tau) = 0$. Owing to the introduced quantities, a system of equations and inclusion (2.5), (2.22), (2.23) is given in the form

$$\ddot{\varphi}(\tau) - a\varphi(\tau) + b\varphi^3(\tau) \in \varepsilon F(\omega_1 - \dot{\varphi})p(\tau), \qquad 0 < \tau < \infty, \tag{2.99}$$

$$\varphi(0) = \varphi^\circ, \qquad \dot{\varphi}(0) = \omega^\circ; \tag{2.100}$$

$$p(\tau) = h_U(\tau) - u^w(\tau), \tag{2.101}$$

$$u^w(\tau) = k^w \int_0^\tau |\omega_1 - \dot{\varphi}(\xi)| p(\xi) d\xi, \tag{2.102}$$

where

$$F(\dot{\varphi}) = f_s \mathrm{Sgn}(\dot{\varphi}) - \alpha\dot{\varphi} + \beta\dot{\varphi}, \tag{2.103}$$

$$\omega_1 = \omega_* + \zeta_k \sin \omega\tau. \tag{2.104}$$

The considered problem is reduced to looking for a solution of the nonlinear system of differential and integral equations with respect to velocity $\dot{\varphi}$ and wear u^w.

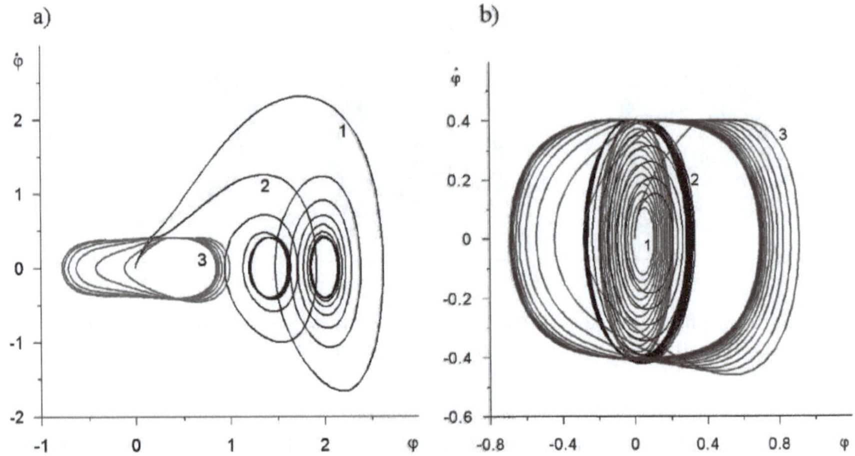

Fig. 2.37: Phase trajectory of bush movement in the conditions of wear of the bush: (a) curve 1, $a = 4$; 2, $a = 2$; 3, $a = 0.1$; (b) curve 1, $a = -4$; 2, $a = -2$; 3, $a = -0.1$.

Numerical analysis. The results of calculations for different values of the parameter a, ζ for the steel cylinder ($E_1 = 19 \cdot 10^{10}$Pa, $\nu_1 = 0.3$) and $\omega = 2$, $\omega_* = 0.4$, $b = 1$, $\alpha = 0.3$, $\beta = 0.3$, $f_s = 0.3$, $\varphi^\circ = 0$, $\omega^\circ = 0$ are shown in Fig. 2.37. Consider the case $h_U(\tau) = H(\tau)$.

The case when the system behaviour depends on control parameter a in the absence of external disturbance $\zeta_k = 0$ has been considered. A numerical analysis of Equations (2.99)–(2.102) has been performed for $\varepsilon = 1$, $k^w = 0.1$. The phase trajectory for positive values of the parameter a is represented in Fig. 2.37a and for negative a in Fig. 2.37b.

Also for the same case a dimensionless wear dependence upon dimensionless time is shown in Fig. 2.38. As can be seen, wear significantly influences stick-slip movements. This kind of movement changes with time into harmonic ones. At the same time, the character of the movements does not play a significant role in wear value.

Furthermore the system movement under the external kinematic disturbance was considered.

In Section 2.3.4.2 Melnikov's method is applied to analyse chaotic motions of the pad driven kinematically. Melnikov's function is derived when $k^w = 0$ (no wear), and the value of k_* is found from $a = 1$.

When Melnikov's function (2.70) changes sign then the system movement is chaotic.

Calculations were performed for $\zeta_k = 3.5$, $\varepsilon = 0.1$. Melnikov's function does not change sign. Consequently according to Melnikov's theory, chaos will not appear. A numerical analysis of Equations (2.99)–(2.102) was performed. Then the system trajectory on the phase plane was drawn. In Fig. 2.39a the dashed curve is typical of systems with friction. Curve 1 in Fig. 2.39a corresponds to the behaviour of the system with wear $k^w = 0.1$.

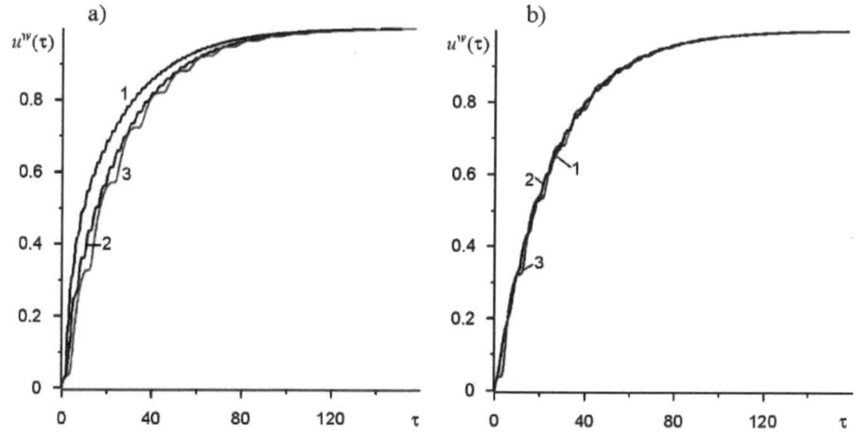

Fig. 2.38: Time evolution of dimensionless wear for different values of the parameter a: (a) curve 1, $a = 4$; 2, $a = 2$; 3, $a = 0.1$; (b) curve 1, $a = -4$; 2, $a = -2$; 3, $a = -0.1$.

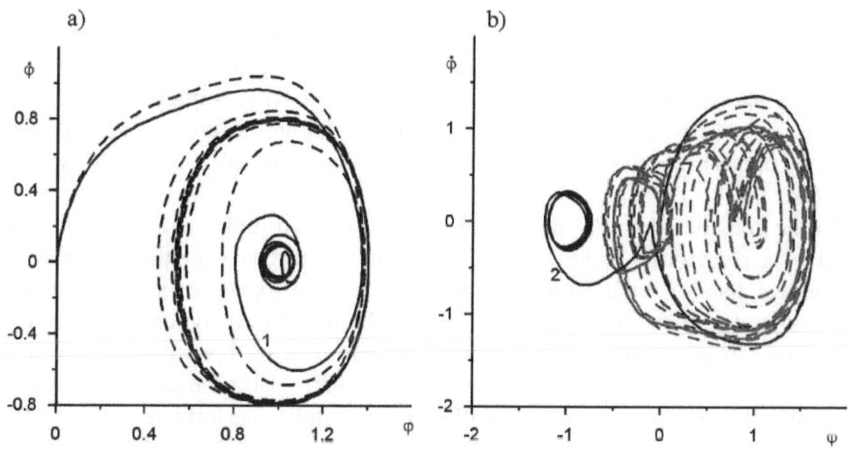

Fig. 2.39: Phase trajectory of bush movement in conditions of wear of the bush: (a) solid curve 1: $\zeta_k = 3.5$, $k^w = 0.1$, dashed curves $\zeta_k = 3.5$, $k^w = 0$; (b) solid curve 2: $\zeta_k = 4.5$, $k^w = 0.1$, dashed curves $\zeta_k = 4.5$, $k^w = 0$.

Furthermore, for calculations we have assumed the following value of the control parameter $\zeta_k = 4.5$. Melnikov's function $M(\tau_0)$ changes sign in this case and according to Melnikov's theory chaos should appear. Also for the same case the numerical analysis of Equations (2.99)–(2.102) was performed. Figure 2.39b shows the phase trajectory of the analysed system. Contrary to the previously analysed movement for $\zeta_k = 3.5$ now we can see that the system oscillates chaotically (see the dashed curve in Fig. 2.39b).

When wear is taken into account, the system will behave as represented by curve 2 in Fig. 2.39b. Chaos is absent.

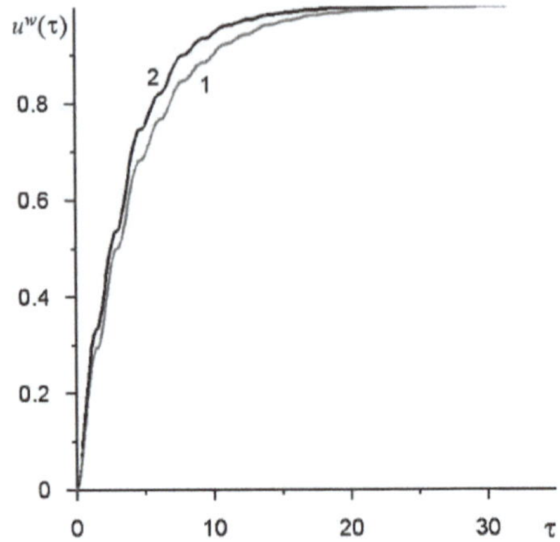

Fig. 2.40: Time evolution of dimensionless wear u^w for different values of the parameter ζ_k. Curve 1: $\zeta_k = 3.5$; 2: $\zeta_k = 4.5$.

Also for this case, the dimensionless wear dependence upon dimensionless time for different values of the dimensionless amplitude of kinematic disturbance is shown in Fig. 2.40. It can be seen that wear sufficiently influences chaotic and harmonic movement. They tend to harmonic movement with time. At the same time the movement character does not play a significant role in wear.

In the considered work, the equation of self-excited movement of a solid shaft fixed in a rigid bush under external kinematic disturbance in the conditions of a classical friction model and abrasive wear is developed. It is shown that for the analysed system in conditions of wear absence the analytical calculation of Melnikov's function is possible; that is, we can choose those parameters of the system that ensure its regular or chaotic movement. Also a numerical calculation of wear influence on the system oscillation character is performed.

2.3.7 *Dynamics with external temperature perturbation*

In this section we assume that in the initial time instant both cylinder (shaft) and bush are in free state (without internal excitation, $U_0 = 0$, $T_* = T_0$, $U_* = 2\alpha_1(1+\nu_1)R_1T_0$, $P_* = 2\alpha_1E_1T_0/(1-2\nu_1)$), and dimensionless bush temperature is governed by the equation $h_T(\tau) = 2(1 - \exp(-\delta\tau^2))$. Owing to heat transfer, the cylinder starts to extend its volume and a contact between cylinder and bush occurs.

The cylinder rotates with angular velocity $\Omega(t) = \Omega_*\omega_1(t)$ such that the centrifugal forces can be neglected. We assume that angular speed of the shaft rotation changes in accordance with $\omega_1 = \omega_* + \zeta_k \sin\omega' t$, where ζ_k is the dimensionless

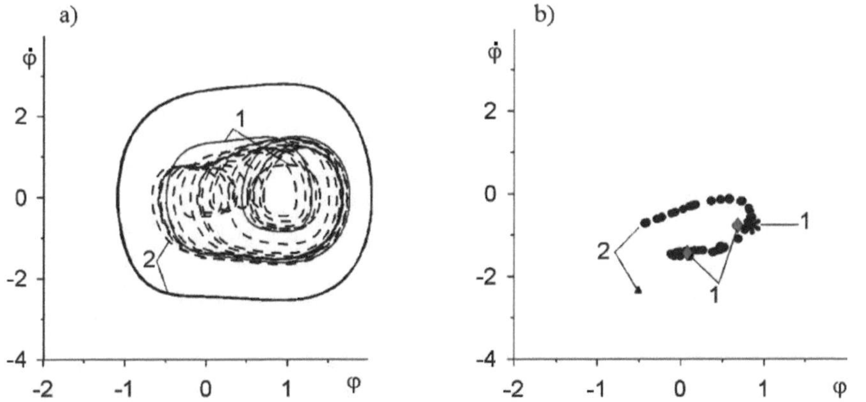

Fig. 2.41: Phase plane (a) and Poincaré section (b) of bush motion (curves 1: $\zeta_k = 3.5$, curves 2: $\zeta_k = 4.5$; solid curves: $\gamma = 1$, dashed curves: $\gamma = 0$).

amplitude of the kinematic excitation. We take $k_* = k_1(l_0/l_1 - 1)(1 + l_1/R_2) - k_2$, and the dependence of friction on relative velocity is approximated by the function (2.61). First, we assume that there is no wear ($k^w = 0$).

In dimensionless form this problem is governed by Equations (2.5), (2.6), (2.22), and (2.31)–(2.33).

In the case when the frictional heat generation is not taken into account ($\gamma = 0$), the contact pressure tends to a constant value ($p(\tau) \to 1$). As we have already mentioned in Section 2.3.4, in this case the bush may vibrate in a chaotic manner. The following parameters are fixed during numerical analysis: $\varepsilon = 0.1$, $\omega = 2$, $\omega_* = 0.4$, $Bi = 10$, $\widetilde{\omega} = 0.1$, $\delta = 10$, $b = 1$, $f_s = 0.3$, $\alpha = 0.3$, and $\beta = 0.3$. The following initial conditions are taken: $x(0) = 0$ and $y(0) = 0$. We are going to investigate the influence of control parameter ζ_k on the system dynamics. Curves 1 in Figs. 2.41–2.44 correspond to the values of $\zeta_k = 3.5$, whereas curves 2 correspond to $\zeta_k = 4.5$. Dashed curves in Figs. 2.41–2.44 correspond to the lack of frictional heat generation ($\gamma = 0$), whereas solid curves are related to frictional heat generation exhibited by coefficient $\gamma = 1$. Figure 2.41a shows trajectories of the analysed system in phase space, whereas Fig. 2.41b presents the Poincaré section.

In the case when the frictional heat generation is not taken into account, the contact friction as well as the contact pressure tend to constant values $p(\tau) \to 1$, $\theta(\tau) \to 2$. Note that our analytical predictions (2.31)–(2.33) give the same values.

Bush motion (as shown in Fig. 2.41) is periodic, which is in agreement with analytical prediction $\zeta_k = 3.5$ (see dashed curves 1 and point 1 in Fig. 2.41), whereas for parameter $\zeta_k = 4.5$ it is chaotic (see dashed curves 2 and points 2 in Fig. 2.41).

The frictional heat generation causes the occurrence of a period-2 bush motion for the parameter $\zeta_k = 3.5$ and period-1 motion for the parameter $\zeta_k = 4.5$

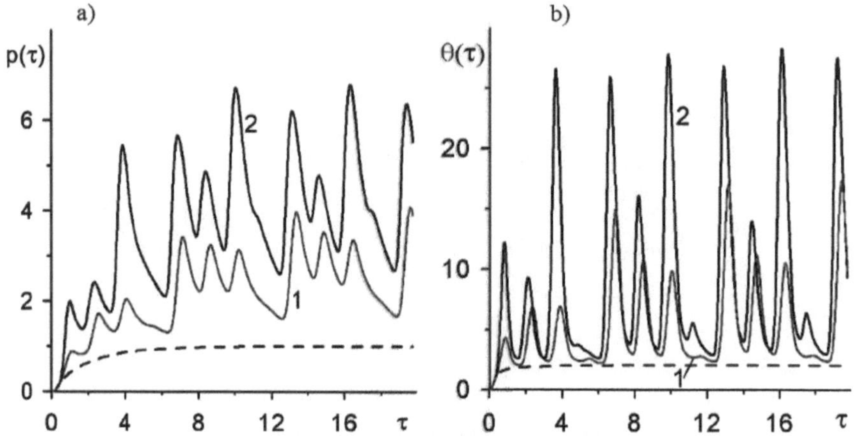

Fig. 2.42: Time histories of dimensionless contact pressure (a) and contact temperature (b) (curves 1: $\zeta_k = 3.5$, curves 2: $\zeta_k = 4.5$; solid curves: $\gamma = 1$, dashed curves: $\gamma = 0$).

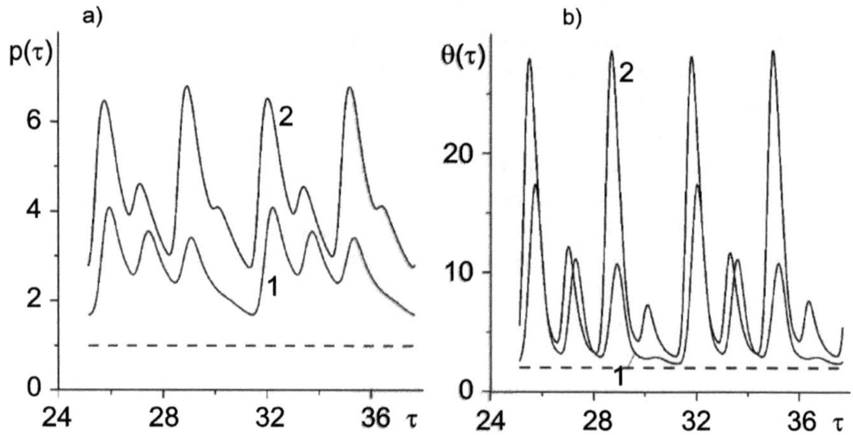

Fig. 2.43: Time histories for $\tau \in (25, 38)$ of dimensionless contact pressure (a) and contact temperature (b) (curves 1: $\zeta_k = 3.5$, curves 2: $\zeta_k = 4.5$; solid curves: $\gamma = 1$, dashed curves: $\gamma = 0$).

(see solid curves 1 and 2 in Fig. 2.41a). The contact pressure and the contact temperature change in time and for $\tau > 32$ and for $\zeta_k = 3.5$ the dimensionless period is defined by the formula $2T_p = 2 \cdot 2\pi/\omega = 2\pi$ (curves 1 in Fig. 2.44). On the other hand, for $\tau > 180$ and for $\zeta_k = 4.5$ the dimensionless period $T_p = 2\pi/\omega = \pi$ (see curves 2 in Fig. 2.44).

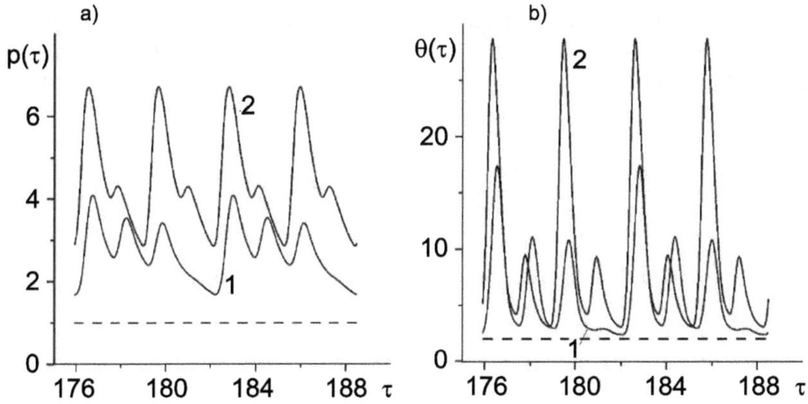

Fig. 2.44: Time histories for $\tau \in (176, 190)$ of dimensionless contact pressure (a) and contact temperature (b) (curves 1: $\zeta_k = 3.5$, curves 2: $\zeta_k = 4.5$; solid curves: $\gamma = 1$, dashed curves: $\gamma = 0$).

2.4. External shaft mechanical excitations

2.4.1 Shaft inertial motion with tribological processes

We analyse a one-dimensional contact model of a thermoelastic inertial shaft with a fixed rigid bush while frictional heat generation and wear take place [Pyryev (2000b)]. The influence of input parameters on the contact characteristics (shaft rotational velocity, contact pressure, temperature, and wear) is studied. A similar model has been examined by [Goryacheva (1988)] (without thermal expansion effects), with reference to the influence of various wear coefficients on the wear amount. A similar problem for the elastic layer [Pyryev (1994)] was analysed by [Olesiak, Pyryev(1998)], and [Pyryev, Grilitskiy (1996)] and without wear in reference [Pyryev, Grilitskiy (1995)]. Such problems were also studied earlier by [Pyryev, Grilitskiy (1996)] (without accounting for body masses) and by [Barber et al. (1985)] in the case of uniform braking. The braking processes, assuming that external loadings are known, have been analysed by [Chichinadze et al. (1979)]. A quasi-stationary approach to the discussed problem was proposed by [Aleksandrov, Annakulova (1990)].

Here, we assume that a contact zone and its geometry enable a sufficient approximation of the problem by a one-dimensional model. Thus, various peculiarities associated with real frictional behaviour can be included in the analysis [Chichinadze et al. (1979), Goriaceva (1988), Aleksandrov, Annakulova (1990)].

Next, we assume that $k_2/k_* \to \infty$ and that the bush is fixed; that is, $\varphi_2 = 0$. Taking $\tilde{\omega} = 1$, $\phi^0 = 0$ we have $t_* = R_1^2/a_1$. Furthermore, we assume that a friction coefficient does not depend on the shaft rotational velocity; that is, $F = f_s =$ const. The shaft rotates at the velocity $\Omega(t) = \Omega_* \omega_1(t)$. It is assumed also that the shaft rotational velocity changes in accordance with the shaft rotational motion due to input of the moment $M_0 h_M(t)$ ($h_M \to 1$, $t \to \infty$) (see Equations (2.24)).

In the dimensionless form, Equations (2.18) to (2.23), (2.27), and (2.28) governing our problem are as follows,

$$\frac{\partial^2\theta(r,\tau)}{\partial r^2} + \frac{1}{r}\frac{\partial\theta(r,\tau)}{\partial r} = \frac{\partial\theta(r,\tau)}{\partial\tau}, \qquad 0<\tau<\tau_c, \qquad 0<r<1; \quad (2.105)$$

$$\frac{\partial\theta(1,\tau)}{\partial r} + Bi[\theta(1,\tau) - h_T(\tau)] = \gamma f_s \omega_r p(\tau) + \gamma f_s \omega_r Q(\tau); \quad (2.106)$$

$$r\frac{\partial\theta(r,\tau)}{\partial r}\bigg|_{r\to 0} = 0, \qquad 0<\tau<\tau_c; \quad (2.107)$$

$$\theta(r,0) = 0, \qquad 0<r<1. \quad (2.108)$$

$$p(\tau) = -u^w(\tau) + \int_0^1 \theta(\eta,\tau)\eta d\eta, \quad 0<\tau<\tau_c; \quad (2.109)$$

$$u^w(\tau) = \gamma f_s \omega_r \xi \int_0^\tau p(\eta) d\eta + \gamma f_s \omega_r \xi \int_0^\tau Q(\eta) d\eta, \qquad 0<\tau<\tau_c; \quad (2.110)$$

$$\dot\phi(\tau) = \omega_1^0 + a_M \left[m_0 \int_0^\tau h_M(\eta) d\eta - f_s \int_0^\tau p(\eta,\dot\phi) d\eta \right], \qquad 0<\tau<\tau_c, \quad (2.111)$$

$$Q(\tau) = (\dot\phi/\omega_r - 1)p(\tau), \quad (2.112)$$

where

$$\xi = \frac{k^w}{\gamma f_s} = \frac{K^w \lambda_1}{2 f_s \alpha_1 a_1 (1+\nu_1)}.$$

Note that in Equations (2.106), (2.110), and (2.111) ω_r denotes an arbitrary constant enabling separation of the linear system part.

Applying the Laplace transformations [Carslaw, Jaeger (1959)], a solution to the problem (2.105)–(2.112) has the following form,

$$\theta(r,\tau) = \psi_\theta(R,\tau) + \tilde{v}\frac{d^2}{d\tau^2}G_1(r,\tau) * Q(\tau), \quad (2.113)$$

$$p(\tau) = \psi_p(\tau) + \tilde{v}\frac{d}{d\tau}F_2(\tau) * Q(\tau), \quad (2.114)$$

$$u^w(\tau) = \psi_u(\tau) + \tilde{v}\xi\frac{d}{d\tau}F_1(\tau) * Q(\tau), \quad (2.115)$$

$$\dot{\phi}(\tau) = \psi_\phi(\tau) - a_M f \tilde{v} F_2(\tau) * Q(\tau), \tag{2.116}$$

where

$$\psi_\theta(r,\tau) = Bi \frac{d}{d\tau} G_2(r,\tau) * h_T(\tau),$$

$$\psi_p(\tau) = Bi \frac{d^2}{d\tau^2} F_3(\tau) * h_T(\tau),$$

$$\psi_u(\tau) = Bi \tilde{v}\xi \frac{d}{d\tau} F_3(\tau) * h_T(\tau),$$

$$\psi_\phi(\tau) = \omega_1^\circ + a_M \left[m_0 h_M(\tau) * H(\tau) - f\, Bi \frac{d}{d\tau} F_3(\tau) * h_T(\tau) \right],$$

$$G_n(r,\tau) = g_n + \sum_{m=1}^{\infty} \frac{D_n(r,s_m)}{s_m \Delta'(s_m)} e^{s_m \tau},$$

$$F_n(\tau) = f_n + \sum_{m=1}^{\infty} \frac{\Delta_n(s_m)}{s_m \Delta'(s_m)} e^{s_m \tau}, \quad n = 1,2,3, \tag{2.117}$$

$$\Delta_1(s_m) = Bi\ C_m + s_m S_m; \quad \Delta_2(s_m) = s_m S_m - \xi \Delta_1(s_m); \quad \Delta_3(s_m) = S_m;$$

$$\Delta(s) = s\Delta_1(s) - \tilde{v}\Delta_2(s); \quad D_1(r,s) = C_m^R; \quad D_2(r,s) = (s_m + \tilde{v}\xi) C_m^R;$$

$$\Delta'(s_m) = 0.5\{(Bi\ S_m + C_m)(s_m + \tilde{v}\xi) + 2\Delta_1(s_m) - \tilde{v} C_m\};$$

$$C_m^R = I_0(r\sqrt{s_m}), \qquad S_m = I_1(\sqrt{s_m})/\sqrt{s_m}, \qquad C_m = I_0(\sqrt{s_m}),$$

$$m = 1,2,3,\ldots$$

$$f_1 = \frac{1}{\tilde{v}\xi}, \; f_2 = -\frac{1}{\tilde{v}}, \; f_3 = \frac{1}{2Bi\ \tilde{v}\xi}, \; g_1 = \frac{1}{Bi\ \tilde{v}\xi}, \; g_2 = \frac{1}{Bi}, \; \tilde{v} = \gamma f_s \omega_r,$$

and s_m are the roots of the characteristic equation $\Delta(s) = 0$ $(m = 1,2,3,\ldots)$. The analysis carried out by [Pyryev, Grilitskiy (1996)] shows that $\mathrm{Im} s_m = 0$, $\mathrm{Re} s_m < 0$ for $m = 3,4,\ldots$, whereas for $m = 1,2$ they lie either in left or right half-planes of the complex plane. For $0 < \xi < \xi_1$ $(\xi_1 = 1/(1+Bi/2))$ and $\tilde{v} \le v_2$ the roots are negative, for $v_2 < \tilde{v} < v_1$ the root is a complex conjugate with negative real part, for $v_1 < \tilde{v} < v_3$ the root is a complex conjugate with positive real part, and for $v_3 < \tilde{v}$ the root is positive. The approximate relations for v_i $(i = 1,2,3)$ are reported by [Pyryev, Grilitskiy (1996)].

The quantities $p(\tau)$ and $\dot{\phi}(\tau)$ that appeared in Equations (2.114) and (2.116) are governed by a system of nonlinear Volterra–Hammerstein integral equations of the second-order convolution type [Verlan, Sizikov (1986)]. The numerical investigations of the problem are carried out using quadrature and the following asymptotic estimations.

$$G_1(1,\tau) = \frac{4\tau}{3}\sqrt{\frac{\tau}{\pi}} + O(\tau^2), \qquad G_2(1,\tau) = 2\sqrt{\frac{\tau}{\pi}} + O(\tau),$$

$$F_1(\tau) = \tau + O(\tau^{1.5}), \qquad \tau \to 0,$$

$$F_2(\tau) = (1-\xi)\tau + O(\tau^{1.5}), \qquad \tau \to 0,$$

$$F_3(\tau) = 0.5\tau^2 + O(\tau^{2.5}), \qquad \tau \to 0. \tag{2.118}$$

2.4.1.1 Solution properties

The behaviour of frictional characteristics of the thermoelastic contact is described by the functions: $h_T(\tau) = (1 - \exp(-\delta\tau^2))H(\tau)$, $h_M(\tau) = H(\tau)$. Depending on the parameters, the following famous system dynamic states can be realized.

Behaviour in the initial time instant. Accounting for the Laplace transform, the following asymptotic terms are found for the thermoelastic contact characteristics in the initial time instant,

$$\theta(1,\tau) = \delta\frac{16Bi}{15}\tau^2\sqrt{\frac{\tau}{\pi}} + O(\tau^{3.5}),$$

$$p(\tau) = \delta\frac{Bi}{3}\tau^3 + O(\tau^{3.5}),$$

$$u^w(\tau) = \delta\frac{Bi}{12}\gamma f\omega_1^\circ\xi\tau^4 + O(\tau^{4.5}),$$

$$\dot{\phi}(\tau) = \omega_1^\circ + a_M\left(m_0\tau - f\delta\frac{Bi}{12}\tau^4 + O(\tau^5)\right), \qquad \tau \to 0. \tag{2.119}$$

Shaft motion at constant velocity. When the shaft rotates at constant velocity $\dot{\phi}(\tau) = \omega_r = \text{const}$, the function $Q(\tau) = 0$ and the considered problem is linear. First terms of (2.113)–(2.116) provide solutions to this problem. Note that for $\tilde{v} \geq v_3$, $\xi < \xi_1$, the so-called frictional instability occurs; that is, the frictional contact characteristics increase exponentially. The critical velocity is

$$v_3 = 2Bi\frac{1 + \frac{\xi}{\xi_1} + \sqrt{\left(1 + \frac{\xi}{\xi_1}\right)^2 - (1-\xi)^2}}{(1-\xi)^2}. \tag{2.120}$$

If there is no wear ($\xi = 0$), the dimensionless critical velocity is equal to $2Bi$.

Behaviour of a solution when it approaches a stationary solution. During wear and conditions $f_* > 1$ ($f_* = f_s/2m_0$) there are formally stationary solutions of nonlinear problem (2.105)–(2.112) defined by (in Equation (2.13), (2.24), and (2.105) differential terms with respect to time are neglected) the parameters: $p_{st} = 0.5f_*^{-1}$, $\theta_{st} = 1$, $\omega_{st} = 0$, and $u^w = 0.5 - p_{st}$. Analysis of roots of the linearized characteristic equation in the vicinity of the stationary solution governed by (2.105)–(2.112) gives $\mathrm{Re}s_1 > 0$; that is, the stationary solutions are unstable and the inequality $\tau_c < \infty$ is satisfied.

In the case without wear ($\xi = 0$), the stationary solution of the nonlinear problem for $f_* < 1$ is

$$p_{st} = 0.5f_*^{-1}, \qquad \theta_{st} = 2p_{st}, \qquad \omega_{st} = 2Bi(1 - f_*)/\gamma f_s. \tag{2.121}$$

The characteristic algebraic equation of the linearized nonlinear problem (2.105)–(2.112) has the form ($\xi = 0$):

$$s\Delta_1(s) + (\gamma f_s a_M m_0 - 2Bi\, s\,(1 - f_1))\Delta_3(s) = 0. \tag{2.122}$$

It can be shown that if $\mathrm{Re}s_m < 0$, $m = 1, 2, 3, \ldots$, then the stationary solution ($\tau \to \infty$) of the nonlinear problem is stable. Behaviour of a nonlinear problem in the neighbourhood of the stationary solution is determined for $\tau \to \infty$ through the roots $s_{1,2}$ of the characteristic equation (2.122) which lie closest to the imaginary axis. Using the characteristic equation (2.122), the following analytic formula defines the first two roots

$$s_{1,2} = (-B \pm \sqrt{D})/A, \tag{2.123}$$

where

$$B = f_*Bi + \gamma f_s a_m m_0/16, \qquad D = Bi^2 f_*^2 - \gamma f_s a_m m_0(B_1 + \gamma f_s a_m m_0/768),$$

$$A = B_1 + f_*Bi/8 + \gamma f_s a_m m_0/192, \qquad B_1 = 1 + Bi/4 + f_*Bi/8.$$

If the following condition is satisfied,

$$\gamma f_s a_M m_0 > 384(\sqrt{D_1} - B_1), \tag{2.124}$$

where $D_1 = B_1^2 + f_*^2 Bi^2/192$, then the roots (2.123) are complex and the contact characteristics exhibit damped oscillation. The imaginary part of the roots defines the period; that is, $T_p = 2\pi/\mathrm{Im}s_1$.

2.4.1.2 Numerical results

In order to verify the theoretical analysis concerning dynamics of the contact characteristics, numerical computations are carried out for the obtained solution for different values of the control parameter ξ. A cylinder made from steel is examined ($\nu_1 = 0.3$, $E_1 = 190\,\text{GPa}$, $\alpha_1 = 14 \cdot 10^{-6\circ}\text{C}^{-1}$, $\lambda_1 = 21\,\text{W m}^{-1\circ}\text{C}^{-1}$, $a_1 = 5.9 \cdot 10^{-6}\,\text{m}^2\,\text{s}^{-1}$) for $R_1 = 30\,\text{mm}$, $\delta = 10^4$, $Bi = 24$, $t_* = 153\,\text{s}$. Depending on the parameters, the considered model can exhibit either braking or acceleration processes further approaching a stationary state.

Acceleration process ($f_* < 1$). The problem is dimensionalised for $T_0 = 5^\circ\,\text{C}$, $\omega_1^\circ = 15.2$, $f_s = 0.1$, $m_0 = 1/\pi$, $a_M = 300\pi$ and for $P_* = 6.65 \cdot 10^{-2}\,\text{MPa}$, $U_* = 5.46 \cdot 10^{-6}\,\text{m}$. Stationary values (without wear) are as follows: $p_{st} = 3.18$, $\theta_{st} = 6.37$, and $\omega_{st} = 108.3$. Both numerical and analytical analyses show that during acceleration and when the contact characteristic approaches a stationary state and in the absence of wear (the conditions (2.124) hold), damped vibrations occur (dashed curves in Fig. 2.45).

Note that the occurrence of even small wear essentially disturbs the contact characteristics (see curves 1–4 in Fig. 2.45).

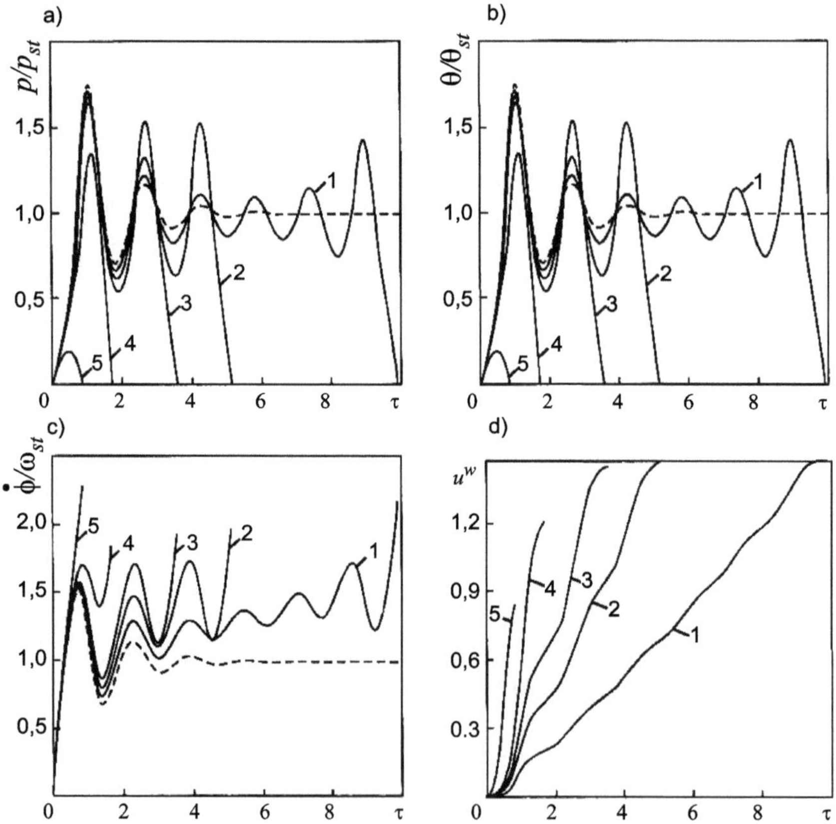

Fig. 2.45: Time histories of the relative pressure $p(\tau)/p_{st}$ (a), relative shaft velocity $\dot\phi(\tau)/\omega_{st}$ (c) and wear $u^w = U^w(\tau)/U_*$ (d) versus dimensionless time during accelerating (curves 1: $\xi = 0.001$, curves 2: $\xi = 0.002$, curves 3: $\xi = 0.003$, curves 4: $\xi = 0.01$, curves 5: $\xi = 0.05$, dashed curves: $\xi = 0$).

Figure 2.45 shows the influence of ξ on relative pressure (Fig. 2.45a), relative temperature (Fig. 2.45b), relative velocity (Fig. 2.45c), and wear (Fig. 2.45d) on dimensionless time.

The period of oscillations of the contact characteristics (curves 1 and 2 in Fig. 2.45) is equal to $T_{os} = 1.6$. The period estimated analytically through the roots (2.123) is of the amount of 1.8. Results are improved when the roots $s_{1,2}$ are close to the imaginary axis. Increasing the shaft inertial moment (decrease of the parameter ε) causes an increase of the oscillation period. An increase of wear stabilises time history (see curves 5 in Fig. 2.45), whereas time contact interval τ_c and wear amount u^w decrease. When acceleration of the periodic motion is zero, then the contact pressure achieves its stationary value.

Braking process ($f_* > 1$). The following parameters are taken: $T_0 = 10°$ C, $m_0 = 0$, $\omega_1^\circ = 76.3 \cdot 10^2$, $f_s = 0.01$, $a_M = 2\pi \cdot 10^6$ and $P_* = 13.3 \cdot 10^{-2}$ MPa, $U_* = 10.9 \cdot 10^{-6}$ m. The numerical analysis shows that when the shaft stops and its temperature achieves the bush temperature, it will not move again (curves 1 and 2 in Fig. 2.46). Owing to an increase of the wear coefficient, the braking time (curves 1 and 2 in Fig. 2.48) and wear increase, whereas contact pressure and temperature decrease. A further increase of the wear coefficient causes disappearance of the contact zone between the shaft and bush (curves 3 in Fig. 2.46). An increase of the shaft inertial moment causes an increase of the braking time.

2.4.2 *Inertialess shaft and bush dynamics and frictional heat generation*

In this section (contrary to 2.3), one of the bodies is subject to mechanical excitation, which makes us refer to the additional equation of this body. The latter new model opens some new essential questions. Does frictional TEI occur in this case, and if not, then how is the system dynamics realized? Is it possible to realize a self-excited motion? What kind of the system motion appears during the braking process? Answers to these questions are given in this section. Note that now, contrary to the case described in Section 2.3, another mechanism of the contact occurrence between the bodies takes place.

Let us remember that in Section 2.3 bush and shaft are in contact owing to the introduced initial shaft compression. In this section we study the contact between bush and shaft which occurs via thermal shaft extension. In real conditions this situation takes place when the shaft thermal extension exceeds the bush heat extension. The mentioned case has been analysed, for instance, in the works of [Pyryev et al. 1995, Pyryev, Mandzyk 1996, Grilitskiy et al. (1997), (1998)] for an infinite two-layer circular cylinder under friction heating. In this work we assume that the shaft transfers heat ideally and hence it is not thermally extended.

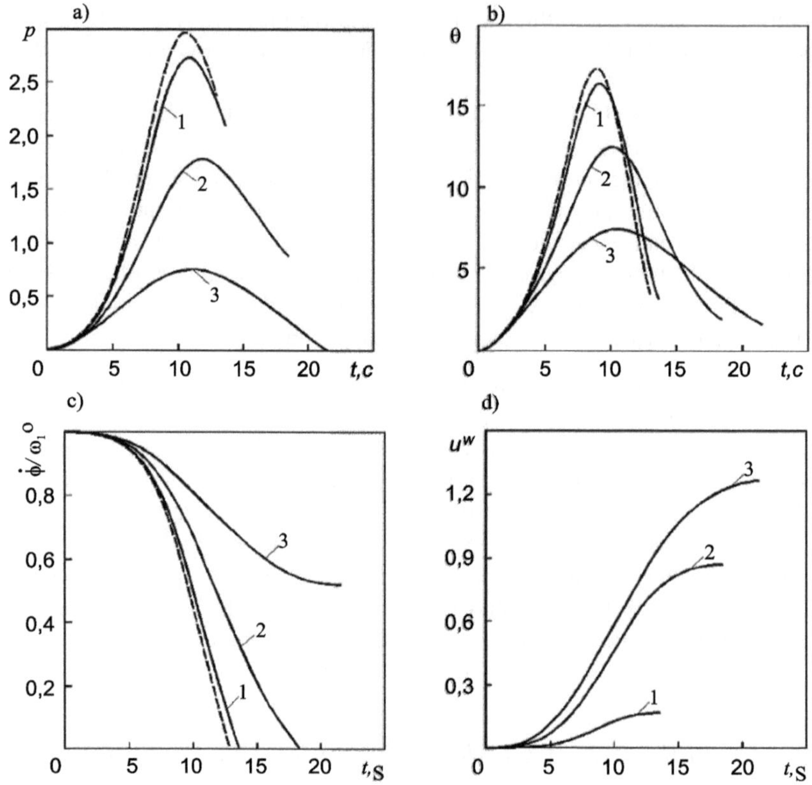

Fig. 2.46: Evolution in time of dimensionless contact pressure $p = P/P_*$ (a), contact temperature $\theta = T/T_0$ (b), rotational shaft velocity $\dot{\phi}/\omega_1^\circ$ (c) and wear $u^w = U^w/U_*$ in breaking instant (curves 1: $\xi = 0.01$, curves 2: $\xi = 0.05$, curves 3: $\xi = 0.1$, dashed curves: $\xi = 0$).

Observe that the system under analysis is described by two characteristic times, that is, the characteristic time related to system oscillations t_D (small) and the characteristic time related to heat transfer t_T (large). If the ratio of these two time-scales is 'small,' then the system can be treated as an uncoupled one.

A numerical verification of the presented analytical estimations of periodic oscillations completes the considerations.

It is assumed that the shaft has the inertial moment B_1 and rotates according to the rotational motion due to input torque action. It is also assumed that at the beginning the shaft and bush are not subjected to the external load ($T_* = T_0$, $U_* = \alpha_1(1+\nu_1)R_1T_0$, $P_* = \alpha_1E_1T_0/(1-2\nu_1)$, $\gamma = \alpha_1E_1a_1/(1-2\nu_1)\lambda_1$, $V_* = R_1/t_T$, $t_D = \sqrt{B_2/k_2}/R_2$, $\tilde{\omega} = t_D/t_T$, $a_M = 2\pi R_1^2 t_T^2 P_*/B_1$, $U_0 = 0$), and the dimensionless bush temperature changes within the rule $h_T(\tau) = (1-\exp(-\delta\tau^2))$. Note that due to heat transfer the rotating shaft starts to expand and a contact between the shaft and bush appears. Consider the case when the bush is coupled with the housing through springs with the stiffness k_2 ($k_1 = k_3 = 0$). Owing to an arbitrary parameter $k_* = k_2$, which gives the constant $a = -1$. It is also assumed that $\chi = 1$ (hence $t_* = \sqrt{B_2/(k_* R_2^2)}$).

The following friction coefficient versus the relative velocity is accounted for in (2.61).

Observe that $F(y) = f(V_* y)$ has a local minimum for $y_{\min} = \sqrt{\alpha/3\beta} = V_{\min}/V_*$. The function $\mathrm{Sgn}(x)$ is approximated by the function $\mathrm{sgn}_{\varepsilon_0}(y)$ of the form (2.58).

2.4.2.1 Application of the Laplace transform

Applying the Laplace transform to Equations (2.18), (2.22), and (2.23) (see our earlier consideration described in Section 2.2.4), the following functions are obtained,

$$p(\tau) = 2Bi \int_0^\tau G_p(\tau - \xi)\dot{h}_T(\xi)d\xi$$

$$+ 2\gamma \int_0^\tau \dot{G}_p(\tau - \xi)F(\dot{\phi} - \dot{\varphi})p(\xi)(\dot{\phi} - \dot{\varphi})d\xi, \tag{2.125}$$

$$\theta(r,\tau) = Bi \int_0^\tau G_\theta(r, \tau - \xi)\dot{h}_T(\xi)d\xi +$$

$$+ \gamma \int_0^\tau \dot{G}_\theta(r, \tau - \xi)F(\dot{\phi} - \dot{\varphi})p(\xi)(\dot{\phi} - \dot{\varphi})d\xi, \tag{2.126}$$

where $G_p(\tau)$, $G_\theta(r,\tau)$ arc the known functions (see (2.33)) of the form

$$\{G_p(\tau), G_\theta(1,\tau)\} = \frac{\{0.5, 1\}}{Bi} - \sum_{m=1}^{\infty} \frac{\{2Bi, 2\mu_m^2\}}{\mu_m^2(Bi^2 + \mu_m^2)} e^{-\mu_m^2 \tau}, \tag{2.127}$$

and μ_m $(m = 1, 2, 3, \dots)$ are the roots of the characteristic Equation (2.33).

Note that the considered problem is governed by the system of nonlinear ordinary differential Equations (2.5) and (2.27),

$$\ddot{\varphi}(\tau) + \tilde{\omega}^{-2}\varphi(\tau) = \varepsilon F(\dot{\phi} - \dot{\varphi})p(\tau),$$

$$\ddot{\phi} = a_M[m_0 h_M(\tau) - F(\dot{\phi} - \dot{\varphi})p(\tau)], \tag{2.128}$$

and the integral equation (2.125) governing the rotational velocities $\dot{\varphi}(\tau)$, $\dot{\phi}$ and the contact pressure $p(\tau)$. The temperature is defined by formulas (2.126).

2.4.2.2 Stationary dynamics

Observe that in Equation (2.128) the variable $\phi(\tau)$ does not appear, and hence the phase space of the bush–shaft system is three-dimensional. Consider the case when after a transitional process the cylinder starts to rotate at constant velocity $\dot{\phi}(\tau) = \omega_{st}$ and the bush does not move $\dot{\varphi} = 0$. Then, the steady-state solution of the problem under consideration (in Equations (2.128) and (2.18) the differential terms are omitted) has the following form,

$$p_{st} = \frac{1}{1-v}, \qquad \theta_{st} = \frac{1}{1-v}, \qquad \varphi_{st} = \tilde{\omega}^2 \varepsilon m_0, \qquad v = \frac{\gamma \omega_{st} F(\omega_{st})}{Bi}, \tag{2.129}$$

where ω_{st} is the solution to the nonlinear equation

$$F(\omega_{st}) = \frac{m_0}{1 + \gamma m_0 \omega_{st}/2Bi}. \tag{2.130}$$

Function F is similar to that defined in (2.61) with (2.130) taken into account, m_0 is the applied dimensionless moment, γ is the parameter related to thermal distortivity ($\gamma = 1.87$ for stainless steel, $\gamma = 1.71$ for aluminium alloy, $\gamma = 1.61$ for copper, and $\gamma = 1.38$ for titanium alloy), and Bi is the Biot number.

As has been already mentioned (Section 2.3.2.2), the stationary solutions (2.129) have physical meaning only if $v < 1$.

Figure 2.47 shows a graphical solution of Equation (2.130) for various parameters m_0 and Bi. For a stainless steel shaft ($\alpha_1 = 14 \cdot 10^{-6}{}^\circ\mathrm{C}^{-1}$, $\lambda_1 = 21\,\mathrm{W\,m^{-1}{}^\circ C^{-1}}$, $\nu = 0.3$, $a_1 = 5.9\,\mathrm{mm^2\,s^{-1}}$, $E_1 = 190\,\mathrm{GPa}$) parameter $\gamma = 1.87$, and for $R_1 = 4 \cdot 10^{-3}\,\mathrm{m}$ the characteristic time $t_T = 2.71\,\mathrm{s}$, and $V_* = 1.47 \cdot 10^{-3}\,\mathrm{m\,s^{-1}}$. Solid curve 1 corresponds to $m_0 = 0.14$, $Bi = 10$, solid curve 2 corresponds to $m_0 = 0.1$, $Bi = 10$, solid curve 3 corresponds to $m_0 = 0.05$, $Bi = 10$, and finally, solid curve 4 corresponds to $m_0 = 0.14$, $Bi = 1$. The dashed curve represents function $F(\omega_{st})$.

Equation (2.130) may have: one solution ω_{st}^3 ($F'(\omega_{st}^3) > 0$) for $m_0 = 0.14$, $Bi = 10$ (first case); three solutions ω_{st}^1, ω_{st}^2, ω_{st}^3 ($F'(\omega_{st}^1) > 0$, $F'(\omega_{st}^2) < 0$, $F'(\omega_{st}^3) > 0$) for $m_0 = 0.1$, $Bi = 10$ (second case); one solution $\omega_{st}^1 = 0$ for $m_0 = 0.05$, $Bi = 10$ and for approximation (2.58) $\omega_{st}^1 \approx \varepsilon_0 m_0/2f_s$, $F'(\omega_{st}^1) \approx 2f_s/\varepsilon_0$ (third case); one solution ω_{st}^2 ($F'(\omega_{st}^2) < 0$) for $m_0 = 0.14$, $Bi = 1$ (fourth case).

For a small value of γ/Bi ($\gamma/Bi \ll 1$) (frictional heat generation is neglected $\gamma = 0$) and for $m_0 \in [0, f_{\min})$, Equation (2.130) can have one solution $\omega_{st}^1 = 0$. It may also have three solutions ω_{st}^1, ω_{st}^2, and ω_{st}^3, for $m_0 \in (f_{\min}, f_s)$, and one solution ω_{st}^3 for $m_0 \in (f_s, \infty)$. Equation (2.130) shows that the stationary solution does not depend on parameter $\tilde{\omega}$. Let us now analyse perturbations of the stationary process (2.130), which are defined by the equation $h_T(\tau) = 1 + h_T^*(\tau)$.

A solution is sought in the form

$$\varphi(\tau) = \varphi_{st} + \varphi^*(\tau), \qquad \theta(r,\tau) = \theta_{st}(r) + \theta^*(r,\tau),$$

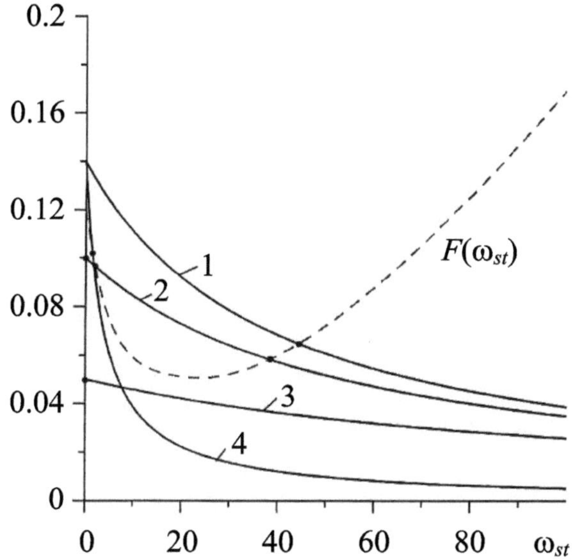

Fig. 2.47: Graphical solution to Equation (2.130). Solid curves 1: $m_0 = 0.14$, $Bi = 10$, 2: $m_0 = 0.1$, $Bi = 10$, 3: $m_0 = 0.05$, $Bi = 10$, 4: $m_0 = 0.14$, $Bi = 1$. The dashed curve corresponds to $F(\omega_{st})$.

$$p(\tau) = p_{st} + p^*(\tau), \qquad \dot{\varphi} = \dot{\varphi}^*(\tau), \qquad \dot{\phi} = \omega_{st} + \dot{\phi}^*(\tau), \tag{2.131}$$

where $|\varphi^*| \ll 1$, $|\theta^*| \ll 1$, $|p^*| \ll 1$, $|\dot{\varphi}^*| \ll 1$, $|\dot{\phi}^*| \ll 1$.

After linearization of the right-hand sides of Equations (2.128) and the boundary condition (2.2.2), the following set of perturbation equations is obtained.

$$\ddot{\varphi}^*(t) + \tilde{\omega}^{-2}\varphi^*(t) = \varepsilon F(\omega_{st})p^*(\tau) + \varepsilon F'(\omega_{st})p_{st}(\dot{\phi}^* - \dot{\varphi}^*), \qquad 0 < \tau < \infty, \tag{2.132}$$

$$\varphi^*(0) = 0, \qquad \dot{\varphi}^*(0) = 0; \tag{2.133}$$

$$p^*(\tau) = 2\int_0^1 \theta^*(\xi, \tau)\xi d\xi, \qquad 0 < \tau < \infty; \tag{2.134}$$

$$\frac{\partial^2\theta^*(r,\tau)}{\partial r^2} + \frac{1}{r}\frac{\partial\theta^*(r,\tau)}{\partial r} = \frac{\partial\theta^*(r,\tau)}{\partial\tau}, \qquad 0 < \tau < \infty, \qquad 0 < r < 1; \tag{2.135}$$

$$\frac{\partial\theta^*(1,\tau)}{\partial r} + Bi\theta^*(1,\tau) = Bi\, h_T^* + \gamma[\omega_{st}F(\omega_{st})p^*(\tau)$$
$$+ p_{st}(\dot{\phi}^* - \dot{\varphi}^*)(F(\omega_{st}) + \omega_{st}F'(\omega_{st}))], \qquad (2.136)$$

$$r\frac{\partial\theta^*(r,\tau)}{\partial r}\bigg|_{r\to 0} = 0, \qquad 0 < \tau < \infty, \qquad \theta^*(r,0) = 0, \qquad 0 < r < 1, \qquad (2.137)$$

$$\ddot{\phi}^*(t) = -a_M[F(\omega_{st})p^* + F'(\omega_{st})p_{st}(\dot{\phi}^* - \dot{\varphi}^*)], \qquad 0 < \tau < \infty, \qquad (2.138)$$

$$\phi^*(0) = 0, \quad \dot{\phi}^*(0) = 0. \qquad (2.139)$$

Application of the Laplace transformation to the linear system (2.132)–(2.139) of the form

$$\{\bar{\theta}^*(r,s), \bar{p}^*(s), \bar{\varphi}^*(s), \bar{h}_T^*(s), \bar{\phi}^*(s)\} = \int_0^\infty \{\theta^*, p^*, \varphi^*, h_T^*, \phi^*\}e^{-s\tau}d\tau,$$

provides a solution in the form of the Laplace transform. The characteristic equation of the linearized problem has the form

$$\Delta_1^*(s) = 0, \qquad (2.140)$$

$$\Delta_1^*(s) = s\Delta^*(s) + a_M p_{st}(\tilde{\omega}^2 s^2 + 1)[\beta_2\Delta_1(s) + 2Bi\, v\beta_1\Delta_2(s)],$$

$$\Delta^*(s) = \Delta_1(s)\Omega_2(s) - 2Bi\, v\Delta_2(s)\Omega_1(s), \qquad (2.141)$$

$$\Omega_1(s) = \tilde{\omega}^2\left(s^2 - \frac{\varepsilon\beta_1}{1-v}s\right) + 1, \qquad \Omega_2(s) = \tilde{\omega}^2\left(s^2 + \frac{\varepsilon\beta_2}{1-v}s\right) + 1,$$

$$\beta_2 = F'(\omega_{st}), \quad \beta_1 = \frac{F(\omega_{st})}{\omega_{st}}.$$

The roots s_m (Re$s_1 >$ Re$s_2 > \cdots >$ Re$s_m > \cdots$, $m = 1, 2, 3, \ldots$) of the characteristic equation (2.140) may lie in the left-hand part (LHP) Re$s < 0$ (a stationary solution is stable) or in the right-hand part (RHP) Re$s > 0$ (a stationary solution is unstable) of the complex plane (s is a complex variable). The parameters separating the two half-planes are called critical.

If frictional heat generation is not taken into account ($\gamma = 0$), the characteristic equation is governed by the following cubic one: $s\Omega_2(s) + \beta_2 a_M(\tilde{\omega}^2 s^2 + 1) = 0$. Its roots lie in the RHP of the complex plane if $\beta_2 < 0$. For a given bush velocity (one may assume $a_M \to 0$), the characteristic equation is reduced to the form $\Delta^*(s) = 0$, whereas for $\gamma = 0$ (the frictional heat generation is not taken into account) it is $\Omega_2(s) = 0$.

Let us analyse the stationary stable solution in more detail. The characteristic function has the form

$$\Delta_1^*(s) = \sum_{m=0}^{\infty} s^m b_m, \tag{2.142}$$

$$b_0 = a_M Bi\, p_{st}(\beta_2 + v\beta_1), \quad d_0 = Bi(1-v),$$

$$b_1 = \frac{d_0 + a_M p_{st}[2(2+Bi)\beta_2 + Bi\, v\beta_1]}{8},$$

$$d_1 = 0.5 + Bi[0.25 - 0.125\, v + \tilde{\omega}^2 p_{st}\varepsilon(\beta_2 + v\,\beta_1)],$$

$$b_m = d_{m-1} + a_M p_{st}[(d_m^{(1)} + \tilde{\omega}^2 d_{m-2}^{(1)})\beta_2 + 2Bi\, v(d_m^{(2)} + \tilde{\omega}^2 d_{m-2}^{(2)})\beta_1],$$

$$d_m = d_m^{(1)} - 2Bi\, v d_m^{(2)} + \tilde{\omega}^2 \varepsilon p_{st}(\beta_2 d_{m-1}^{(1)} + 2Bi\, v\beta_1 d_{m-1}^{(2)})$$
$$+ \tilde{\omega}^2(d_{m-2}^{(1)} - 2Bi\, v d_{m-2}^{(2)}), \qquad m = 2, 3, \ldots$$

$$d_m^{(1)} = \frac{Bi + 2m}{2^{2m}(m!)^2}, \qquad d_m^{(2)} = \frac{1}{2^{2m+1} m!(1+m)!}, \qquad m = 0, 1, \ldots.$$

Observe that for small value of a_M the roots being sought can be found in Equation (2.142) for $m = 2$:

$$s_{1,2} = 0.5(-b_1 \pm \sqrt{b_1^2 - 4b_0 b_2})/b_2. \tag{2.143}$$

It should be emphasised that for a given shaft velocity ($a_M \to 0$), Equation (2.142) provides the condition $v > 1$ of a frictional TEI. However, when a moment of inertia of the shaft ($a_M > 0$) is taken into account, then the frictional TEI does not occur. In the latter case, the system itself controls the rotational shaft velocity by always keeping $v < 1$. Note that the shaft dynamics significantly influences the values of the characteristic equation roots only if the parameter $\tilde{\omega}^2(a_M + \varepsilon)$ cannot be considered a 'small' one.

Under the assumption of the temperature increase of the surrounding medium up to $T_* = 5°\text{C}$, the stainless steel shaft of the radius $R_1 = 4 \cdot 10^{-3}\,\text{m}$ is further investigated ($P_* = 33.2\,\text{MPa}$ and $t_T = 2.71\,\text{s}$). Furthermore, it is assumed that either the moment $M_0 = 334\,\text{N}$ ($m_0 = 0.1$) or the moment $M_0 = 468\,\text{N}$ ($m_0 = 0.14$) is applied to the shaft. Let either $\alpha_T = 5.25 \cdot 10^4\,\text{W m}^{-2}{}°\text{C}^{-1}$ ($Bi = 10$) or $\alpha_T = 5.25 \cdot 10^3\,\text{W m}^{-2}{}°\text{C}^{-1}$ ($Bi = 1$). It is also assumed that the shaft moment of inertia $B_1 = 245.8\,\text{kg} \cdot \text{m}$ ($a_M = 100$). On the other hand, it is assumed that the bush moment of inertia $B_2 = 245.8\,\text{kg} \cdot \text{m}$ (the internal radius $R_2 = 4 \cdot 10^{-2}\,\text{m}$), and the stiffness of the springs $k_2 = 2.1 \cdot 10^6\,\text{N m}^{-2}$ (which gives $t_D = 0.271\,\text{s}$ and $\varepsilon = 100$). The ratio of small t_D and large t_T time scales is $\tilde{\omega} = 0.1$.

In the first case ($m_0 = 0.14$, $Bi = 10$) there is one solution $\omega^3_{st} = 44.40$, $p^3_{st} = \theta^3_{st} = 2.16$, $\varphi^3_{st} = 0.14$ ($\beta_1 = 0.14 \cdot 10^{-2}$, $\beta_2 = 0.12 \cdot 10^{-2}$, $v = 0.54$), which is stable (the first roots of Equation (2.140) $s_{1,2} = -0.18 \pm 10.07i$ lie in the LHP). The expected 'period' of damped oscillations is equal to $2\pi/\mathrm{Im}s_1 = 0.62$.

In the second case ($m_0 = 0.1$, $Bi = 10$) there are three solutions. The solution $\omega^3_{st} = 38.40$, $p^3_{st} = \theta^3_{st} = 1.72$, $\varphi^3_{st} = 0.1$ ($\beta_1 = 0.15 \cdot 10^{-2}$, $\beta_2 = 0.1 \cdot 10^{-2}$, $v = 0.41$) is stable (the first roots of Equation (2.140) $s_{1,2} = -0.11 \pm 10.04i$, $s_3 = -0.53$ lie in the LHP). The solution $\omega^2_{st} = 1.99$ ($\beta_1 = 4.85 \cdot 10^{-2}$, $\beta_2 = -0.96 \cdot 10^{-2}$, $v = 3.58 \cdot 10^{-2}$) is unstable (the roots of Equation (2.140) $s_{1,2} = 0.47 \pm 9.97i$, $s_3 = 0.87$ lie in the RHP). The solution with approximation (2.58) ($\beta_1 = 2.4 \cdot 10^3$, $\beta_2 = 2.4 \cdot 10^3$, $v = 7.8 \cdot 10^{-7}$), which corresponds to a periodic motion of the bush and shaft being in a stick ($\omega^1_{st} \approx 0$) (the roots of Equation (2.140) $s_{1,2} = -0.52 \cdot 10^{-4} \pm 7.07i$ lie on the imaginary axis). Observe that in the last case the roots may be found directly from the characteristic equation $s^2 + \omega_0^2 = 0$, where $\omega_0 = 1/(\tilde{\omega}\sqrt{1 + \varepsilon/a_M})$.

In the third case ($m_0 = 0.05$, $Bi = 10$) there is one solution. The solution with approximation (2.58) $\omega^1_{st} = 0.21 \cdot 10^{-4}$, $p^1_{st} = \theta^1_{st} = 1.0$, $\varphi^1_{st} = 0.05$ ($\beta_1 = 2.4 \cdot 10^3$, $\beta_2 = 2.4 \cdot 10^3$, $v = 1.95 \cdot 10^{-7}$) corresponds to a periodic motion (the roots of Equation (2.140) $s_{1,2} = -0.52 \cdot 10^{-4} \pm 7.07i$ are purely imaginary). As in the second case, in the case considered now the roots may also be found directly from the characteristic equation $s^2 + \omega_0^2 = 0$.

The steady-state solutions ω^1_{st} correspond to stick conditions with rigid body torsional vibrations. Eigenvalues obtained in these cases have a small real part due to regularization of the step function in the Stribeck approximation. These real parts are spurious effects of the regularization and the real physical behaviour does not involve any slip.

In the fourth case ($m_0 = 0.14$, $Bi = 1$) there is again only one solution. The solution $\omega^2_{st} = 1.41$, $p^2_{st} = \theta^2_{st} = 1.37$, $\varphi^2_{st} = 0.14$ ($\beta_1 = 7.28 \cdot 10^{-2}$, $\beta_2 = -1.08 \cdot 10^{-2}$, $v = 0.27$) is unstable (the roots of Equation (2.140) $s_{1,2} = 0.745 \pm 10.06i$, $s_{3,4} = 0.13 \pm 1.36i$ lie in the RHP). Observe that the roots $s_{1,2}$ crucially affect the self-excited oscillations as they have the largest real parts and give the oscillations with the estimated period of $T_p = 2\pi/\mathrm{Im}s_1 = 0.62$, which has been successfully verified in a numerical way. If there is one unstable solution, the corresponding unsteady-state solution approaches a stick-slip periodic solution.

The most interesting fourth case is characterized by the following limiting cases. For $m_0 = 0.14$, $Bi = 1$, and when the bush vibrations $\tilde{\omega} = 0$ (fifth case) are neglected, only one solution occurs. The solution $\omega^2_{st} = 1.41$, $p^2_{st} = \theta^2_{st} = 1.37$, $\varphi^2_{st} = 0.14$ ($\beta_1 = 7.28 \cdot 10^{-2}$, $\beta_2 = -1.08 \cdot 10^{-2}$, $v = 0.27$) is unstable (the roots of Equation (2.140) $s_{1,2} = 0.145 \pm 1.37i$ lie in the RHP). Note that the roots $s_{1,2}$ can be well approximated via Equation (2.140) taking into account only the three first terms of (2.142) ($b_0 = 1.2$, $b_1 = -0.043$, $b_2 = 0.615$, $b_3 = 0.0725$). Again, they govern the self-excited oscillations with the period $T_p = 2\pi/\mathrm{Im}s_1 = 4.58$.

For $m_0 = 0.14$, $Bi = 1$, and with no frictional heat generation $\gamma = 0$, there is one solution $\omega^3_{st} = 86.71$, $p^3_{st} = \theta^3_{st} = 1$, $\varphi^3_{st} = 0.14$ ($\beta_1 = 0.16 \cdot 10^{-2}$, $\beta_2 = 0.21 \cdot 10^{-2}$, $v = 0$), which is stable (the first roots of Equation (2.140)

$s_{1,2} = -0.11 \pm 10.0i$, $s_3 = -0.216$ lie in the LHP). An expected motion has the character of damped oscillations with the period $2\pi/\mathrm{Im}s_1 = 0.629$ and with the logarithmic decrement $2\pi\mathrm{Re}s_1/\mathrm{Im}s_1 = 0.068$.

When comparing the roots of the last and the first case it becomes clear that if the frictional heat generation ($|\mathrm{Re}s_1|$) is taken into account, then the logarithmic oscillation decrement increases. However, the "period" of the damped oscillations increases only slightly.

2.4.2.3 Stick-slip process

Let us analyse the system for $t \to \infty$ ($h_M(\tau) = 1$), when a stick $\omega_r = \dot{\phi} - \dot{\varphi} = 0$ for $\tau \in t_{st}$ ($t_{st} = (\tau_1, \tau_2) \cup \cdots (\tau_{2i-1}, \tau_{2i}) \cup \cdots$), or a slip for $\tau \in t_{sl}$ ($t_{sl} \in (0, \tau_1) \cup \cdots (\tau_{2i}, \tau_{2i+1}) \cup \cdots$) occurs. In the first case $\tau \in t_{st}$, the bush and cylinder move together, and the governing equation has the form

$$\ddot{\varphi}(\tau) + \omega_0^2 \varphi(\tau) = \varepsilon m_0 \tilde{\omega}^2 \omega_0^2, \qquad \omega_0 = 1/(\tilde{\omega}\sqrt{1 + \varepsilon/a_M}), \qquad \tau \in t_{st}. \quad (2.144)$$

The solution of (2.144) describes a periodic motion

$$\varphi(\tau) = \varepsilon m_0 \tilde{\omega}^2 + C_1 \cos(\omega_0 \tau) + C_2 \sin(\omega_0 \tau), \qquad \tau \in t_{st}, \quad (2.145)$$

with the period $2\pi\tilde{\omega}\sqrt{1 + \varepsilon/a_M}$. The cylinder–bush system oscillates periodically, and the bush is subjected to an action of the friction force

$$F(0)p(\tau) = m_0 + \tilde{\omega}^2 (C_1 \cos(\omega_0 \tau) + C_2 \sin(\omega_0 \tau))/(a_M + \varepsilon). \quad (2.146)$$

The contact pressure is estimated by the formula

$$p(\tau) = 2Bi \int_0^{\tau} G_p(\tau - \xi) h_T(\xi) d\xi$$

$$+ 2\gamma \sum_{m=1}^{i} \int_{\tau_{2m-2}}^{\tau_{2m-1}} G_p(\tau - \xi) F(\dot{\phi} - \dot{\varphi}) p(\xi) (\dot{\phi} - \dot{\varphi}) d\xi, \qquad \tau \in (\tau_{2i}, \tau_{2i+1}). \quad (2.147)$$

2.4.2.4 Numerical analysis

The numerical analysis of the problem has been carried out using the Runge–Kutta method for Equations (2.128) and the quadrature method for Equations (2.125) and (2.126) applying the asymptotic estimations

$$G_\theta(1, \tau) \approx 2\sqrt{\tau/\pi}, \qquad G_p(\tau) \approx \tau, \qquad \tau \to 0. \quad (2.148)$$

Formula (2.61) has been used to approximate the dependence of the friction kinematic coefficient on the relative velocity. Numerical calculations have been executed for various values of parameters m_0 and Bi, for which an analytical analysis has been performed as well.

2.4.2.5 Acceleration process

Assume that at the initial time instant the force moment $h_M(\tau) = 1 - \exp(-\delta\tau^2)$ acts on the shaft ($\delta = 100$). This moment forces the shaft to rotate with acceleration. The dimensionless environment temperature is governed by the equation $h_T(\tau) = 1 - \exp(-\delta\tau^2)$. Due to the heat transfer the rotating cylinder begins to expand and eventually comes into contact with the bush. Figs. 2.48–2.51 show the outcomes of the calculations carried out for ten values of m_0 and Bi, and for the initial conditions $\varphi^\circ = 0$, $\omega^\circ = 0$, $\phi^\circ = 0$, $\dot{\phi}^\circ = 0$. Curve 1 illustrates the first case ($m_0 = 0.14$, $Bi = 10$, $\tilde{\omega} = 0.1$, $\gamma = 1.87$), curve 2 represents the second case ($m_0 = 0.1$, $Bi = 10$, $\tilde{\omega} = 0.1$, $\gamma = 1.87$), curve 4 corresponds to the fourth case ($m_0 = 0.14$, $Bi = 1$, $\tilde{\omega} = 0.1$, $\gamma = 1.87$), and curve 5 presents the fifth case ($m_0 = 0.14$, $Bi = 1$, $\tilde{\omega} = 0$, $\gamma = 1.87$).

Figure 2.48a shows the dependence of the dimensionless angular velocity $\dot{\phi}$ of the cylinder (dashed curve) and the bush $\dot{\varphi}$ (solid curve) on the dimensionless time τ for the first and the second of the considered cases is reported. It can be seen that in all cases the system behaviour is in agreement with the analytical predictions. In the first case, after certain transitional processes the shaft starts to rotate with constant velocity $\omega_{st} = 44.4$. The bush displays damped oscillations with the period $T_p = 0.62$. In the second and third case (already for a small force moment), the cylinder and bush come into contact and start oscillating periodically as one body with the period $T_p = 0.89$. The stick type oscillations are periodic.

Figures 2.48b,c illustrate the dependence of the dimensionless angular velocity $\dot{\phi}$ of the cylinder (dashed curve) and the bush $\dot{\varphi}$ (solid curve) on the dimensionless time τ for the fourth and the fifth case. In the fourth case, the system shows stick-slip oscillation ($T_p = 0.642$), and in the fifth case the oscillation exhibited by the system is of the thermal stick-slip type ($T_p = 4.49$). Let us recall that the root of the characteristic equation responsible for instability gives the approximated period $T_p = 2\pi/\mathrm{Im}s_1 = 4.58$.

In Figs. 2.49a,b, time histories of both the contact pressure and temperature for the considered cases are shown using solid curves 1–5. In cases 4 and 5, the contact characteristics undergo changes in time.

2.4.2.6 Braking process

It is assumed that at the initial state the shaft rotates at angular velocity $\dot{\phi}^\circ$ ($h_M(\tau) = 0$). The dimensionless temperature of the bush changes in agreement with the formula $h_T(\tau) = 1 - \exp(-\delta\tau^2)$. Owing to heat transfer, the rotating

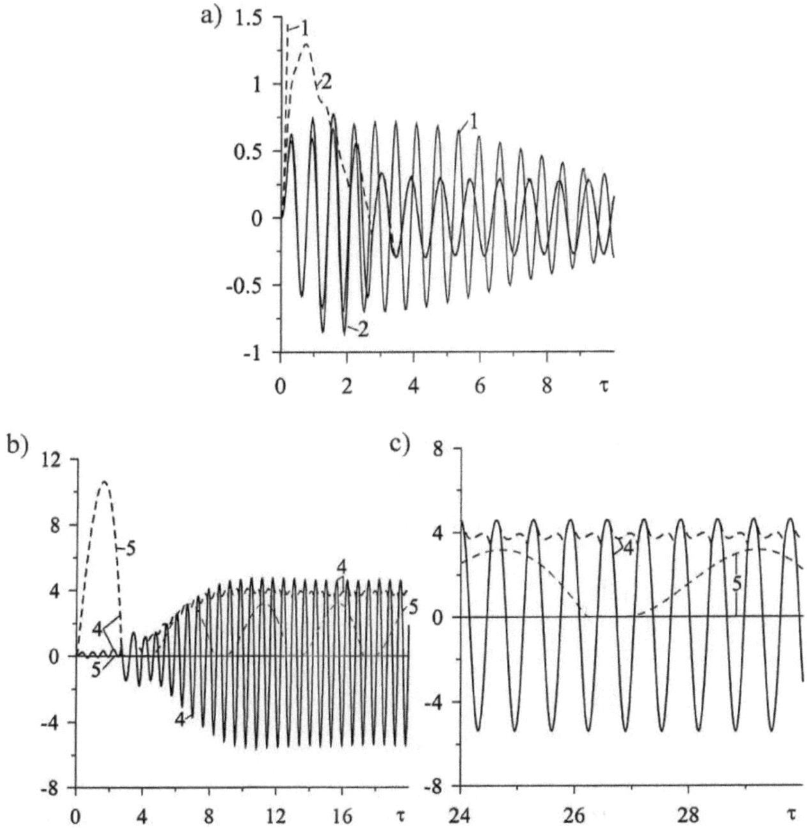

Fig. 2.48: Bush dimensionless velocity $\dot{\varphi}$ (solid curves) and shaft dimensionless velocity $\dot{\phi}$ (dashed curves) versus dimensionless time τ during acceleration for different values of m_0 and $\tilde{\omega}$. Curve 1: $m_0 = 0.14$, $Bi = 10$, $\tilde{\omega} = 0.1$, 2: $m_0 = 0.1$, $Bi = 10$, $\tilde{\omega} = 0.1$, 4: $m_0 = 0.14$, $Bi = 1$, $\tilde{\omega} = 0.1$, 5: $m_0 = 0.14$, $Bi = 1$, $\tilde{\omega} = 0$.

cylinder expands and comes into contact with the bush; that is, it starts to brake. The initial conditions are as follows: $\varphi^\circ = 0$, $\omega^\circ = 0$, $\phi^\circ = 0$, $\dot{\phi}^\circ = 100$. Computational examples for some values of γ are shown in Figs. 2.50 and 2.51. Curve 1 corresponds to the case $Bi = 10$, $\tilde{\omega} = 0.1$, $\gamma = 1.87$, whereas curve 2 represents the case $Bi = 10$, $\tilde{\omega} = 0$, $\gamma = 1.87$ (the bush does not oscillate).

Figure 2.50 shows the dimensionless time histories of the dimensionless angular velocity of the shaft $\dot{\phi}$ (dashed curves) and the bush $\dot{\varphi}$ (solid curves) during the braking process. It is seen (curve 1) that the shaft angular velocity decreases, and the bush undergoes oscillations until the two angular velocities reach the same value. Because there is no driving moment and the damping has not been introduced, the bush and shaft start to oscillate as one body with the period $T_p = 2\pi/\omega_0$. When the bush dynamics is not taken into account ($\tilde{\omega} = 0$; see curve 2), the shaft velocity also decreases and finally the shaft stops. A comparison of the results

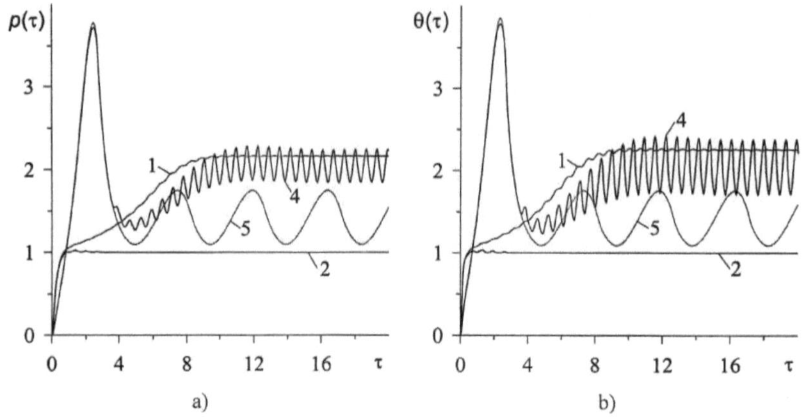

Fig. 2.49: Dimensionless contact pressure $p(\tau)$ (a) and contact temperature $\theta(\tau) = \theta(1,\tau)$ (b) versus dimensionless time τ during acceleration for different values of m_0, Bi, and $\tilde{\omega}$. Curve 1: $m_0 = 0.14$, $Bi = 10$, $\tilde{\omega} = 0.1$, 2: $m_0 = 0.1$, $Bi = 10$, $\tilde{\omega} = 0.1$, 4: $m_0 = 0.14$, $Bi = 1$, $\tilde{\omega} = 0.1$, 5: $m_0 = 0.14$, $Bi = 1$, $\tilde{\omega} = 0$.

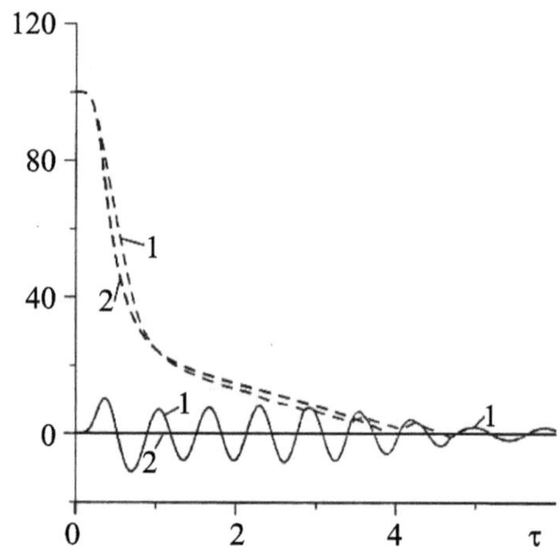

Fig. 2.50: Time history of the braking pad angular speed $\dot{\varphi}$ (solid curves) and the shaft speed $\dot{\phi}$ (dashed curves) during braking ($m_0 = 0$) for various values of $\tilde{\omega}$. Curves 1: $\tilde{\omega} = 0.1$, 2: $\tilde{\omega} = 0$.

represented by curves 1 and 2 leads to the conclusion that the shaft braking time for $\tilde{\omega} = 0$ is smaller than the time interval needed for the shaft and bush to achieve a fixed contact (stick) with each other ($\tilde{\omega} = 0.1$). In Figs. 2.51a,b, time histories of the contact pressure and the temperature are given. Both of the characteristics increase in the beginning when $\gamma > 0$ (see curves 1 and 2). It can be concluded that during the braking process the maximal values of the pressure and of the contact temperature become smaller when the dynamics of the bush is taken into account.

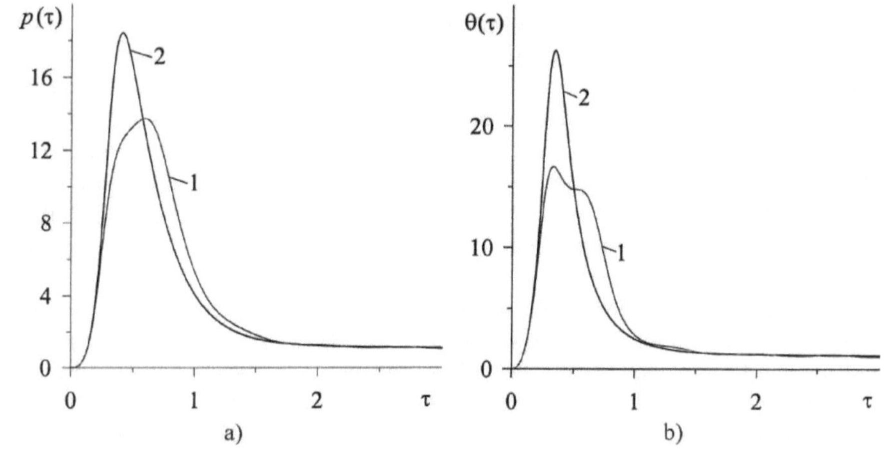

Fig. 2.51: Time history of contact pressure during shaft braking ($m_0 = 0$) for various values of $\tilde{\omega}$. Curves 1: $\tilde{\omega} = 0.1$, 2: $\tilde{\omega} = 0$.

To conclude, one of the classical models of an elastically supported braking pad being in frictional contact with a rotating shaft during frictional heat generation and heat expansion is analysed with the use of the Stribeck friction model. The stability analysis of the stationary solutions is followed by the numerically verified analytical estimation of the periodic stick-slip occurrence. It is detected and illustrated that the stick-slip motion appears in the presence of the driving moment, heat transfer, and thermal expansion of the shaft materials. In addition, numerical calculations illustrating the influence of the parameters on the dynamics and contact characteristics of the investigated model in the acceleration and braking processes have been performed.

It should be emphasized that, contrary to the results reported in [Pyryev, Grilitskiy (1996)] where already at small wear the frictional TEI occurs (contact parameters increase exponentially after the relative speed exceeds the critical value, which is found from the condition $v = 1$) and either overheating [Pyryev et al. (1995)] or brake heating [Aleksandrov, Annakulova (1990)] may appear, the considered system can never be overheated. An increase of the friction force (or moment of friction force), frictional heat generation, and system overheating is caused by an increase of the contact pressure while one of the contacting bodies moves at constant speed and heat expansion is bounded. In order to keep the motion speed constant, the friction force (or moment of friction force) increases, and consequently, energy is supplied to the system. Although the system heat expansion is bounded, the contact pressure may increase, which yields an increase of both the friction force (moment of friction force) and the frictional heat generation. However, the system will not be overheated as the moving body starts to brake. The heat balance leads to a cooling process, and hence the contact pressure and friction force (moment of friction force) start to decrease. This, however, again brings an increase in the relative speed and in the frictional heat generation.

The described process will repeat again and again, which physically means that the system controls itself to avoid overheating. This type of control is called passive. Additionally, it is suggested that the phenomena referred to as 'stick-slip' in the classical terminology be called 'thermal stick-slip' instead. This new terminology seems to be more adequate because a slip relates to a heating process caused by an independent movement of two bodies, whereas a stick corresponds to the process of cooling that takes place when the relative velocity of two contacting bodies equals zero. It is worth noting that in the latter case although the bush does not move, the shaft exhibits thermal stick-slip dynamics.

When the frictional heat generation is not taken into account ($\gamma = 0$), the stick-slip oscillations cannot appear. In this case there are two stable stationary solutions, and one unstable, and hence any trajectory is always attracted by one of the stable critical points (equilibrium). For $\gamma > 0$, thermal stick-slip oscillation can appear, and then either a short bush oscillation period ($T_p = 0.624$ for case 4) or a long period, being the one of shaft oscillations ($T_p = 4.49$ for case 5), is achieved. Owing to the dynamics of the considered system, the maximal pressure and temperature values decrease during the braking process.

In other words, in the system where heat expansion of the contacting bodies is bounded, for the sake of stability it is better to apply the mechanical external excitation (considered here) than the kinematic excitation.

2.4.3 *Inertial damped dynamics of cylinder and bush and tribological processes*

In this section (contrary to Section 2.4.2) we assume that damping of bush vibrations occurs ($h \neq 0$) and its wear is ($k^w \neq 0$). We assume that the cylinder has the moment of inertia B_1 and rotates according to the applied moment. We also assume that in the initial time instant, the cylinder and bush are not driven internally ($U_0 = 0$, $T_* = T_0$, $U_* = 2\alpha_1(1+\nu_1)R_1T_0$, $P_* = 2\alpha_1 E_1 T_0/(1-2\nu_1)$), and that dimensionless bush (pad) temperature is changed according to the law $h_T(\tau) = 2(1-\exp(-\delta\tau^2))$. Owing to heat exchange the rotating cylinder begins to increase up to a contact with the bush, and then the wear process begins. Let us consider the case when the bush (bearing) is linked to a basis (frame) by springs characterized by stiffness k_2 ($k_1 = k_3 = 0$). Because an arbitrary parameter $k_* = k_2$, it gives the following value $a = -1$. In addition, we assume that $\chi = 1$ (it provides $t_* = \sqrt{B_2/(k_* R_2^2)}$).

Note that the considered problem is manifested by the system of nonlinear differential Equations (2.5) and (2.27) and integral equation (2.125) governing angular velocities $\dot{\varphi}(\tau)$, $\dot{\phi}(\tau)$ and contact pressure $p(\tau)$ values. Temperature is defined by (2.126), whereas wear by formulas (2.23).

2.4.3.1 Stick-slip process

Consider now how the system behaves for $t \to \infty$ ($h_M(\tau) = 1$), when either a stick motion occurs $\omega_r = \dot{\phi} - \dot{\varphi} = 0$ for $\tau \in t_{st}$ ($t_{st} = (\tau_1, \tau_2) \cup \cdots \ (\tau_{2i-1}, \tau_{2i}) \cup \cdots$), or a slip motion takes place for $\tau \in t_{sl}$ ($t_{sl} \in (0, \tau_1) \cup \cdots \ (\tau_{2i}, \tau_{2i+1}) \cup \cdots$). Both bush and shaft for $\tau \in t_{st}$ do not move relatively to each other. In this case the process is governed by the equation

$$\ddot{\varphi}(\tau) + 2h\omega_0^2\dot{\varphi}(\tau) + \omega_0^2\varphi(\tau) = \varepsilon m_0\omega_0^2, \omega_0 = 1/\sqrt{1+\varepsilon/a_M}, \qquad \tau \in t_{st}. \tag{2.149}$$

Its solution gives free vibration at frequency $\omega_h = \omega_0\sqrt{1-\omega_0^2h^2}$ and with the period of damped vibrations $2\pi/\omega_h$:

$$\varphi(\tau) = \varepsilon m_0 + e^{-h\omega_0^2\tau}(C_1\cos(\omega_h\tau) + C_2\sin(\omega_h\tau)), \qquad \tau \in t_{st}. \tag{2.150}$$

2.4.3.2 Numerical analysis

Numerical analysis of the stated problem has been carried out using the fourth-order Runge–Kutta method applied to Equations (2.5), (2.6), (2.27), (2.28) and quadrature applied to Equations (2.125) and (2.126) accounting for the asymptotic estimation (2.148).

The dependence of kinematic friction versus the relative velocity (see Fig. 2.2) has been estimated by formula (2.61). The numerical computations are performed for various parameters k^w and γ.

2.4.3.3 Cylinder and bush dynamics during acceleration

Assume that in the initial time instant the force moment $h_M(\tau) = 1 - \exp(-\delta\tau^2)$ acts on the shaft. This moment forces the shaft to rotate with an acceleration. Dimensionless temperature of the bush is governed by the equation $h_T(\tau) = 2(1 - \exp(-\delta\tau^2))$. We take zero as initial conditions: $\varphi^\circ = 0$, $\omega^\circ = 0$, $\phi^\circ = 0$, $\dot{\phi}^\circ = 0$. Results of computations are given in Figs. 2.52a, 2.53 and 2.54 for a few values of parameter k^w and γ.

The following values are taken for a computational purpose: $\varepsilon = 1$, $a_M = 1$, $Bi = 1$, $\gamma = 20$, $h = 0.05$, $\tilde{\omega} = 0.1$, $\delta = 10$. In Fig. 2.52a dimensionless angular velocity of the cylinder $\dot{\phi}$ (dashed curves) and bush $\dot{\varphi}$ (solid curves) versus dimensionless time τ during acceleration ($m_0 = 0.5$) for some parameters k^w characterizing bush wear are given.

Figure 2.53 shows time evolutions of contact pressure (a) and temperature (b) during acceleration for the same parameter values. In Fig. 2.54 time evolutions of dimensionless friction force (Fig. 2.54a) and dimensionless wear (Fig. 2.54b) are presented. In the mentioned figures curves 1 correspond to the case $k^w = 0.01$, whereas curves 2 correspond to $k^w = 0.1$.

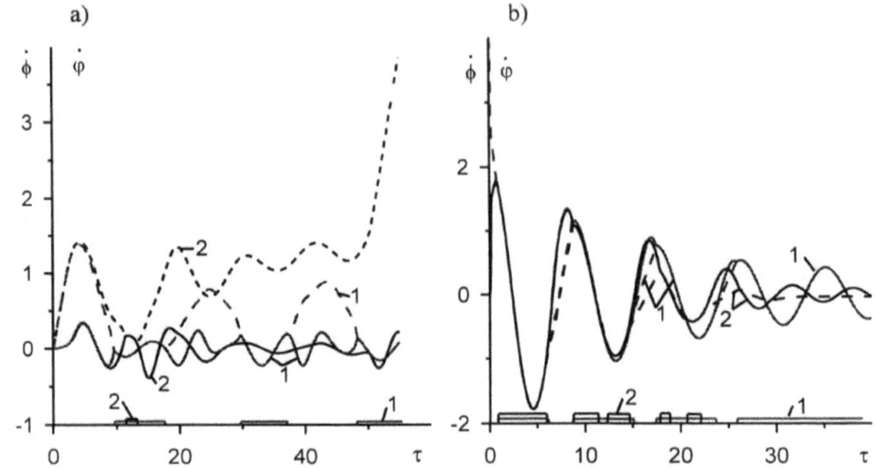

Fig. 2.52: Time histories of angular speed of the braking pad $\dot{\varphi}$ (solid curves) and shaft $\dot{\phi}$ (dashed curves) during acceleration (a) and braking (b) for different values of the parameter k^w (curves 1: $k^w = 0.01$; curves 2: $k^w = 0.1$).

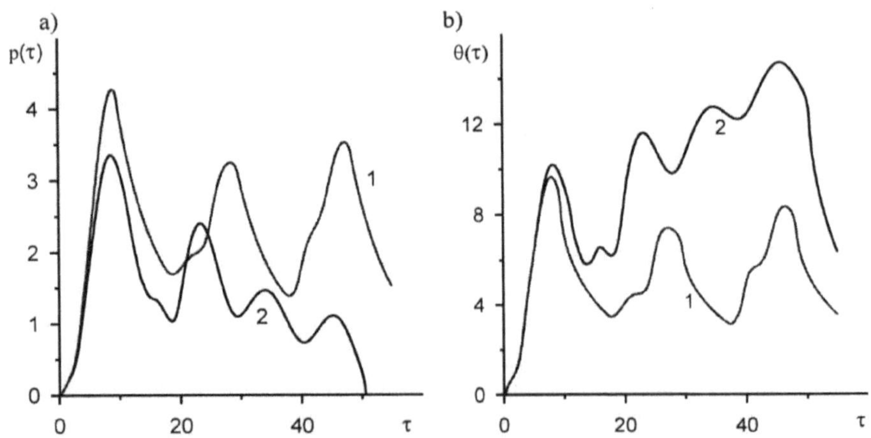

Fig. 2.53: Time histories of contact pressure (a) and contact temperature (b) during acceleration for different values of the parameter k^w (curves 1: $k^w = 0.01$; curves 2: $k^w = 0.1$).

Consider first dynamics for a small value of the wear coefficient $k^w = 0.01$ (curve 1). In response to a driven moment action, the shaft starts to rotate (solid curves). Owing to thermal shaft radial expansion, the contact pressure p increases. To conclude, both dimensionless contact pressure $p(\tau)$ (Fig. 2.53a) and dimensionless friction force $F(\dot{\phi}_1 - \dot{\phi}_2)p(\tau)$ (right-hand side of Equation (2.5)) increase giving an increase of bush velocity and contact temperature $\theta(\tau)$ and wear bush $u^w(\tau)$. For example, in time instant $\tau = 4.27$ ($\tau = 4.76$) the shaft (bush) velocity

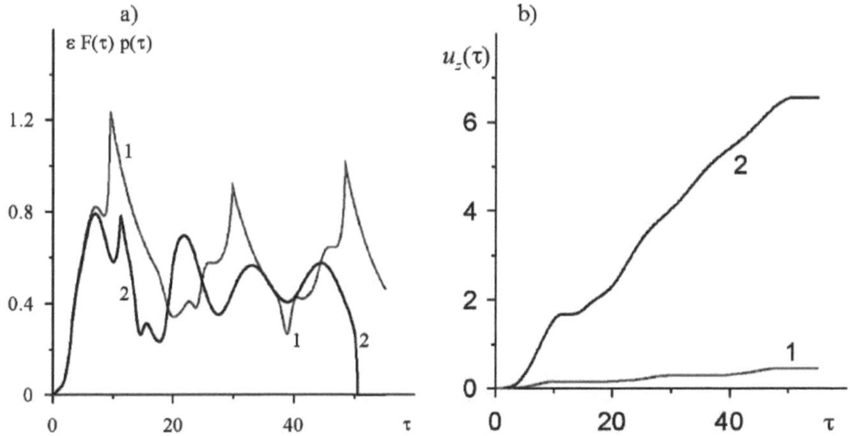

Fig. 2.54: Time histories of friction force (a) and wear (b) during acceleration for different values of the parameter k^w (curves 1: $k^w = 0.01$; curves 2: $k^w = 0.1$).

starts to decrease (Fig. 2.53a). The maximal values of contact pressure are achieved for dimensionless time units $\tau = 8.95$ (see curve 1 in Fig. 2.53a). For time instant $\tau_1 = 9.54$ the relative sliding velocity of both bodies is equal to zero, and a stick phase begins, which ends for $\tau_2 = 17.5$.

In the stick state for $\tau \in (\tau_1, \tau_2)$ the shaft temperature decreases owing to heat exchange, and therefore both contact pressure (see curve 1 in Fig. 2.53a) and friction decrease, but wear does not undergo any changes. Beginning from $\tau_2 = 17.5$, a sliding phase appears within the interval of $\tau \in (\tau_2, \tau_3)$, where $\tau_3 = 29.7$. In this phase, both an increase and decrease of the shaft velocity are observed, the bush vibrates, and also contact temperature and pressure exhibit oscillating character (see curves 1 in Fig. 2.53a). Friction accompanied by vibrations also increases the bush wear (Fig. 2.53a). For $\tau_3 = 29.7$ the next stick phase occurs for $\tau \in (\tau_3, \tau_4)$, where $\tau_4 = 37.3$. It is worth noting that during stick phases $\tau \in \tau_{st} = (\tau_1, \tau_2) \cup \cdots (\tau_{2i-1}, \tau_{2i}) \cup \cdots$ the system velocity oscillates periodically with the period $2\pi(1 + \varepsilon/a_M)/\sqrt{1 + \varepsilon/a_M - h^2} = 8.89$.

Consider now the system dynamics for a larger value of the wear coefficient $k^w = 0.1$ (curve 2). In the beginning of the sliding phase for $\tau \in (0, \tau_1)$, where $\tau_1 = 11.7$, all earlier mentioned characteristics of two contacting bodies are similar to the previous case associated with small wear (see curves 2 in Fig. 2.52a). Only one sliding phase $\tau \in (\tau_1, \tau_2)$, where $\tau_2 = 13.3$, is exhibited. After $\tau_2 = 13.3$ the bush starts to vibrate and the shaft rotation velocity, as well as contact temperature and wear, is increased. The contact pressure approaches zero $\tau_c = 50.6$ in an oscillatory manner (Fig. 2.53a). Beginning from this time instant, the contact between two bodies is lost. The shaft starts to rotate with an acceleration, whereas the bush vibrates with the period $2\pi/\sqrt{1 - h^2}$. Zones of sticks are shown in Fig. 2.52a and are marked by horizontal intervals 1 and 2. One may conclude from the figure that during acceleration and increase of the dimensionless wear

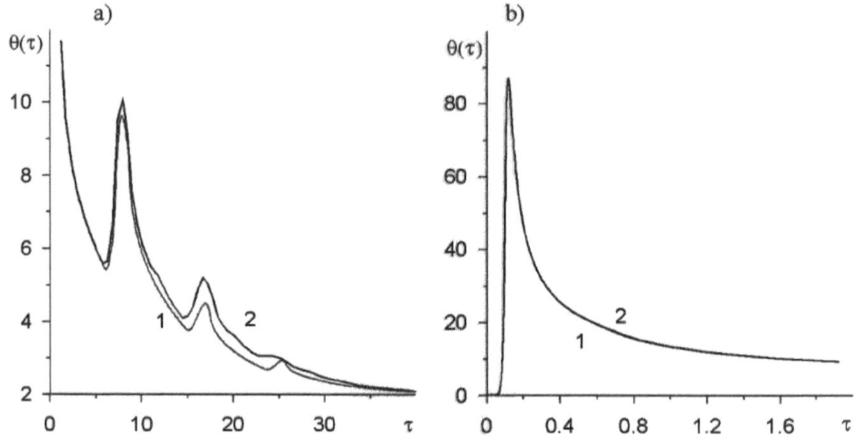

Fig. 2.55: Time histories of contact temperature for $\tau \in (2, 40)$ (a) and contact temperature for $\tau \in (0, 2)$ (b) during braking ($m_0 = 0$) for different values of parameter k^w (curves 1: $k^w = 0.01$; curves 2: $k^w = 0.1$).

coefficient the previously observed periodic stick-slip motion is substituted by a motion of no contacting bodies ($\tau > 50.6$).

2.4.3.4 Cylinder and bush dynamics during braking

It is assumed that during the initial moment the cylinder rotates at velocity ϕ° ($h_M(\tau) = 0$, $m_0 = 0$). Dimensionless temperature of the bush (pad) changes in accordance with $h_T(\tau) = 2(1 - \exp(-\delta\tau^2))$. Owing to heat transfer, the shaft starts to extend and begins to keep contact with the bush, and hence braking occurs. Initial conditions are as follows: $\varphi^\circ = 0$, ω°, $\phi^\circ = 0$, $\dot{\phi}^\circ = 4$. Computational results are shown in Figs. 2.52b, 2.55 and 2.56 for some values of parameter k^w. Curves 1 correspond to the case $k^w = 0.01$, curves 2 correspond to $k^w = 0.1$. In Fig. 2.52b dimensionless dependence of cylinder angular velocity $\dot{\phi}$ (dashed curves) and bush $\dot{\varphi}$ (solid curves) on time τ during braking is presented. Figure 2.55b shows temperature time evolution for $\tau \in (0, 1.8)$ and Fig. 2.55a for $\tau \in (1.8, 40)$. Figures 2.56a and 2.56b illustrate time evolutions of contact pressure and wear, respectively.

Consider first dynamics for a small value of the wear coefficient $k^w = 0.01$ (curve 1 in Figs. 2.52b, 2.55, 2.56). As a result of temperature increase in a surrounding medium, temperature of the shaft rotating at dimensionless velocity $\dot{\phi}_1^\circ = 4$ also increases. The shaft and bush start to touch each other, a contact pressure increases and achieves its maximal value $p = 6.72$ for time instant $\tau = 0.26$, and both friction force and bush velocity increase. At $\tau_1 = 0.75$ the velocity of two bodies (sliding velocity) will be equal to zero, and a stick phase begin (it ends at $\tau_2 = 5.9$). As in the previous case, due to heat exchange the shaft

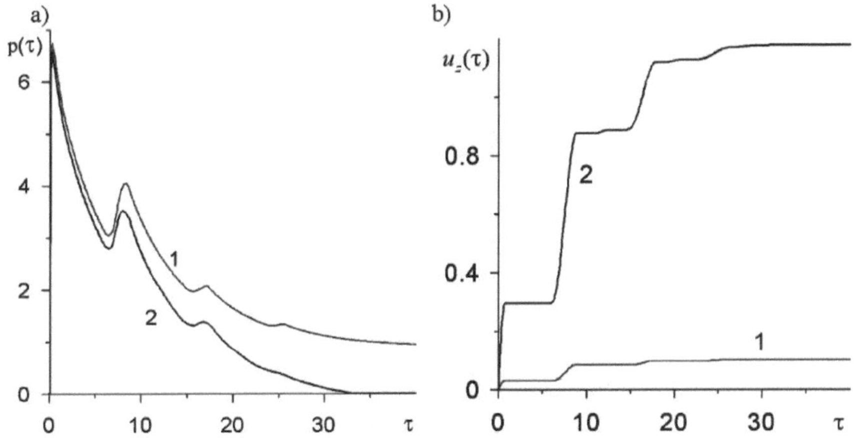

Fig. 2.56: Time histories of contact pressure (a) and wear (b) during braking ($m_0 = 0$) for different values of parameter k^w (curves 1: $k^w = 0.01$; curves 2: $k^w = 0.1$).

temperature starts to decrease in the stick phase $\tau \in (\tau_1, \tau_2)$, which causes a decrease of both contact pressure (curve 1 in Fig. 2.56a) and friction force. The latter one changes its sign rapidly at $\tau = 4.57$. Wear process is constant during the stick phase (Fig. 2.56b), that is, at $\tau \in (\tau_1, \tau_2)$. Zones of sticks are shown in Fig. 2.52b and marked by horizontal intervals 1. Beginning from $\tau_7 = 26.0$ the stick phase is exhibited, which is observed until damped oscillations (with the period $2\pi(1+\varepsilon/a_M)/\sqrt{1+\varepsilon/a_M-h^2} = 8.89$) vanish.

Finally, let us consider the braking process for the largest value of the wear coefficient $k^w = 0.1$ (curve 2). The corresponding stick phases are shown in Fig. 2.52b and denoted by horizontal intervals 2. For $\tau_c = 33.3$ the contact pressure is equal to zero (Fig. 2.56a). The shaft stops, whereas damped bush vibrations are observed with the period $2\pi/\sqrt{1-h^2}$.

Note that when the shaft displacement achieves its extreme values ($\ddot{\psi} = 0$, see dashed curves), the friction force changes its sign. In the stick phases $\tau \in \tau_{st} = (\tau_1, \tau_2) \cup \cdots (\tau_{2i-1}, \tau_{2i}) \cup \cdots$ the wear process is not observed.

2.5. Dynamics of contacting bodies with impacts

In this section the model of a contact system with heat and wear generated by friction and/or impacts is studied. The methods and mathematical models of such systems applied thus far by others contribute only partially to the description of complex dynamics. First, the analysis of contacting dynamic models omit tribological processes on a contact body surface. Second, the mentioned models do not include either the body inertia or impact phenomena usually appearing within the body clearance. We contribute to the problem by matching both phenomena, which

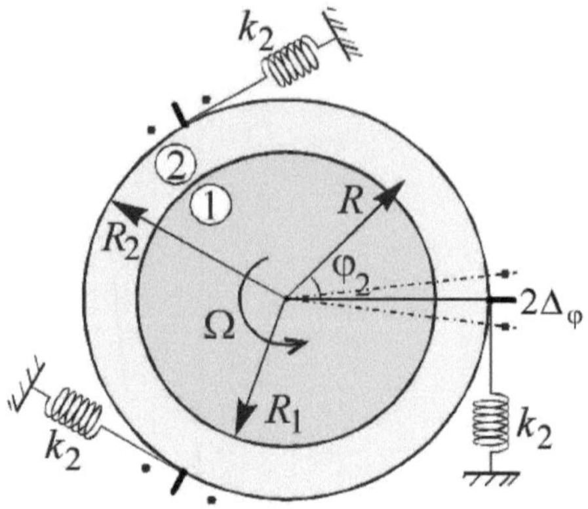

Fig. 2.57: Analysed system.

improves modelling of dynamic behaviour of contacting bodies. Analysis of both stick-slip and slip-slip motion exhibited by the system is performed (impactless behaviour of this model has already been studied by the authors [Awrejcewicz, Pyryev (2002), (2004a), (2004b)]), among others. Analytically predicted vibro-impact stick-slip and slip-slip dynamics have also been verified numerically.

2.5.1 *Mathematical modelling*

Attention is focused on modelling of nonlinear dynamics of two bodies consisting of a stiff bush with clearance $2\Delta_\varphi$ (see Fig. 2.57). The bush is coupled with housing by springs with stiffness k_2 and is mounted on the rotating thermoelastic shaft 1. The following assumptions are taken: (i) the shaft rotates with enough small angular velocity Ω such that centrifugal forces can be omitted; (ii) nonlinear kinetic friction occurs between the bush and the shaft; (iii) heat is generated on the contacting surface $R = R_1$ due to friction; and (iv) heat transfer between contacting bodies is governed by Newton's law.

2.5.1.1 Equations for shaft rotational movement of an absolutely rigid bush

Let axis Z be a cylinder axis. The equilibrium state of the moments of forces with respect to the shaft axis gives

$$B_2\ddot{\varphi}_2(t) + k_2R_2^2\varphi_2(t) = f(V_r)2\pi R_1^2 P(t), \qquad |\varphi_2(t)| < \Delta_\varphi, \qquad \dot{\varphi}_2(t) \neq \Omega, \tag{2.151}$$

$$\ddot{\varphi}_2(t) = 0, \qquad |\varphi_2(t)| < \Delta_\varphi, \qquad \dot{\varphi}_2(t) = \Omega, \tag{2.152}$$

$$\dot{\varphi}_2^+ = -k\dot{\varphi}_2^-, \qquad |\varphi_2| = \Delta_\varphi, \qquad \dot{\varphi}_2^-\varphi_2 > 0, \tag{2.153}$$

where: $V_r = R_1\Omega - R_1\dot{\varphi}_2(t)$ relative velocity of the contact bodies, k is the coefficient of restitution, $\dot{\varphi}_2^-$ ($\dot{\varphi}_2^+$) is the bush velocity just before (after) impact, B_2 is the moment of inertia of the bush per length unit, $f(V_r)$ is the kinetic friction coefficient depending on relative velocity, and $P(t)$ is the contact pressure. The initial value problem is defined in the following way.

$$\varphi_2(0) = \varphi_2^\circ, \qquad \dot{\varphi}_2(0) = \omega_2^\circ. \tag{2.154}$$

Relation-approximating curve $f(V_r)$ has the following form,

$$f(V_r) = \mathrm{sgn}(V_r)F(|V_r|),$$

$$F(V_r) = \begin{cases} f_s - \kappa V_r, & 0 < V_r \le V_{\min} \\ f_s - \kappa V_{\min}, & V_{\min} < V_r \end{cases} \tag{2.155}$$

where f_s, κ, $V_{\min}$ are constant coefficients.

2.5.1.2 Thermoelastic shaft

Inertial terms occurring in the equation of motion are omitted in our study and the problem may be considered as a quasi-static one. In the case of axially symmetric shaft stresses, the governing equations can be derived using the theory of thermal stresses for an isotropic body [Nowacki (1962)]. Applying cylindrical coordinates one gets the following set of Equations ((2.7),(2.8)) with the attached mechanical (2.9) and thermal boundary conditions

$$\lambda_1 \frac{\partial T_1(R_1,t)}{\partial R} + \alpha_T T_1(R_1,t) = (1-\eta)f(V_r)V_r P(t), \tag{2.156}$$

$$\lambda_1 2\pi R \frac{\partial T_1(R,t)}{\partial R}\bigg|_{R\to 0} = 0, \qquad 0 < t < t_c, \tag{2.157}$$

and with the following initial conditions

$$T_1(R,0) = 0, \qquad 0 < R < R_1. \tag{2.158}$$

Velocity of the bush wear is proportional to a certain power of friction force (2.13). Shaft radial stresses $\sigma_R(R,t)$ may be found using radial displacement $U(R,t)$ and temperature $T_1(R,t)$ from the following formula (2.14).

Upon integration of Equation (2.7) and taking into account (2.9) and (2.14), the contact pressure is

$$P(t) = \frac{2E_1\alpha_1}{1-2\nu_1}\frac{1}{R_1^2}\int_0^{R_1} T_1(\xi,t)\xi d\xi$$

$$+\frac{E_1}{(1-2\nu_1)(1+\nu_1)R_1}[U_0 h_U(t) - U^w(t)]. \qquad (2.159)$$

2.5.2 *Algorithm of solution*

Let us introduce the following dimensionless parameters:

$$\tau = \frac{t}{t_*},\ \tau_c = \frac{t_c}{t_*},\ r = \frac{R}{R_1},\ \varphi(\tau) = \frac{\varphi_2}{\Delta_\varphi},\ p = \frac{P}{P_*},\ \theta = \frac{T_1}{T_*},\ u^w = \frac{U^w}{U_0},$$

$$\alpha_0 = \frac{f_0}{f_s},\ \omega_1 = \frac{\Omega t_*}{\Delta_\varphi} = \frac{1}{\gamma\sqrt{\alpha_0}},\ \mu_0 = \frac{1-\alpha_0}{\alpha_0},\ \eta_0 = \frac{V_0 - V_{\min}}{V_0},$$

$$\varepsilon = \mu_0\gamma\sqrt{\alpha_0} = \frac{\mu_0}{\omega_1},\ \gamma = \sqrt{\frac{2\pi R_1^2 P_* f_s \Delta_\varphi}{B_2\Omega^2}},\ \omega_0^2 = \frac{k_2 R_2^2 t_*^2}{B_2},\ Bi = \frac{\alpha_T R_1}{\lambda_1},$$

$$k^w = \frac{K^w\Delta_\varphi E_1}{(1-2\nu_1)(1+\nu_1)},\ \gamma_1 = \frac{(1-\eta)E_1\alpha_1 R_1^2 f_0\Delta_\varphi}{\lambda_1(1-2\nu_1)t_T},\ \tilde{\omega} = \frac{t_*}{t_T},\ \Psi = \frac{F}{f_0},$$

$$x = \frac{\varphi_2^\circ}{\Delta_\varphi},\ y = \frac{\omega_2^\circ t_*}{\Delta_\varphi},\ f(V_0) = f_0,\ h_U(\tau) = h_U(t_*\tau), \qquad (2.160)$$

where

$$t_* = \sqrt{\frac{B_2\Delta_\varphi}{f_0 2\pi R_1^2 P_*}},\qquad V_* = \frac{R_1\Delta_\varphi}{t_*},\qquad P_* = \frac{E_1 U_0}{(1-2\nu_1)(1+\nu_1)R_1},$$

$$T_* = \frac{U_0}{\alpha_1(1+\nu_1)R_1},\qquad t_T = \frac{R_1^2}{a_1},\quad V_0 = \Omega R_1.$$

The dimensionless equations governing dynamics of the analysed system have the form

$$\ddot{\varphi}(\tau) + \omega_0^2\varphi(\tau) = \mathrm{sgn}(\omega_1 - \dot{\varphi})\Psi(\dot{\varphi})p(\tau),\qquad |\varphi(\tau)| < 1,\qquad \dot{\varphi}(\tau) \neq \omega_1,$$

$$\ddot{\varphi}(\tau) = 0,\qquad |\varphi(\tau)| < 1,\qquad \dot{\varphi}(\tau) = \omega_1,$$

$$\dot{\varphi}^+ = -k\dot{\varphi}^-,\qquad |\varphi| = 1,\qquad \dot{\varphi}^-\varphi > 0,$$

$$\varphi(0) = x,\qquad \dot{\varphi}(0) = y, \qquad (2.161)$$

where

$$\Psi(\dot{\varphi}) = \begin{cases} 1 + \varepsilon\omega_1\eta_0, & \dot{\varphi} < \omega_1\eta_0, \omega_1(2-\eta_0) < \dot{\varphi} \\ 1 + \varepsilon\dot{\varphi}, & \omega_1\eta_0 < \dot{\varphi} < \omega_1 \\ 1 + 2\varepsilon\omega_1 - \varepsilon\dot{\varphi}, & \omega_1 < \dot{\varphi} < \omega_1(2-\eta_0) \end{cases} . \tag{2.162}$$

In order to solve the motion Equations (2.161) one needs to know contact pressure (2.22) $p(\tau)$ and wear (2.23):

$$p(\tau) = h_U(\tau) - u^w(\tau) + \int_0^1 \theta(\xi,\tau)\xi d\xi, \tag{2.163}$$

$$u^w(\tau) = k^w \int_0^\tau |\omega_1 - \dot{\varphi}(\tau)| p(\tau) d\tau. \tag{2.164}$$

The one-dimensional transient heat conduction equation under consideration takes the following dimensionless form (2.18), whereas the thermal boundary conditions are

$$\left[\frac{\partial\theta(r,\tau)}{\partial r} + Bi\theta(r,\tau)\right]_{r=1} = \gamma_1\tilde{\omega}^{-1}\Psi(\dot{\varphi}(\tau))|\omega_1 - \dot{\varphi}(\tau)|p(\tau),$$

$$2\pi r \frac{\partial\theta(r,\tau)}{\partial r}\bigg|_{r\to 0} = 0, \tag{2.165}$$

and initial conditions are as follows,

$$\theta(r,0) = 0. \tag{2.166}$$

Applying an inverse Laplace transform [Carslaw, Jaeger (1959)], the nonlinear problem governed by Equations (2.18), (2.165), and (2.166) is reduced to the following integral equation of the second kind of Volterra type,

$$p(\tau) = h_U(\tau) - u^w(\tau)$$

$$+ 2\gamma_1\tilde{\omega}^{-1} \int_0^\tau \dot{G}_p(\tau-\xi)\Psi(\dot{\varphi}(\xi))|\omega_1 - \dot{\varphi}(\xi)|p(\xi)d\xi. \tag{2.167}$$

Then the problem is reduced to consideration of Equations (2.161) and (2.167), which yield both dimensionless pressure $p(\tau)$ and velocity $\dot{\varphi}(\tau)$. The temperature is defined by the following formula,

$$\theta(r,\tau) = \gamma_1 \tilde{\omega}^{-1} \int_0^{\tau} \dot{G}_\theta(r,\tau-\xi)\Psi(\dot{\varphi}(\xi))|\omega_1 - \dot{\varphi}(\xi)|p(\xi)d\xi, \tag{2.168}$$

where $G_p(\tau)$ and $G_\theta(1,\tau)$ are defined by (2.33).

Temperature is defined by (2.168). Numerical analysis of the problem is carried out using the Runge–Kutta method for (2.161) and the quadrature method for (2.167) and (2.168), and taking into account the following asymptotes,

$$G_\theta(1,\tau) \approx 2\sqrt{\tau\tilde{\omega}/\pi}, \qquad G_p(\tau) \approx \tilde{\omega}\tau, \qquad \tau \to 0. \tag{2.169}$$

2.5.3 On the periodic motion with impacts

First the case of bush vibrations without tribological processes is studied ($\gamma_1 = 0$, $k^w = 0$). For this case we have $p(\tau) = h_U(\tau)$. Our system governed by Equations (2.161) may exhibit four different periodic motions. Namely:

(i) Periodic orbit with one impact, where a stick does not appear (Fig. 2.58)
(ii) Periodic orbit with one impact, where a stick-slip occurs (Fig. 2.60)
(iii) Periodic orbit with two impacts, where a slip of the contacting bodies occurs (Fig. 2.59)
(iv) Periodic orbit with two impacts, where a stick-slip appears (Fig. 2.61).

In what follows we assume that $\varepsilon \ll 1$, $\omega_0^2 \ll 1$, and $\eta_0 \leq -1$. It means that the system dynamics is exhibited in the interval $(0 < V_0 < V_{\min})$, where a decreasing slope of the kinetic friction coefficient is observed.

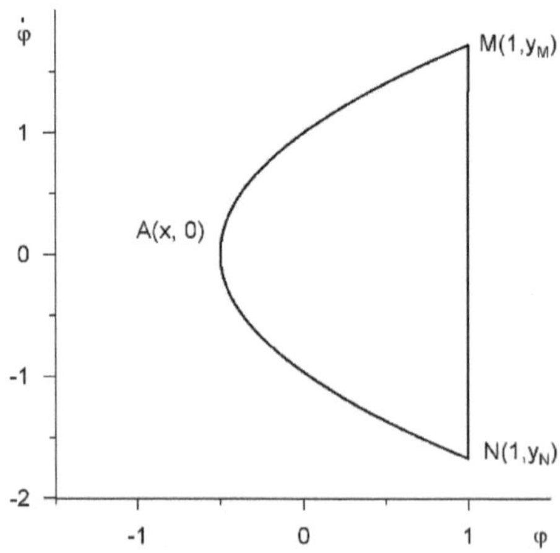

Fig. 2.58: Phase curve of the bush dynamics with one impact per motion period ($\omega_1 = 2.5$, $\delta = 2$, $\varepsilon = 0.01$, $x = -0.5$, $k = 0.9884$).

Because $\eta_0 \le -1$, then $(V_0 - V_{\min})/V_0 \le -1$, and hence $V_0 \le 0.5V_{\min}$.

Observe that if inequality $\omega_0^2 \ll 1$ is violated, then the secular terms occur while solving Equation (2.170).

Let us consider the conditions of the first type periodic motion with one impact per period (Fig. 2.58). Our aim is to find an appropriate value of the parameter k (restitution coefficient) to realize the mentioned periodic motion (phase trajectory $AMNA$). It is assumed that the phase trajectory part AM starts in the point $A(x,0)$, $x \in (-1,1)$. The Cauchy problem associated with Equation (2.161) is as follows.

$$\ddot{\varphi}(\tau) = 1 + \varepsilon\dot{\varphi}(\tau) - \omega_0^2\varphi(\tau), \qquad \varphi(0) = x, \qquad \dot{\varphi}(0) = 0, \qquad x \in (-1,1). \tag{2.170}$$

A solution being sought is given in the form $\varphi(\tau) = \varphi_{AM}(\tau, x, 0)$, $\dot{\varphi}(\tau) = \omega_{AM}(\tau, x, 0)$. Let the point M have coordinates $M(1, y_M)$ being achieved by moving point $(\varphi, \dot{\varphi})$ at time instant τ_{AM}. Therefore our problem is reduced to the consideration of the following system of nonlinear algebraic equations

$$1 = \varphi_{AM}(\tau_{AM}, x, 0), \qquad y_M = \omega_{AM}(\tau_{AM}, x, 0) \tag{2.171}$$

to be solved with respect to τ_{AM} and y_M.

Because we have assumed $\varepsilon \ll 1$ and $\omega_0^2 \ll 1$ and assuming $\omega_0^2 = \varepsilon\delta$, we begin with a solution to the Cauchy problem (2.170) keeping the accuracy of $O(\varepsilon^2)$ in the form

$$\varphi_{AM}(\tau, x, 0) = x + 0.5\tau^2 + (1/6)\tau^3\varepsilon - (1/24)\tau^2(12x + \tau^2)\delta\varepsilon + O(\varepsilon^2), \tag{2.172}$$

$$\omega_{AM}(\tau, x, 0) = \tau + 0.5\tau^2\varepsilon - (1/6)\tau(6x + \tau^2)\delta\varepsilon + O(\varepsilon^2). \tag{2.173}$$

Therefore, Equations (2.171) take the form

$$1 = x + 0.5\tau_{AM}^2 + (1/6)\tau_{AM}^3\varepsilon - (1/24)\tau_{AM}^2(12x + \tau_{AM}^2)\delta\varepsilon + O(\varepsilon^2), \tag{2.174}$$

$$y_M = \tau_{AM} + 0.5\tau_{AM}^2\varepsilon - (1/6)\tau_{AM}(6x + \tau_{AM}^2)\delta\varepsilon + O(\varepsilon^2). \tag{2.175}$$

A solution to nonlinear Equation (2.174) is sought in the form $\tau_{AM} = A_{AM} + B_{AM}\varepsilon + O(\varepsilon^2)$. Then $\tau_{AM}^2 = A_{AM}^2 + 2A_{AM}B_{AM}\varepsilon + O(\varepsilon^2)$, $\tau_{AM}^3 = A_{AM}^3 + 3A_{AM}^2 B_{AM}\varepsilon + O(\varepsilon^2)$, $\tau_{AM}^4 = A_{AM}^4 + 4A_{AM}^3 B_{AM}\varepsilon + O(\varepsilon^2)$, and the obtained results are substituted to (2.174) to yield

$$\begin{aligned} 1 = x &+ 0.5(A_{AM}^2 + 2A_{AM}B_{AM}\varepsilon) + (1/6)A_{AM}^3\varepsilon \\ &- (1/24)12xA_{AM}^2\delta\varepsilon - (1/24)A_{AM}^4\delta\varepsilon + O(\varepsilon^2). \end{aligned}$$

Comparing terms standing by the same power of ε one gets

$$2(1-x) = A_{AM}^2, \qquad 0 = A_{AM}(B_{AM}+(1/6)A_{AM}^2)-(1/24)A_{AM}^2(12x+A_{AM}^2)\delta.$$

Taking into account the first of the latter two equations one obtains

$$A_{AM} \equiv \tau_1 = \sqrt{2(1-x)},$$

whereas the second equation yields

$$B_{AM} = -(1/6)A_{AM}^2 - (1/24)A_{AM}(-12x - A_{AM}^2)\delta$$
$$= -(1/6)\tau_1^2 - (1/24)\tau_1(5\tau_1^2 - 12)\delta.$$

The solution to Equation (2.174) is

$$\tau_{AM} = \tau_1 - (1/6)\tau_1^2\varepsilon - (1/24)\tau_1(5\tau_1^2 - 12)\delta\varepsilon + O(\varepsilon^2). \tag{2.176}$$

Substituting (2.176) into (2.175) one obtains

$$y_M = \tau_1 + (1/3)\tau_1^2\varepsilon + (1/8)\tau_1(\tau_1^2 - 4)\delta\varepsilon + O(\varepsilon^2). \tag{2.177}$$

On the other hand, the phase trajectory part NA begins in point $N(1, y_N)$. The Cauchy problem associated with Equation (2.161) has the following form,

$$\ddot{\varphi}(\tau) = 1 + \varepsilon\dot{\varphi}(\tau) - \omega_0^2\varphi(\tau), \qquad \varphi(0) = 1, \qquad \dot{\varphi}(0) = y_N, \tag{2.178}$$

where $y_N = -k\,y_M$. Problem (2.178) has the following solution, $\varphi(\tau) = \varphi_{NA}(\tau, 1, y_N)$, $\dot{\varphi}(\tau) = \omega_{NA}(\tau, 1, y_N)$. Because point A has coordinates $A(x, 0)$, it is achieved by moving point $(\varphi, \dot{\varphi})$ at time instant τ_{NA}. Therefore, the following nonlinear equations hold

$$x = \varphi_{NA}(\tau_{NA}, 1, y_N), \qquad 0 = \omega_{NA}(\tau_{NA}, 1, y_N), \tag{2.179}$$

which define the values τ_{NA} and y_N being sought.

Recall that $\varepsilon \ll 1$ and $\omega_0^2 \ll 1$, and the Cauchy problem associated with Equation (2.178) is

$$\varphi_{NA}(\tau, 1, y_N) = 1 + y_N\tau + 0.5\tau^2 + (1/6)\tau^2(\tau + 3y_N)\varepsilon$$
$$- (1/24)\tau^2(12 + \tau^2 + 4y_N\tau)\delta\varepsilon + O(\varepsilon^2), \tag{2.180}$$

$$\omega_{NA}(\tau, 1, y_N) = y_N + \tau + 0.5\tau(\tau + 2y_N)\varepsilon$$
$$- (1/6)\tau(6 + \tau^2 + 3y_N\tau)\delta\varepsilon + O(\varepsilon^2). \tag{2.181}$$

Equations (2.179) take the following form,

$$x = 1 + y_N \tau_{NA} + 0.5\tau_{NA}^2 + (1/6)\tau_{NA}^2(\tau_{NA} + 3y_N)\varepsilon$$

$$- (1/24)\tau_{NA}^2(12 + \tau_{NA}^2 + 4y_N\tau_{NA})\delta\varepsilon + O(\varepsilon^2), \tag{2.182}$$

$$0 = y_N + \tau_{NA} + 0.5\tau_{NA}(\tau_{NA} + 2y_N)\varepsilon$$

$$- (1/6)\tau_{NA}(6 + \tau_{NA}^2 + 3y_N\tau_{NA})\delta\varepsilon + O(\varepsilon^2). \tag{2.183}$$

Using Equation (2.183) we express τ_{NA} by y_N as follows.

$$\tau_{NA} = -y_N + 0.5y_N^2\varepsilon - (y_N - (1/3)y_N^3)\delta\varepsilon + O(\varepsilon^2). \tag{2.184}$$

Substitution of (2.184) into (2.182) yields

$$y_N = -\tau_1 + (1/3)\tau_1^2\varepsilon + (1/8)\tau_1(4 - \tau_1^2)\delta\varepsilon + O(\varepsilon^2), \qquad \tau_1 = \sqrt{2(1-x)}. \tag{2.185}$$

Now, substituting (2.185) to (2.184) the quantity τ_{NA} is expressed through τ_1 in the following way,

$$\tau_{NA} = \tau_1 + (1/6)\tau_1^2\varepsilon - (1/24)\tau_1(5\tau_1^2 - 12)\delta\varepsilon + O(\varepsilon^2). \tag{2.186}$$

Taking into account the obtained values (2.177) and (2.185), and equation $y_N = -k\,y_M$, the restitution coefficient being sought is given explicitly in the following form,

$$k_{AMNA} = 1 - (2/3)\tau_1\varepsilon + O(\varepsilon^2). \tag{2.187}$$

Summation of both time intervals τ_{AM} and τ_{NA} gives the period of case (i):

$$\tau_{AMNA} = 2\tau_1 - (1/12)\tau_1(5\tau_1^2 - 12)\delta\varepsilon + O(\varepsilon^2). \tag{2.188}$$

Note that the periodic motion (i) takes place for $y_M < \omega_1$. Furthermore, the use of Equation (2.177) provides the estimation $x_0 < x < 1$, where

$$x_0(\omega_1) = 1 - (1/2)\omega_1^2 + (1/3)\omega_1^3\varepsilon + (1/8)\omega_1^2(\omega_1^2 - 4)\delta\varepsilon + O(\varepsilon^2). \tag{2.189}$$

Let us consider conditions of the occurrence of third-type dynamics (iii) associated with two impacts per motion period (Fig. 2.59). Our aim is to define the value of parameter k for which the required periodic motion appears (phase trajectory $BMNDB$). Let us find the coordinates of point $B(-1, y_B)$, where the phase trajectory part BM begins.

We assume that the phase trajectory goes from point $A(x.0)$, $x \in (-\infty, -1)$ into point B if there is no boundary at $x = -1$. In this case the Cauchy problem associated with differential Equation (2.161) takes the form

$$\ddot{\varphi}(\tau) = 1 + \varepsilon\dot{\varphi}(\tau) - \omega_0^2\varphi(\tau), \qquad \varphi(0) = x, \qquad \dot{\varphi}(0) = 0, x \in (-\infty, -1). \tag{2.190}$$

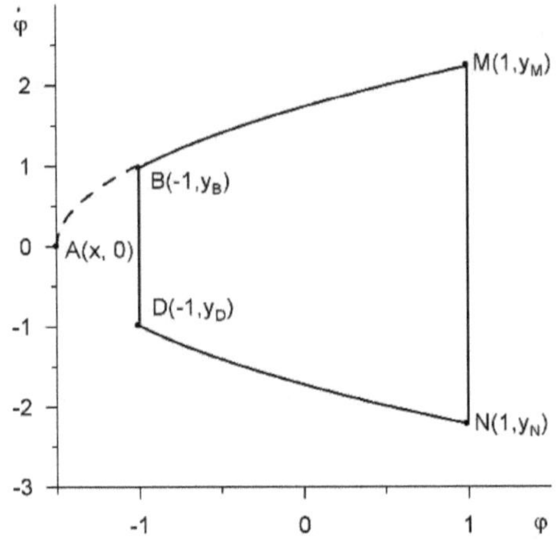

Fig. 2.59: Periodic bush phase orbit with two impacts ($\omega_1 = 2.5$, $\delta = 2$, $\varepsilon = 0.01$, $x = -1.5$, $k = 0.9887$).

A solution to problem (2.190) is as follows: $\varphi(\tau) = \varphi_{AB}(\tau, x, 0)$, $\dot{\varphi}(\tau) = \omega_{AB}(\tau, x, 0)$. Because point B is achieved by the moving point $(\varphi, \dot{\varphi})$ at time instant τ_{AB}, one obtains the two following nonlinear equations,

$$-1 = \varphi_{AB}(\tau_{AB}, x, 0), \qquad y_B = \omega_{AB}(\tau_{AB}, x, 0) \tag{2.191}$$

yielding τ_{AB} and y_B.

For $\varepsilon \ll 1$ and $\omega_0^2 \ll 1$ the Cauchy problem associated with (2.190), keeping the accuracy order of $O(\varepsilon^2)$, has a form analogous to that of (2.172) and (2.173). The system of Equations (2.191) takes the following form,

$$-1 = x + 0.5\,\tau_{AB}^2 + (1/6)\tau_{AB}^3\varepsilon - (1/24)\tau_{AB}^2(12x + \tau_{AB}^2)\delta\varepsilon + O(\varepsilon^2), \tag{2.192}$$

$$y_B = \tau_{AB} + 0.5\tau_{AB}^2\varepsilon - (1/6)\tau_{AB}(6x + \tau_{AB}^2)\delta\varepsilon + O(\varepsilon^2). \tag{2.193}$$

The solution to Equation (2.192) is as follows.

$$\tau_{AB} = \tau_2 - (1/6)\tau_2^2\varepsilon - (1/24)\tau_2(5\tau_2^2 + 12)\delta\varepsilon + O(\varepsilon^2),$$

$$\tau_2 = \sqrt{-2(1+x)}. \tag{2.194}$$

Substituting (2.194) to (2.193) the following coordinate is found.

$$y_B = \tau_2 + (1/3)\tau_2^2\varepsilon + (1/8)\tau_2(4 + \tau_2^2)\delta\varepsilon + O(\varepsilon^2). \tag{2.195}$$

Phase trajectory part BM begins at point $B(-1, y_B)$. The Cauchy problem of Equation (2.161) takes the form

$$\ddot{\varphi}(\tau) = 1 + \varepsilon\dot{\varphi}(\tau) - \omega_0^2\varphi(\tau), \qquad \varphi(0) = -1, \qquad \dot{\varphi}(0) = y_B, \tag{2.196}$$

where y_B is defined by (2.195). A solution to problem (2.196) is as follows, $\varphi(\tau) = \varphi_{BM}(\tau, -1, y_B)$, and $\dot{\varphi}(\tau) = \omega_{BM}(\tau, -1, y_B)$. Point M has coordinates $M(1, y_M)$ and it is achieved by moving point $(\varphi, \dot{\varphi})$ at time instant τ_{BM}, which defines the following set of equations,

$$1 = \varphi_{BM}(\tau_{BM}, -1, y_B), \qquad y_M = \omega_{BM}(\tau_{BM}, -1, y_B). \tag{2.197}$$

A solution to Equations (2.197) gives

$$\tau_{BM} = \tau_1 - \tau_2 - (2/3)\varepsilon + (1/24)(12\tau_2 + 5\tau_2^3$$

$$- (32 + 28\tau_2^2 + 5\tau_2^4)/\tau_1)\delta\varepsilon + O(\varepsilon^2), \tag{2.198}$$

$$y_M = \tau_1 + (1/3)\tau_1^2\varepsilon + (1/8)\tau_2^2\tau_1\delta\varepsilon + O(\varepsilon^2). \tag{2.199}$$

Phase trajectory part ND begins in point $N(1, y_N)$, and the Cauchy problem associated with differential Equation (2.161) takes the form

$$\ddot{\varphi}(\tau) = 1 + \varepsilon\dot{\varphi}(\tau) - \omega_0^2\varphi(\tau), \qquad \varphi(0) = 1, \qquad \dot{\varphi}(0) = y_N, \tag{2.200}$$

where $y_N = -k\, y_M$. Thus, we have the following solution of problem (2.200), $\varphi(\tau) = \varphi_{ND}(\tau, 1, y_N)$, $\dot{\varphi}(\tau) = \omega_{ND}(\tau, 1, y_N)$. Point D has coordinates $D(-1, y_D)$, where $y_D = -y_B/k$ and it will be achieved by moving point $(\varphi, \dot{\varphi})$ at time instant τ_{ND}. One obtains the following equations

$$-1 = \varphi_{ND}(\tau_{ND}, 1, y_N), \qquad y_D = \omega_{ND}(\tau_{ND}, 1, y_N) \tag{2.201}$$

yielding the τ_{ND} and y_D being sought. The second equation of (2.201) gives

$$\tau_{ND} = y_D - y_N + (1/2)(y_N^2 - y_D^2)\varepsilon$$

$$+ (1/6)(y_N - y_D)(-6 + 2y_N^2 - y_D^2 - y_N y_D)\delta\varepsilon + O(\varepsilon^2). \tag{2.202}$$

Substituting (2.202) to the first Equation (2.201) and taking into account $y_N = -k\, y_M$ and $y_B = -k\, y_D$ one gets

$$k_{BMNDB} = 1 + \frac{\tau_2^3 - \tau_1^3}{3(2 + \tau_2^2)}\varepsilon + O(\varepsilon^2). \tag{2.203}$$

Time interval τ_{ND} is defined by Equation (2.202) taking into account (2.203), $y_N = -k\, y_M$, and $y_B = -k\, y_D$, as well as (2.193) and (2.199):

$$\tau_{ND} = \tau_1 - \tau_2 + \frac{2}{3(2+\tau_2^2)}(2 - 2\tau_1\tau_2 + \tau_2^2)\varepsilon$$
$$+ \frac{1}{24(2+\tau_2^2)}(22\tau_2^3 + 5\tau_2^5 - 16\tau_1 - 5\tau_2^4\tau_1 + 24\tau_2 - 18\tau_2^2\tau_1)\delta\varepsilon + O(\varepsilon^2). \tag{2.204}$$

Now summing both time intervals τ_{BM} and τ_{ND} one obtains the period associated with case (iii) of the form

$$\tau_{BMNDB} = 2(\tau_1 - \tau_2) - \frac{4\tau_2\tau_1}{3(2+\tau_2^2)}\varepsilon$$
$$+ \frac{(5\tau_2^2 + 12)\tau_2\tau_1 - (5\tau_2^4 + 28\tau_2^2 + 32)}{12\tau_1}\delta\varepsilon + O(\varepsilon^2). \tag{2.205}$$

Note that periodic motion in case (iii) occurs only if $y_M < \omega_1$. Therefore, taking (2.199) into account, one obtains the following inequality $x_0 < x < -1$, where x_0 is defined by (2.189).

Let us define conditions in which the second type motion (ii) occurs; that is, with one impact per period (Fig. 2.60) and with a stick-slip behaviour. Following previous approaches we define the value of parameter k responsible for realization of this motion type (phase trajectory $ACMNA$). Assuming that the trajectory part AC begins at $A(x, 0)$, $x \in (-1, 1)$, the Cauchy problem for the differential equation (2.161) takes the form

$$\ddot{\varphi}(\tau) = 1 + \varepsilon\dot{\varphi}(\tau) - \omega_0^2\varphi(\tau), \qquad \varphi(0) = x, \qquad \dot{\varphi}(0) = 0, \qquad x \in (-1, 1). \tag{2.206}$$

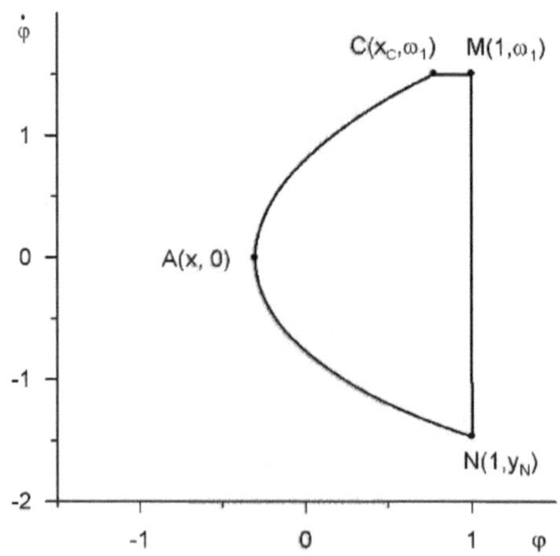

Fig. 2.60: Bush phase curve with one impact per motion period in stick-slip conditions ($\omega_1 = 1.5$, $\delta = 2$, $\varepsilon = 0.1$, $x = -0.3$, $k = 0.9796$, $x_1 = -0.3359$, $x_0 = -0.1109$).

We take the following solution $\varphi(\tau) = \varphi_{AC}(\tau, x, 0)$, $\dot{\varphi}(\tau) = \omega_{AC}(\tau, x, 0)$. Let point C have coordinates $C(x_C, \omega_1)$ and let it be achieved by moving point $(\varphi, \dot{\varphi})$ at time instant τ_{AC}. The following nonlinear equations

$$x_C = \varphi_{AC}(\tau_{AC}, x, 0), \qquad \omega_1 = \omega_{AC}(\tau_{AC}, x, 0) \tag{2.207}$$

are used to determine τ_{AC} and x_C.

Proceeding in a way similar to the previous analysis we take $\varepsilon \ll 1$ and $\omega_0^2 \ll 1$, and $\omega_0^2 = \varepsilon\delta$ to find

$$\tau_{AC} = \omega_1 - 0.5\omega_1^2\varepsilon + (1/6)\omega_1(\omega_1^2 - 3\tau_1^2 + 6)\delta\varepsilon + O(\varepsilon^2), \tag{2.208}$$

$$x_C = 1 - 0.5(\tau_1^2 - \omega_1^2) - (1/3)\omega_1^3\varepsilon + (1/8)\omega_1^2(\omega_1^2 - 2\tau_1^2 + 4)\delta\varepsilon + O(\varepsilon^2). \tag{2.209}$$

Condition $x_C \le 1$ gives the boundary $x \le x_0$ ($x_C|_{x=x_0} = 1$), where x_0 is defined by (2.189).

Phase trajectory NA begins at $N(1, y_N)$, and the Cauchy problem for differential equation (2.161) takes the form

$$\ddot{\varphi}(\tau) = 1 + \varepsilon\dot{\varphi}(\tau) - \omega_0^2\varphi(\tau), \qquad \varphi(0) = 1, \qquad \dot{\varphi}(0) = y_N, \tag{2.210}$$

where $y_N = -k\omega_1$. The solution to problem (2.210) is: $\varphi(\tau) = \varphi_{NA}(\tau, 1, y_N)$, $\dot{\varphi}(\tau) = \omega_{NA}(\tau, 1, y_N)$. Point A has coordinates $A(x, 0)$ and it is achieved by moving point $(\varphi, \dot{\varphi})$ at time instant τ_{NA}. The following nonlinear set of equations

$$x = \varphi_{NA}(\tau_{NA}, 1, y_N), \qquad 0 = \omega_{NA}(\tau_{NA}, 1, y_N) \tag{2.211}$$

gives

$$\tau_{NA} = \tau_1 + (1/6)\tau_1^2\varepsilon - (1/24)\tau_1(5\tau_1^2 - 12)\delta\varepsilon + O(\varepsilon^2), \tag{2.212}$$

$$k_{ACMNA} = \tau_1/\omega_1 - (1/3)(\tau_1^2/\omega_1)\varepsilon - (1/8)(\tau_1/\omega_1)(4 - \tau_1^2)\delta\varepsilon + O(\varepsilon^2). \tag{2.213}$$

A natural limitation $k \le 1$ yields the inequality $x \ge x_1$, where

$$x_1(\omega_1) = 1 - 0.5\omega_1^2 - (1/3)\omega_1^3\varepsilon + (1/8)\omega_1^2(\omega_1^2 - 4)\delta\varepsilon + O(\varepsilon^2). \tag{2.214}$$

Summation of τ_{AC} and τ_{NA} as well as of $\tau_{CM} = (1 - x_C)/\omega_1$ gives the estimation of a period in case (ii):

$$\begin{aligned}\tau_{ACMNA} &= (\tau_1 + \omega_1)^2/(2\omega_1) + (1/6)(\tau_1^2 - \omega_1^2)\varepsilon \\ &\quad + (1/24)(\tau_1 + \omega_1)(12 - 5\tau_1^2 + \omega_1^2 - \omega_1\tau_1)\delta\varepsilon + O(\varepsilon^2).\end{aligned} \tag{2.215}$$

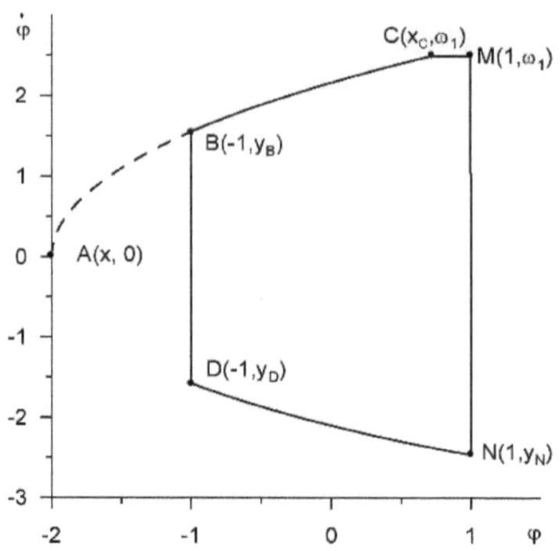

Fig. 2.61: Bush phase stick-slip curve with two impacts per period ($\omega_1 = 2.5$, $\delta = 2$, $\varepsilon = 0.05$, $x = -2.0$, $k = 0.9853$, $x_3 = -2.097$, $x_0 = -1.689$).

In the second case periodic motion ($-1 \leq x \leq 1$) occurs when $x_1 \leq x \leq x_0$, where x_0 and x_1 are defined by formulas (2.189) and (2.214).

Let us consider now the conditions of appearance of the fourth type of periodic stick-slip motion with two impacts per period (Fig. 2.61). Also in this case we establish a value of the parameter k being responsible for a stick-slip periodic motion occurrence (phase trajectory $BCMNDB$). Let us first find coordinates of point $B(-1, y_B)$, where the trajectory BC begins.

We assume that point $A(x, 0)$, $x \in (-\infty, -1)$ achieves point B if there is no limit $x = -1$. The Cauchy problem for Equation (2.161) has the form of (2.190). A solution to problem (2.190) has the form $\varphi(\tau) = \varphi_{AB}(\tau, x, 0)$, $\dot{\varphi}(\tau) = \omega_{AB}(\tau, x, 0)$. Point B is achieved by moving point $(\varphi, \dot{\varphi})$ in time instant τ_{AB}. This observation enables us to derive nonlinear Equations (2.191) with solutions τ_{AB} and y_B defined by (2.194) and (2.195), respectively.

Phase trajectory part BC begins at point $B(-1, y_B)$. The Cauchy problem for differential Equation (2.161) has the form of (2.196), where y_B is defined by (2.195).

A solution to problem (2.196) has the form: $\varphi(\tau) = \varphi_{BC}(\tau, -1, y_B)$, $\dot{\varphi}(\tau) = \omega_{BC}(\tau, -1, y_B)$. Point C has coordinates $C(x_C, \omega_1)$ and it is achieved by moving point $(\varphi, \dot{\varphi})$ at time instant τ_{BC}. Therefore, one gets the following system of equations

$$x_C = \varphi_{BC}(\tau_{BC}, -1, y_B), \qquad \omega_1 = \omega_{BC}(\tau_{BC}, -1, y_B), \tag{2.216}$$

whose solutions follow

$$\tau_{BC} = \omega_1 - \tau_2 + (1/6)(\tau_2^2 - 3\omega_1^2)\varepsilon + (1/24)(12\tau_2 + 5\tau_2^3 - 12\tau_2^2\omega_1 - 24\omega_1 + 4\omega_1^3)\delta\varepsilon + O(\varepsilon^2), \quad (2.217)$$

$$x_C = -1 - 0.5(\tau_2^2 - \omega_1^2) - (1/3)\omega_1^3\varepsilon + (1/8)\omega_1^2(-4 - 2\tau_2^2 + \omega_1^2)\delta\varepsilon + O(\varepsilon^2). \quad (2.218)$$

On the other hand, the phase trajectory ND begins at point $N(1, y_N)$. The Cauchy problem for differential equation (2.161) has the form

$$\ddot{\varphi}(\tau) = 1 + \varepsilon\dot{\varphi}(\tau) - \omega_0^2\varphi(\tau), \qquad \varphi(0) = 1, \qquad \dot{\varphi}(0) = y_N, \quad (2.219)$$

where $y_N = -k\,\omega_1$. The solution of problem (2.219) is: $\varphi(\tau) = \varphi_{ND}(\tau, 1, y_N)$, $\dot{\varphi}(\tau) = \omega_{ND}(\tau, 1, y_N)$. Point D has coordinates $D(-1, y_D)$, where $y_D = -y_B/k$ and it is achieved by moving point $(\varphi, \dot{\varphi})$ at time instant τ_{ND}. Hence, we have the following equations

$$-1 = \varphi_{ND}(\tau_{ND}, 1, y_N), \qquad y_D = \omega_{ND}(\tau_{ND}, 1, y_N) \quad (2.220)$$

yielding k and τ_{ND}. Expressing τ_{ND} by y_D and y_N through the formula

$$\tau_{ND} = y_D - y_N - 0.5(y_D^2 - y_N^2)\varepsilon - (1/6)(y_D - y_N)(2y_N^2 - y_D^2 - y_N y_D - 6)\delta\varepsilon + O(\varepsilon^2), \quad (2.221)$$

and substituting (2.221) to the first equation of (2.220) and taking into account $y_N = -k\,\omega_1$ and $y_D = -y_B/k$, one gets

$$k_{BCMNDB} = \frac{\tau_0}{\omega_1} + \frac{-16\tau_0^2 + \tau_2^3\omega_1^2(\tau_0 + \omega_1) - \tau_2^2\omega_1^2(4 + \tau_0^2)}{6\omega_1(\tau_2^2\omega_1^2 + 2\tau_0^2)}\varepsilon + \frac{\tau_2^2\tau_0\omega_1(\tau_2^2 + 4)}{16(\tau_2^2\omega_1^2 + 2\tau_0^2)}\delta\varepsilon + O(\varepsilon^2), \qquad \tau_0 = \sqrt{2 + \sqrt{4 + \tau_2^2\omega_1^2}}, \quad (2.222)$$

and also the following dimensionless time is estimated

$$\tau_{ND} = \tau_1 - \tau_2 + (2/3)(2 - 2\tau_2\tau_1 + \tau_2^2)\varepsilon/(2 + \tau_2^2) + \frac{\tau_2(22\tau_2^2 + 5\tau_2^4 - 5\tau_2^3\tau_1 + 24 - 18\tau_2\tau_1) - 16\tau_1}{24(2 + \tau_2^2)}\delta\varepsilon + O(\varepsilon^2). \quad (2.223)$$

The natural limitation introduced on the restitution coefficient ($k \leq 1$) yields $x \geq x_2$, where

$$x_2(\omega_1) = x_1(\omega_1) + (2/3)(\omega_1^2 - 4)^{3/2}\varepsilon. \tag{2.224}$$

Now, summing both τ_{BC} and τ_{ND}, as well as $\tau_{CM} = (1 - x_C)/\omega_1$ one gets the period of stick-slip periodic orbit for case (iv):

$$\begin{aligned}\tau_{BCMNDB} &= (\tau_1^2 - 4\tau_2\omega_1 + 2\tau_1\omega_1 + \omega_1^2)/(2\omega_1)\\ &+ (1/6)(\tau_2^4 - 8\tau_2\tau_1 - \tau_2^2\omega_1^2 + 6\tau_2^2 - 2\omega_1^2 + 8)\varepsilon/(2 + \tau_2^2)\\ &+ (1/24)(24\tau_2 + 10\tau_2^3 - 8\tau_1 - 12\omega_1 + \omega_1^3 - 5\tau_2^2\tau_1 - 6\tau_2^2\omega_1)\delta\varepsilon + O(\varepsilon^2).\end{aligned} \tag{2.225}$$

A periodic motion in case (iv) ($-\infty \le x \le -1$) takes place if $x_2 \le x \le x_0$, where x_0 and x_2 are defined by formulas (2.189) and (2.224).

Results of our consideration allow us to give formulas for the coefficient of restitution k for a general case of the following case.

$$k(x,\omega_1) = \begin{cases} k_{ACMNA}, & 0 < \omega_1 < 2 - (4/3)\varepsilon, \quad x_1 < x < x_0\\ k_{AMNA}, & 0 < \omega_1 < 2 - (4/3)\varepsilon, \quad x_0 < x < 1\\ \\ k_{BCMNDB}, & 2 - (4/3)\varepsilon < \omega_1 < 2, \quad x_1 < x < -1\\ k_{ACMNA}, & 2 - (4/3)\varepsilon < \omega_1 < 2, \quad -1 < x < x_0\\ k_{AMNA}, & 2 - (4/3)\varepsilon < \omega_1 < 2, \quad x_0 < x < 1\\ \\ k_{BCMNDB}, & 2 < \omega_1 < 2 + (4/3)\varepsilon, \quad x_2 < x < -1\\ k_{ACMNA}, & 2 < \omega_1 < 2 + (4/3)\varepsilon, \quad -1 < x < x_0\\ k_{AMNA}, & 2 < \omega_1 < 2 + (4/3)\varepsilon, \quad x_0 < x < 1\\ \\ k_{BCMNDB}, & 2 + (4/3)\varepsilon < \omega_1 < \infty, \quad x_2 < x < x_0\\ k_{BMNDB}, & 2 + (4/3)\varepsilon < \omega_1 < \infty, \quad x_0 < x < -1\\ k_{AMNA}, & 2 + (4/3)\varepsilon < \omega_1 < \infty, \quad -1 < x < 1 \end{cases} \tag{2.226}$$

The obtained results are graphically presented in Fig. 2.62.

Observe that the function $k(x,\omega_1)$ possesses the following values $k(x_1,\omega_1) = 1$, $0 < \omega_1 < 2$, $k(x_2,\omega_1) = 1, 2 < \omega_1 < \infty$, $k(1,\omega_1) = 1$ at the boundaries, whereas inside the considered interval it has the following minima

$$\min_{x\in[x_1,1]} k(x,\omega_1) = k(x_0,\omega_1) = 1 - (2/3)\omega_1\varepsilon, \quad 0 < \omega_1 < 2 + (4/3)\varepsilon, \tag{2.227}$$

$$\min_{x\in[x_2,1]} k(x,\omega_1) = k(-1,\omega_1) = 1 - (4/3)\varepsilon, \quad 2 + (4/3)\varepsilon < \omega_1 < \infty, \tag{2.228}$$

which can be presented in the form

$$k_{\min} = \begin{cases} 1 - (2/3)\omega_1\varepsilon, & 0 < \omega_1 < 2 + (4/3)\varepsilon\\ 1 - (4/3)\varepsilon, & 2 + (4/3)\varepsilon < \omega_1 < \infty \end{cases} \tag{2.229}$$

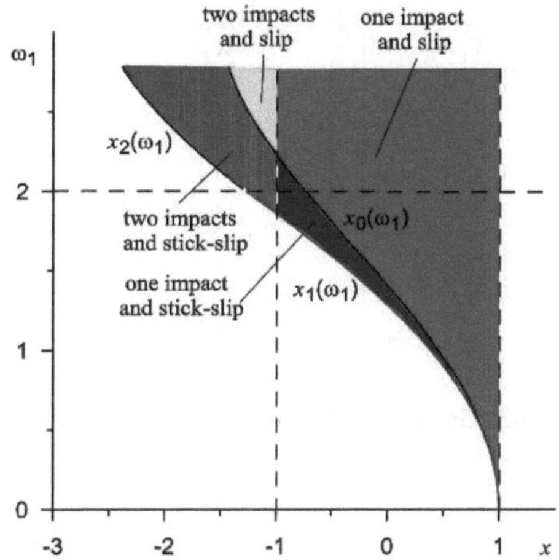

Fig. 2.62: Zones of different periodic impact motions ($\omega_0^2 = 0.2$, $\delta = 2$, $\varepsilon = 0.1$).

Notice that for an arbitrary $k^* \in (k_{\min}, 1)$ there are two values of x_1^*, x_2^* $(k(x_1^*, \omega_1) = k(x_2^*, \omega_1) = k^*)$. Let us introduce the following intervals

$$x_1 < x_1^* < x_0, \qquad x_0 < x_2^* < 1, \quad \text{for } 0 < \omega_1 < 2, \tag{2.230}$$

$$x_2 < x_1^* < x_0, \qquad x_0 < x_2^* < 1, \quad \text{for } 2 < \omega_1 < 2 + (4/3)\varepsilon, \tag{2.231}$$

$$x_2 < x_1^* < -1, \qquad -1 < x_2^* < 1, \quad \text{for } 2 - (4/3)\varepsilon < \omega_1 < \infty. \tag{2.232}$$

It is not difficult to check that a periodic orbit associated with x_1^* (decreasing part of the coefficient $k(x)$) is stable, whereas a periodic orbit associated with x_2^* (increasing part of the coefficient $k(x)$) is unstable.

2.5.4 Numerical analysis

A numerical analysis has been carried out for the following dimensionless parameters: $\varepsilon = 0.1$, $\omega_1 = 1.6$, $\delta = 2$, $\omega_0^2 = 0.2$. Formula (2.229) gives $k_{\min} = 0.89$. If $k^* = 0.93$ ($k^* \in (k_{\min}, 1)$), then our system exhibits two periodic orbits defined by $x_1^* = -0.33$ (stable) case (ii) and $x_2^* = 0.44$ (unstable) case (i) (see Figs. 2.63 and 2.64).

Curves 1 and 2 approach a stable periodic orbit, whereas curve 3 tends to the stable point (1,0). Note that the dashed curve is associated with an unstable orbit (see Fig. 2.64).

Next, a numerical analysis has been carried out for the following dimensionless parameters: $\varepsilon = 0.1$, $\omega_1 = 1.6$, $\omega_0^2 = 0.2$, $\eta_0 = -2$, $\tilde{\omega} = 0.1$. For $x = -0.9$ and

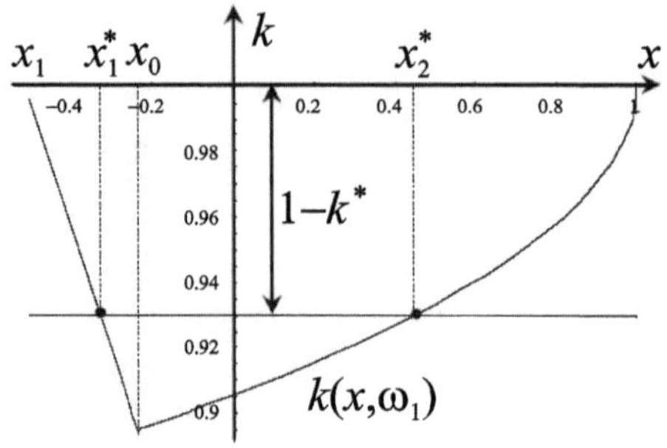

Fig. 2.63: Graphical solution of equation $k(x, \omega_1) = k^*$.

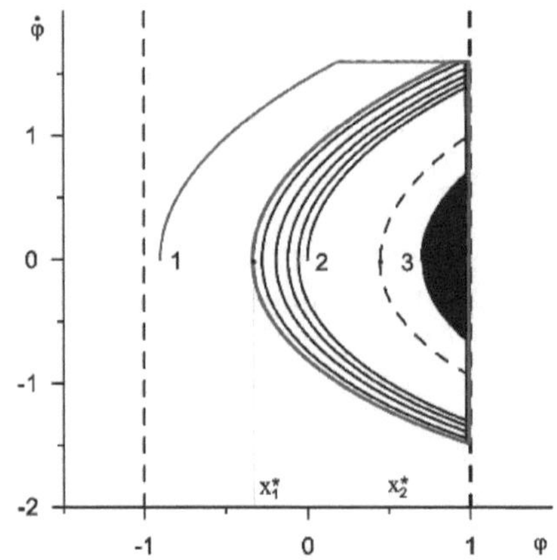

Fig. 2.64: Phase trajectory of the bush movement for different values of x: curves 1: $x = -0.9$ $(0 \le x < x_1^*)$; curves 2: $x = 0$ $(x_1^* < x < x_2^*)$; curves 3: $x = 0.7$ $(x_2^* < x \le 1)$.

$k = 0.93$ the corresponding bush phase trajectory for various parameters γ_1 and k^w is shown in Fig. 2.65.

Time histories of contact pressure, temperature on surface contact, and wear are reported in Fig. 2.66. Curves 1 correspond to the case when $\gamma_1 = 0$ (lack of heat extension), $k^w = 0$ (lack of bush wear). Curves 2 correspond to the case of heat transfer lack ($\gamma_1 = 0$) and $k^w = 0.02$. Curves 3 correspond to the case where the shaft heat expansion is taken into account ($\gamma_1 = 0.1$), but the bush wear is neglected ($k^w = 0$). Curves 4 correspond to the case where both mentioned parameters are taken into account ($\gamma_1 = 0.5$, $k^w = 0.02$).

In the first case ($\gamma_1 = 0$, $k^w = 0$), where the tribological processes are not taken into account the phase curve approaches a stable orbit (curve 1 in Fig. 2.65a). In this case the contact pressure is exhibited by curve 1 in Fig. 2.66a. The bush wear ($\gamma_1 = 0$, $k^w = 0.02$) occurrence decreases the contact pressure (curve 2 in Fig. 2.66a), which tends to zero value (the corresponding phase curve is shown in Fig. 2.65b). Note that after the wear process, the bush moves in a periodic manner.

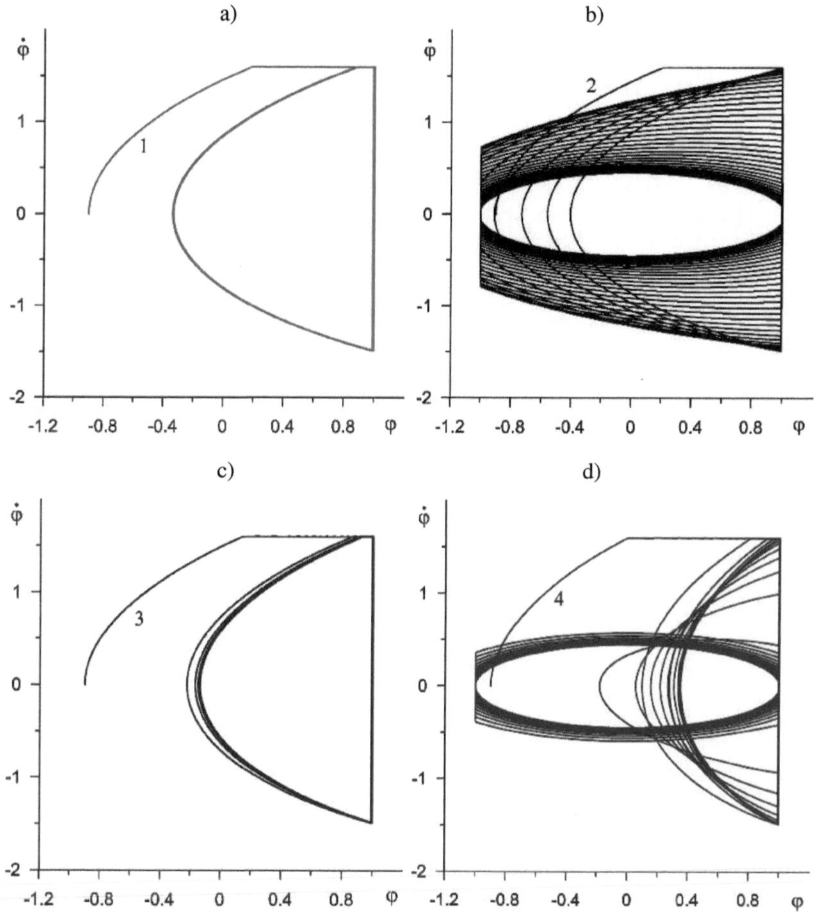

Fig. 2.65: Phase trajectory of the bush movement for different values of tribologic parameters: curves 1: $\gamma_1 = 0$, $k^w = 0$ (without tribologic processes); curves 2: $\gamma_1 = 0$, $k^w = 0.02$; curves 3: $\gamma_1 = 0.1$, $k^w = 0$ (with heat generation); curves 4: $\gamma_1 = 0.5$, $k^w = 0.02$ (with tribologic processes).

Bush wear kinematics is shown in Fig. 2.66c (curve 2). An inclusion of the shaft heat expansion ($\gamma_1 = 0.1$) within the given heat transfer conditions ($Bi = 10$) yields a periodic change of both contact pressure (curve 3 in Fig. 2.66a) and temperature (curve 3 in Fig. 2.66b). The phase curve after a transitional process tends to a new stable periodic orbit (curve 3 in Fig. 2.65c). For a general case, (i.e., where the tribological processes are taken into account ($\gamma_1 = 0.5$, $k^w = 0.02$)) and for the given heat transfer conditions ($Bi = 10$) the obtained results are exhibited by curves 4 in Figs. 2.65 and 2.66. In this case the bush wear increases owing to the shaft heat extension, and the contact pressure first increases and then it tends to zero (curve 4 in Fig. 2.66a). The contact temperature being changed in an oscillatory manner first increases, but then decreases with decrease of the contact

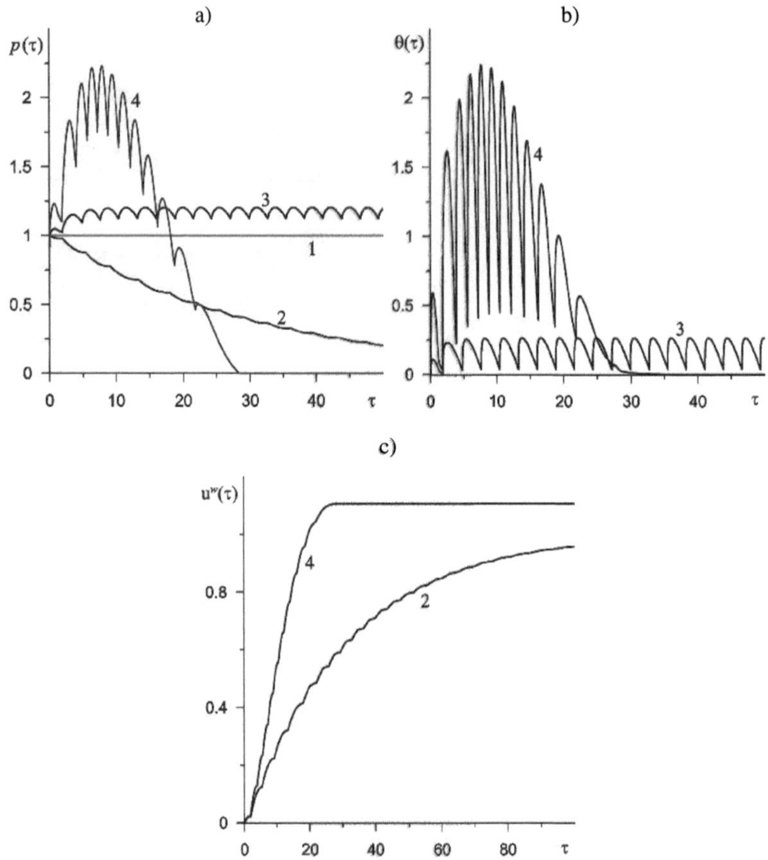

Fig. 2.66: Time histories of dimensionless contact pressure $p(\tau)$ (a), dimensionless contact temperature $\theta(1,\tau) \equiv \theta(\tau)$ (b) and wear $u^w(\tau)$ (c), versus dimensionless time τ for different values of γ_1 and k^w; curves 1: $\gamma_1 = 0$, $k^w = 0$ (without tribologic processes); curves 2: $\gamma_1 = 0$, $k^w = 0.02$ (heat generation included); curves 3: $\gamma_1 = 0.1$, $k^w = 0$ (heat generation included, wear omitted); curves 4: $\gamma_1 = 0.5$, $k^w = 0.02$ (with tribologic processes).

pressure. The bush wear kinematics is exhibited by curve 4 in Fig. 2.66c. Observe that the final wear amount is larger than the initial shaft compression. In this case the phase curve (after the bush wear) approaches a stable periodic orbit (curve 4 in Fig. 2.65d).

In the case when the bush wear is less than the shaft thermal expansion (for instance, in the case of $\gamma_1 = 1$, $Bi = 10$), the contact characteristics increase in an exponential manner with time increase. In the latter case the shaft cannot succeed in making cooling in time.

We have proposed a novel model of vibrations of the bush–shaft system with inclusion of both impacts and tribological processes occurring on the contact surface. A similar system, however, without impacts, has been studied earlier by the authors and it has been described in references [Awrejcewicz, Pyryev (2002), (2004b)]. The occurrence of self-excited vibrations in a more simplified system with a gap (without tribological processes and springs) has also been analysed in reference [Balandin (1993), Balandin et al. (2001)].

Applying the Laplace transform, our problem has been reduced to that of the system of one nonlinear differential equation and one second-order Volterra integral equation with respect to the contact pressure. A kernel of the latter equation is the function of the sliding velocity. We have estimated analytically the restitution coefficient for which a periodic orbit occurs assuming small slope of friction characteristics. We, among others, have shown various periodic motions exhibited by the analysed system and we have verified numerically our theoretical considerations and predictions.

For an arbitrary restitution coefficient $k^* \in (k_{\text{min}}, 1)$ two periodic orbits (stable and unstable) appear on the phase plane. Increase of the parameter k^* from k_{min} to 1 yields an increase (decrease) of the stable (unstable) periodic orbit. For $k^* = 1$ the unstable periodic orbit is reduced to the point $(1, 0)$. Decrease of the parameter k^* causes the approach of both stable and unstable periodic trajectories. For $k^* = k_{\text{min}}$ a bifurcation occurs and a half-stable periodic orbit is born substituting two previous stable and unstable orbits. In other words, for $k < k_{\text{min}}$ a periodic motion is not exhibited by the studied system.

The final conclusion follows. Tribologic processes have an important impact on the studied system dynamics, because they may change it even qualitatively.

Observe that our aim was not to investigate either a real engineering system or construction. We are rather focused on periodic and chaotic nonsmooth dynamics with friction and impacts exhibited by relatively simple models shown in Figs. 2.1 and 2.57. However, the methodology based on a study of the mentioned simple models by analytical (or semi-analytical) methods and numerical methods (necessary for the prediction and monitoring of chaotic vibrations) seems to be powerful for analysing and understanding the phenomena as well as for designing engineering systems. Even if detailed computations from partial differential equations could be a complementary task for applications, the presented approach is profitable for scientists and engineers. Furthermore, the obtained analytical formulas can be used for identification of the parameters of the models, or can serve as verification and validation of the real engineering objects studied via the finite difference method, finite element method, Bubnov–Galerkin method, and others.

The solid friction occurs during a sliding contact between the moving bodies, and the further taken contact geometry allows us to study body dynamics models being close to their real behaviour. Although the introduced approach is not directly applied to a real engineering object, the studied geometry of joints could be applied in the analysis of the clamping and interference joints, as well as in various joints with friction and impact phenomena. The mentioned simple models can be directly applied to study the so-called brake pad as well as Pronny's brake [Andronov et al. (1966)].

2.6. Stick-slip vibrations continuous friction model

In this section an alternative (novel) continuous friction model is presented and studied. Advantages of the proposed model with respect to the other friction models are illustrated and discussed. As an example of a mechanical system a one degree-of-freedom model with dry friction is considered. The analysed system exhibits both regular and nonregular dynamics. The dynamics of the mentioned mechanical model is monitored via standard characteristics (trajectories of motion in the system's phase space, and bifurcation diagrams, as well as the Lyapunov exponents).

2.6.1 Introduction

Dry friction belongs to one of the most known phenomena exhibited by mechanical systems. Its proper mathematical modelling is not an easy task, because in the general case friction force between sliding surfaces is a complex process and depends on several parameters such as relative velocity of sliding surfaces, normal load, time, and temperature. Several formulations describing dry friction force were presented in Section 1.1.5. In a signum model the friction force F_{fr} is defined as a function of the slip velocity ω_r of sliding surfaces in the slip phase and as a function of the force $\Phi(\varphi, \dot{\varphi}, \tau)$ in the stick phase. This model describes the dry friction phenomenon in a correct and accurate way. Note that an exact value of zero is rarely computed during numerical simulation and for this reason the mentioned 'signum model' is equivalent, from a numerical point of view, to the classical Coulomb model. The dependence of friction force on the sliding relative velocity based on the 'signum model' is not a continuous function for $\omega_r = 0$ and standard numerical procedures devoted to solving differential continuous equations cannot be used. The friction curve is therefore often approximated by a continuous or smooth function (smoothing method) but for $\omega_r = 0$ the value of the friction force is always equal to zero and the friction force depends on ω_r but not on $\Phi(\varphi, \dot{\varphi}, \tau)$ in the stick phase, thus equations in the stick mode (static equations) are not correct.

The other so-called 'switch model' from a mathematical point of view treats the system as three different sets of ordinary differential equations: one for the slip phase, a second for the stick phase, and a third for the transition from stick to slip. At each timestep the state vector is inspected to determine whether the analysed system is in the slip, in the stick, or in the transition phase. The conditions for changing to the stick phase or the slip phase operate as switches between the systems and a region of stepness parameter ε_ω is very small in comparison to relative sliding velocity. The finite region is necessary for digital computation, because an exact value of zero will rarely be computed. If the relative velocity lies within the stick band (phase), friction force is equal to the force $\Phi(\varphi, \dot{\varphi}, \tau)$. In the switch model the acceleration of the body during the stick mode is proportional to

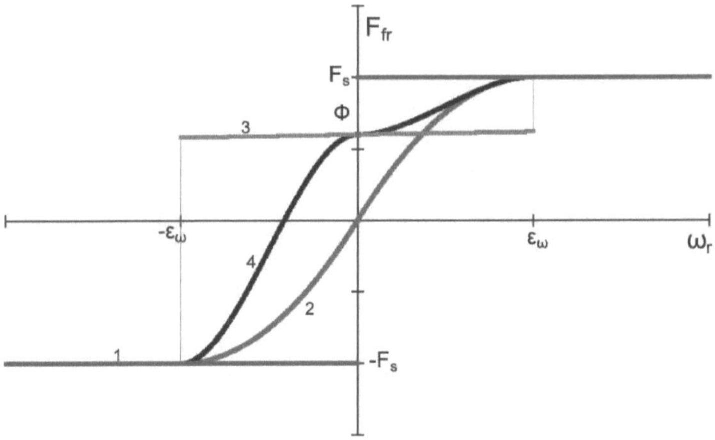

Fig. 2.67: Different approximations of friction force in the near-zero relative velocity: signum model (curve 1), smoothing method (curve 2), switch model (curve 3), and continuous friction model (curve 4).

the relative velocity (with ω_d coefficient). The relative velocity, which is directly dependent on the state of the system, cannot be set to zero in the stick phase as the state vector has to be continuous for ordinary differential equations.

We present a novel continuous friction model taking into account some elements of the mentioned switch model. A continuous friction model uses friction force on the basis of the switch model. In other words, in a continuous friction model the space (Φ, ω_r) is divided into regions (as in the switch model) and friction force is a continuous function on ω_r (as in the smoothing method), and for $\omega_r = 0$ the friction force is equal to the force $\Phi(\varphi, \dot{\varphi}, \tau)$ (as in the signum model). For this reason this continuous friction model can be treated as an approximating friction force appearing in the switch model using the continuous functions and describing the friction force for $\omega_r = 0$ in the correct and accurate way. The curves describing the friction force on the signum model, smoothing model (smoothing method), switch model, and continuous friction model in the region $|\omega_r| \leq \varepsilon_\omega$ are plotted in Fig. 2.67.

In Fig. 2.68 the dependencies of friction force as a function of two variables $F_{fr}(\omega_r, \Phi)$ are plotted for dimensionless parameters $F_s = 1$ and $\varepsilon_\omega = 0.001$.

2.6.2 *One degree-of-freedom model*

To demonstrate the above continuous friction model and compare results with other friction models a single degree-of-freedom (1-dof) mechanical system with dry friction is considered. This model possesses stick-slip periodic and nonperiodic solutions. The analysed system is shown in Fig. 2.69.

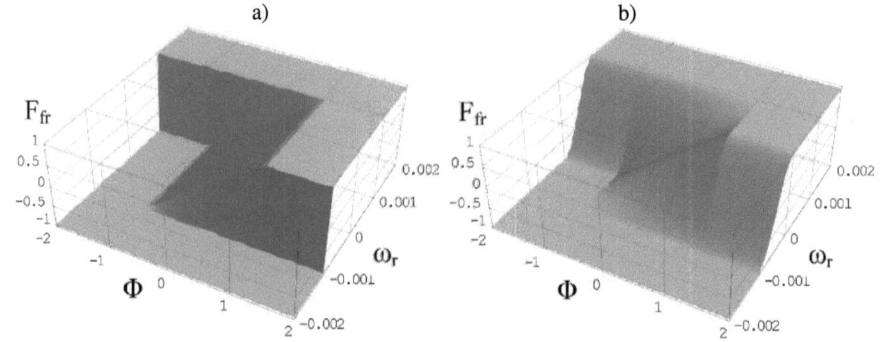

Fig. 2.68: Friction force $F_{fr}(\omega_r, \Phi)$ as a function of two variables: (a) switch model, (b) continuous friction model.

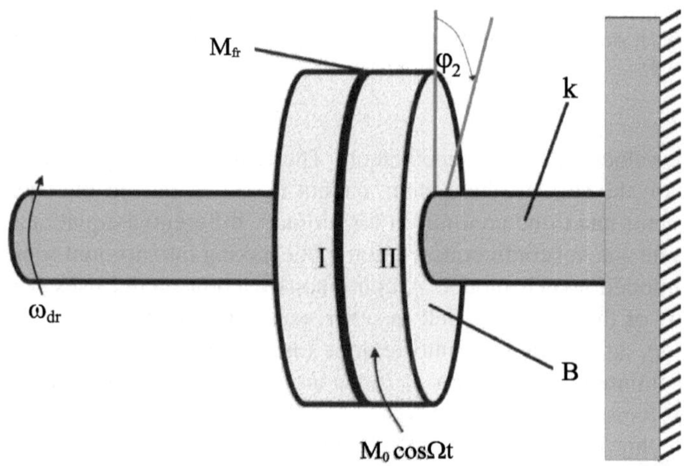

Fig. 2.69: The analysed one degree-of-freedom model with dry friction.

The second disc (II) is characterized by linear stiffness k and the mass moment of inertia B and it is fixed to the frame. The second disc is coupled with the driving first disc (I) that is moving with angular velocity ω_{dr}. Between the discs dry friction occurs which generates a moment of friction force M_{fr} (with maximum static moment of friction force M_s). In addition, harmonic excitation with amplitude M_0 and circular frequency Ω is added to our model.

The following second-order differential equation of this system is

$$B\ddot{\varphi}_2 + k\varphi_2 = M_{fr} + M_0 \cos \Omega t. \tag{2.233}$$

Let us introduce the following similarity coefficients t_*, φ_* and the following dimensionless parameters

$$\tau = \frac{t}{t_*}, \qquad \varphi = \frac{\varphi_2}{\varphi_*}, \qquad \omega = \dot{\varphi}_2 t_* = \omega_2 t_*,$$

$$\omega_d^2 = \frac{kt_*^2}{B}, \qquad \zeta = \frac{M_0 t_*^2}{B\varphi_*}, \qquad F_s = \frac{M_s t_*^2}{B\varphi_*}, \qquad \omega_0 = \Omega t_*, \qquad \omega_1 = \omega_{dr} t_*.$$

Then, in the dimensionless form, the mathematical model reads

$$\ddot{\varphi} = -\omega_d^2 \varphi + \zeta \cos \omega_0 \tau + F_{fr}, \tag{2.234}$$

with the friction model (1.68), where

$$F(|\omega_r|)\mathrm{sgn}(\omega_r)p(\tau) = \frac{F_s \mathrm{sgn}(\omega_r)}{1 + \delta(|\omega_r| - \varepsilon)}, \tag{2.235}$$

$$\Phi(\varphi, \dot{\varphi}, \tau) = \omega_d^2 \varphi - \zeta \cos \omega_0 \tau, \tag{2.236}$$

and the relative angular velocity of the second disc with respect to the first disc is denoted by $\omega_r = \omega_1 - \dot{\varphi}$.

This dimensionless dynamical system can be expressed as a set of first-order ordinary differential equations. The governing equations read

$$\begin{cases} \dot{\varphi} = \omega, \\ \dot{\omega} = -\omega_d^2 \varphi + \zeta \cos \Psi + F_{fr}, \\ \dot{\Psi} = \omega_0, \end{cases} \tag{2.237}$$

where the dot denotes differentiation with respect to dimensionless time τ.

The initial parameters of this model taken from the work [Awrejcewicz et al. (2007)] are: $B = 2[\mathrm{kg} \cdot \mathrm{m}^2]$, $k = 10[\mathrm{N} \cdot \mathrm{m} \cdot \mathrm{rad}^{-1}]$, $M_s = 2[\mathrm{N}]$, $M_0 = 0.5[\mathrm{N}]$, $\Omega = 2[\mathrm{rad} \cdot \mathrm{s}^{-1}]$, $\omega_{dr} = 0.3[\mathrm{rad} \cdot \mathrm{s}^{-1}]$. For $t_* = 1[\mathrm{s}]$, $\varphi_* = 1[\mathrm{rad}]$ the initial dimensionless parameters of this model are: $\omega_d^2 = 5$, $F_s = 1$, $\zeta = 0.25$, $\omega_0 = 2$, $\omega_1 = 0.3$. The dimensionless parameter $\delta = 3$ and numerical dimensionless parameters are: time step $h = 0.001$ and steepness parameter $\varepsilon_\omega = 0.001$.

2.6.3 Numerical calculations

In the signum model and switch model friction force is a noncontinuous function of relative slip velocity and therefore the methods commonly used to compute the Lyapunov exponents cannot be applied. A continuous friction model as is studied in this section, does not possess this disadvantage and can be used during analysis of the systems, where the Lyapunov exponents are computed by standard procedures (Equations (2.87)). Note, that while computing Lyapunov exponents, Equations (2.237) and three additional systems of equations with respect to perturbations are solved.

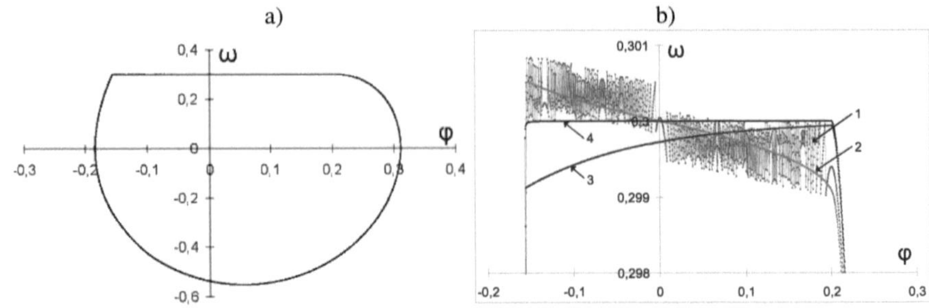

Fig. 2.70: Phase portraits of the analysed system without harmonic excitation for different models of friction: signum model (curve 1), smoothing method (curve 2), switch model (curve 3) and continuous friction model (curve 4).

In order to simulate the stick-slip vibrations using other friction models (signum model, smoothing method ((1.64) approximation), switch model, and continuous friction model) Equations (2.237) are solved. In what follows in order to solve the derived ordinary differential equations the standard numerical algorithms often used to study dynamics of lumped mechanical systems can be directly applied. In our case the thus-far mentioned differential equations are solved via the fourth-order Runge–Kutta method (RK4) and the Gram–Schmidt orthonormalisation method. The dynamics of the system is monitored via standard characteristics such as time histories in the system's phase space and bifurcation diagrams as well as the Lyapunov exponents.

2.6.4 Results

Let us consider first dynamics of the system for $\zeta = 0$, that is, without harmonic excitation. For this case, the phase portraits obtained with the signum model, smoothing method, switch model, and continuous friction model are shown in Fig. 2.70.

The periodic stick-slip oscillations have almost the same sliding velocity at each model (Fig. 2.70a). It is visible too, that in the sticking phase some differences are observed (Fig. 2.70b). The differences occur in the result of another approximating friction force application in the near-zero relative velocity neighbourhood.

The studied mechanical system (with harmonic excitation, $\zeta = 0.25$) possesses both periodic and nonperiodic solutions. Figures 2.71a and 2.71b present different behaviours of the analysed mechanical system in the time interval $\tau \in (200, 500)$.

It was shown above that in comparison with the other friction models better results are obtained for that case using the continuous friction model. It allows us to obtain the same accuracy as in the switch model, but for larger time-step h and steepness parameter ε_ω. Consequently, the switch model is more expensive than the continuous friction model for this case, from the computational point of view.

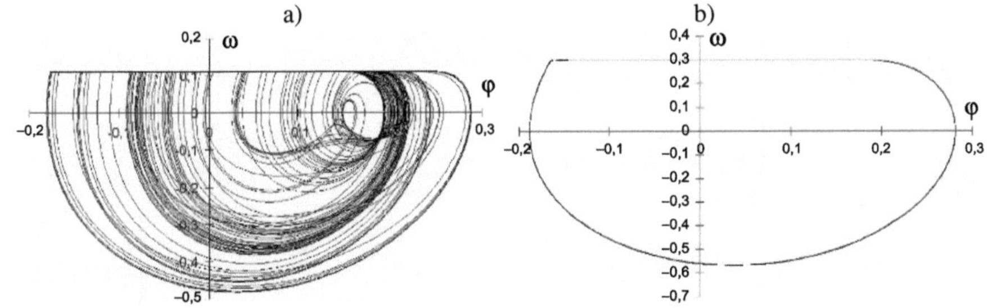

Fig. 2.71: Phase portraits of the analysed system with harmonic excitation for various angular velocities: (a) $\omega_1 = 0.11$; (b) $\omega_1 = 0.3$.

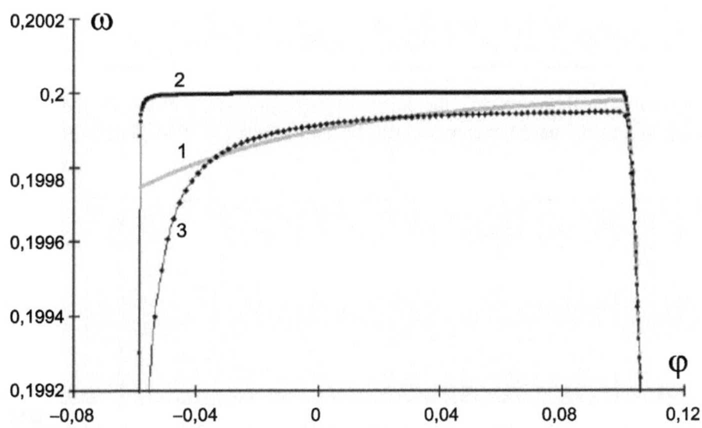

Fig. 2.72: Phase portraits: switch model (curve 1); continuous friction model (curve 2); continuous friction model for ε_ω = 0.01 and h = 0.01 (curve 3).

Figure 2.72 shows a phase portrait of the analysed system for changed parameters $\omega_d^2 = 10$, $\zeta = 0.5$, and $w_1 = 0.2$.

In this case, the sliding velocity is almost the same (for both switch and continuous friction models), but in the sticking phase some differences are observed, too. Contrary to the switch model results, results using the applied continuous friction model are better. Namely, we have obtained almost exact (high-precision) numerical computations even for larger (10 times) parameter ε_ω and time-step h (curve 3). In addition, the switch model is more expensive than the continuous friction model from the computational point of view.

Below, the periodic and nonperiodic solutions are detected using a bifurcation diagram and the Lyapunov exponent identification tools. The bifurcation diagram of the system is shown in Fig. 2.73 with the velocity ω_1 as a control parameter and the angle φ on the vertical axis. In Fig. 2.74 the bifurcation diagram of the system is shown with the ζ as a control parameter and the angle φ on the vertical axis.

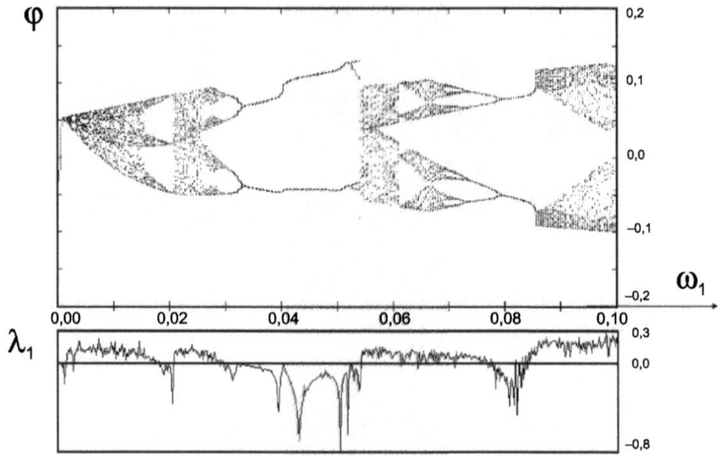

Fig. 2.73: Bifurcation diagram and the largest Lyapunov exponent λ_1 of the analysed 1-dof model with ω_1 as control parameter and φ on the vertical axis ($\omega_d^2 = 10, \zeta = 0.5$).

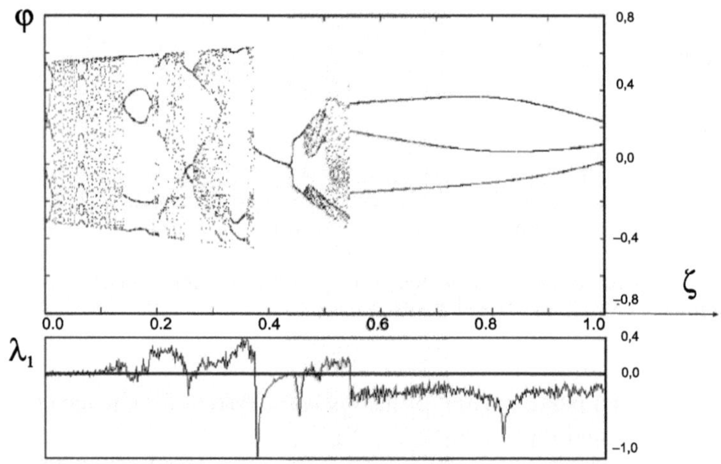

Fig. 2.74: Bifurcation diagram and the largest Lyapunov exponent λ_1 of the analysed 1-dof model with ζ as control parameter and φ on the vertical axis ($\omega_d^2 = 5, \omega_1 = 0.3$).

In the same plots dependencies of the largest Lyapunov exponent λ_1 versus the control parameters are also reported. A study of this bifurcation diagram implies that chaos occurs when the exponent λ_1 is positive. One of the computed Lyapunov exponents is always negative and second is always equal to zero (not shown in Fig. 2.73 and Fig. 2.74), because the studied system of equations is autonomous.

2.6.5 Conclusions

The continuous friction model suitable for simulation of the stick-slip vibrations has been studied and validated using the one degree-of-freedom mechanical system with dry friction. It has been observed that the continuous friction model yields engineering-accepted results and possesses some advantages in comparison to the switch model. Interesting dynamics of the analysed system are reported and analysed, including stick-slip periodic and chaotic behaviours. One of the important advantages of the novel continuous friction model is associated with direct application of the standard numerical procedures devoted to either solving differential equations or to computation of the Lyapunov exponents. The obtained results have been compared with those given by the switch model application, and they indicate better numerical accuracy of the continuous proposed continuous model. This friction model is validated and it gives correct results, even if the numerical steepness parameter ε_ω is extremely large. It allows obtaining the same accuracy as in the switch model for larger time-step h and steepness parameter ε_ω. The continuous friction model may also be force in any other mechanical systems.

Chapter 3
Thermoelastic Contact of Parallelepiped Moving Along Walls

In this chapter four models of thermoelastic contact of parallelepiped bodies are studied. The bodies move vertically between two walls. In the first model (section 3.1), a plate (one body) hangs on a spring and can exhibit vibrations. Walls compress this parallelepiped-like body and move at constant velocity V. In the second model (section 3.2), a contact between the plate and nonmovable walls occurs owing to their heating and thermal extension. Dynamics of a two degrees-of-freedom system with an account of friction and heating processes is studied (section 3.3). A dynamic 2-dof damper with dry friction and heat generation has been modelled mathematically (section 3.4).

Due to friction occurring on the contacting surface, self-excited vibrations appear and heat is generated. This section contributes to unification of applied separate models (i.e., to match both self-excited and thermoelastic contact models).

If frictional heat generation is neglected, then both models exhibit either smooth frictional self-vibrations [Kragelsky, Gitis (1987)] or a stick-slip vibration process [see, e.g., Ibrahim (1994)].

In self-excited systems loss of energy is supplied by energy generated from a source with nonvibrating properties. Energy transfer is controlled by dynamics of the system.

Models of frictional self-excited vibration types play an important role in machine tool construction. They have an essential impact on two important quality criteria of exploitation of various machine tools: the accuracy and uniformity of slow displacements. Note that nonuniformities of machine tool elements have a harmful impact on manufactured elements and the smoothness of their motion is highly required. The behaviour of self-excited vibrations may occur during braking processes while switching on various mechanical junctions (e.g., frictional clutches), during cutting processes, and so on. Note that during displacements of heavy bodies, frictional self-excited vibration excitation decreases the velocity of moving bodies, and hence also decreases energy requirements needed for carrying out the required work.

Both experimental and theoretical investigations of self-excited vibrations provided the following observations: (1) for $V_r < V_{\min}$ stick-slip motion may occur,

J. Awrejcewicz and Yu. Pyryev, *Nonsmooth Dynamics of Contacting Thermoelastic Bodies*, DOI: 10.1007/978-0-387-09653-7_3, © Springer Science+Business Media, LLC 2009 whereas for $V_r > V_{\min}$ self-excited vibrations do not appear ($V_{\min}$ denotes the velocity for which the kinematic friction coefficient is minimal); (2) length of stick intervals increases with sliding velocity decrease; (3) length of jump duration approaches half of the period of vibrations of a corresponding linear system; (4) stick and slip intervals appear periodically; (5) the increase of sliding velocity induces a decrease of both jump amplitudes and maximal value of the first displacement; (6) the increase of a normal load (contact pressure) causes an increase of stick interval duration; (7) increase of the system elasticity leads to an increase of the period of vibrations and decrease of their amplitude [Kragelsky, Gitis (1987)].

Owing to increase of spring stiffness the considered model can be smoothly transformed to that of thermoelastic contact of a body with moving walls [Pyryev (1994), Olesiak, Pyryev (1998)]. Similar models have also been studied in references [Olesiak, Pyryev (2000), Aleksandrov, Annakulova (1990)]. These investigations reveal that when the velocity of wall movement V is larger than a certain threshold value V_{cr}, the so-called thermoelastic instability, or thermal bang may occur [Aleksandrov, Annakulova (1990)]. Thermoelastic instability is manifested by an exponential increase of the system parameters (temperature and contact pressure). Energy produced on a contact surface cannot be absorbed by a surrounding medium and the body is heated.

3.1. Kinematically driven parallelepiped-type rigid plate

Consider the following one-dimensional model of thermoelastic contact of the body with a movable surrounding medium. Below, we consider a parallelepiped-like plate ($b_1 \times b_2 \times L$) that moves vertically and is fixed to a base by a spring of length l_1 and stiffness k_1 (Fig. 3.1).

The plate has mass M_1 and it may move vertically along the walls in the z_1 direction of a rectangular system of coordinates $0x_1y_1z_1$. In the beginning, the body is situated at distance Z_0 and it has velocity $\dot{Z}_0$. The walls moving at velocity V play the role of a surrounding medium. The initial value of the distance between the walls is equal to the plate thickness L. Then, this distance is decreased according to the formula $\varepsilon(t) = U_0 h_U(t)$. As a result of this process, dry friction occurs on the parallelepiped walls $X = \pm L/2$. It is defined by the function $F_{fr}(V_r)$, where V_r is the relative velocity of the plate and walls; that is, $V_r = V - \dot{Z}$. According to Amonton's assumption, friction force F_{fr} is equal to the product of the normal reaction component $N(t)$ and friction coefficient $F_{fr} = f(V_r)N(t)$. It characterizes resistance of the bodies during their relative displacement; $f(V_r)$ denotes the kinetic friction coefficient ($f(-V_r) = -f(V_r)$) (Fig. 2.2).

The action of friction on the contact surface $X = -L/2$, $X = L/2$ generates heat and wear. We follow the generally accepted assumption [see Ling (1959)] that friction work is transformed into heat energy. Furthermore, we assume that walls ideally transform heat and that between the plate and walls heat transfer is governed

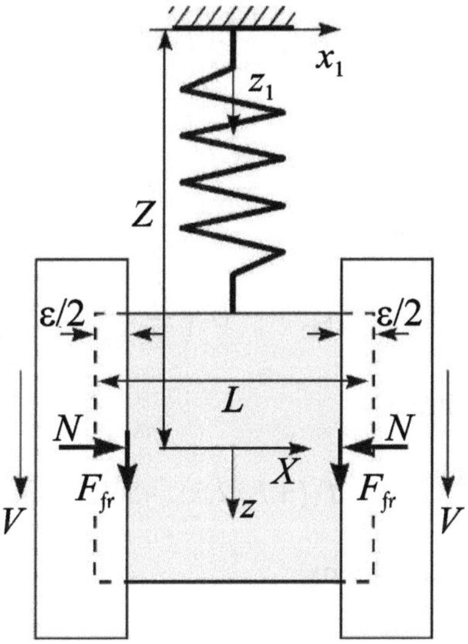

Fig. 3.1: Scheme of the plate moving between two movable walls.

by Newton's law [Carslaw, Jaeger (1959)], and a surrounding medium temperature is T_{sm}. Plate surfaces not being in contact with movable walls are thermally isolated and they have dimension L/b_1, L/b_2. The mentioned assumptions allow us to introduce a one-dimensional model.

The formulated problem is limited to the determination of plate center movement $Z(t)$, plate velocity $\dot{Z}(t)$, contact pressure

$$P(t) = N(t)/b_1 b_2 = -\sigma_{XX}(-L/2, t) = -\sigma_{XX}(L/2, t),$$

plate temperature $T_1(X, t)$, displacement $U(X, t)$ in the X direction, and wear $U^w(t)$.

3.1.1 Mathematical modelling

In the considered case, the mathematical model is governed by the dynamics of a plate center of the form [Ulitko (1990)]

$$m\frac{d^2}{dt^2}Z(t) + c(Z(t) - l_1) - mg = 2f(V_r)P(t), \tag{3.1}$$

and the equation governing the theory of thermal stresses for an isotropic body of the form [Nowacki (1962)]

$$\frac{\partial}{\partial X}\left[\frac{\partial}{\partial X}U(X,t)-\alpha_1\frac{1+\nu_1}{1-\nu_1}T_1(X,t)\right]=0, \tag{3.2}$$

$$\frac{\partial^2}{\partial X^2}T_1(X,t)=\frac{1}{a_1}\frac{\partial}{\partial t}T_1(X,t), \qquad X\in(-L/2,L/2), \tag{3.3}$$

with the attached mechanical

$$U\left(-\frac{L}{2},t\right)=\frac{U_0}{2}h_U(t)-U^w(t), \qquad U\left(\frac{L}{2},t\right)=-\frac{U_0}{2}h_U(t)+U^w(t), \tag{3.4}$$

and heat boundary conditions

$$\pm\lambda_1\frac{\partial T_1(\pm L/2,t)}{\partial X}+\alpha_T(T_1(\pm L/2,t)-T_{sm})=f(V_r)V_rP(t), \tag{3.5}$$

and the following initial conditions

$$T(X,0)=T_{sm}, \qquad X\in(-L/2,L/2), \qquad Z(0)=Z_0, \qquad \frac{d}{dt}Z(0)=\dot{Z}_0. \tag{3.6}$$

In our further considerations, we assume that velocity of the plate wear is proportional to a certain power of the friction force. According to the Archard assumption [Archard (1953)] one gets

$$\dot{U}^w(t)=K^w\left|V_r(t)\right|P(t). \tag{3.7}$$

Normal stress occurring in the plate is defined by the following formula,

$$\sigma_{XX}=\frac{E_1}{1-2\nu_1}\left[\frac{1-\nu_1}{1+\nu_1}\frac{\partial U}{\partial X}-\alpha_1T_1\right]. \tag{3.8}$$

In the above, the following notation is taken: E_1 is the elasticity modulus; ν_1, λ_1, a_1, α_1, α_T are Poisson coefficients, heat transfer, thermal diffusion, linear heat extension of the plate, and heat transfer from a wall to the plate, respectively; $m=M_1/b_1b_2$, $c=C/b_1b_2$, $P(t)=N(t)/b_1b_2$ are the contact pressure.

Integration of Equation (3.2) with Equation (3.8) and boundary conditions (3.4) gives the contact pressure $P(t)=-\sigma_{XX}(-L/2,t)=-\sigma_{XX}(L/2,t)$:

$$P(t)=P_0h_U(t)-2\frac{P_0}{U_0}U^w(t)+\frac{\tilde{E}\tilde{\alpha}}{L}\int\limits_{-L/2}^{L/2}T_1(\xi,t)d\xi, \tag{3.9}$$

where

$$P_0=\tilde{E}\frac{U_0}{L}, \qquad \tilde{\alpha}=\alpha_1\frac{1+\nu_1}{1-\nu_1}, \qquad \tilde{E}=\frac{E_1(1-\nu_1)}{(1+\nu_1)(1-2\nu_1)}. \tag{3.10}$$

Let us introduce the following coefficients of similarity

$$t_* = 1/\omega \text{ [s]}, \qquad L_* = V/\omega \text{ [m]}, \qquad N_* = P_0 L_*^2 \text{ [N]}, \qquad T_* = R_0 V P_0 [^\circ\text{C}],$$

and the following dimensionless parameters

$$x = \frac{X}{L_*}, \quad z' = \frac{Z}{L_*}, \quad \tau = \frac{t}{t_*}, \quad v = \frac{\dot{Z}(t)}{V}, \quad z_0' = \frac{Z_0}{L_*}, \quad \dot{z}_0' = \frac{\dot{Z}_0}{V},$$

$$u = \frac{U}{L_*}, \quad u^w = \frac{U^w}{U_0}, \quad p = \frac{PL_*^2}{N_*}, \quad p_{st} = \frac{P_{st}L_*^2}{N_*}, \quad \theta = \frac{T_1 - T_{sm}}{T_*}, \quad \tilde{\omega} = \frac{\omega_*}{\omega},$$

$$l = \frac{L}{L_*}, \quad l' = \frac{l_1}{L_*}, \quad Bi = R_0 \alpha_T, \quad \Omega = \tilde{\omega}\frac{V_2}{V}, \quad \gamma = \frac{\tilde{E}\tilde{\alpha}V}{\alpha_T}, \quad k^w = \frac{K^w \tilde{E} V}{\omega L}, \tag{3.11}$$

where

$$R_0 = \frac{L}{\lambda_1}, \qquad \omega = \sqrt{\frac{c}{m}}, \qquad \omega_* = \frac{a_1}{L^2}, \qquad V_2 = \frac{P_0}{m\omega_*}. \tag{3.12}$$

In the above, the following notation is applied. P_0 is the contact pressure assuming a lack of heat extension; R_0 is the heat resistance generated by a heat transfer; and ω is the free frequency of mass center vibrations with a lack of a contact pressure.

In the dimensionless form, the considered mathematical model is

$$\ddot{z}'(\tau) + z'(\tau) = 2\Omega f(V(1 - \dot{z}'(\tau)))p(\tau) + l' + gt_*/V, \tag{3.13}$$

$$\frac{\partial^2}{\partial x^2}\theta(x,\tau) = \frac{1}{\tilde{\omega} l^2}\frac{\partial}{\partial \tau}\theta(x,\tau), \qquad x \in \left(-\frac{l}{2}, \frac{l}{2}\right), \tag{3.14}$$

$$\pm l \frac{\partial \theta(\pm l/2, \tau)}{\partial x} + Bi\theta(\pm l/2, \tau) = f(V(1 - \dot{z}'(\tau)))(1 - \dot{z}'(\tau))p(\tau), \tag{3.15}$$

$$\theta(x, 0) = 0, \qquad z(0) = z_0', \qquad \dot{z}(0) = \dot{z}_0', \tag{3.16}$$

where the dimensionless contact pressure is described by the formula:

$$p(\tau) = h_U(\tau) - 2u^w(\tau) + \frac{\gamma Bi}{l} \int_{-l/2}^{l/2} \theta(\eta, \tau) d\eta. \tag{3.17}$$

3.1.2 Laplace transform

In order to solve problem (3.13)–(3.15), the Laplace transform is applied with respect to time τ. The theorem on 'convolution' is applied [Abramowitz, Stegun (1965), Carslaw, Jaeger (1959)] while finding the inverse transform. Finally, we get:

$$\ddot{z}(\tau) + z(\tau) = 2\Omega[f(V(1-\dot{z}(\tau)))p(\tau) - f(V)p_{st}], \tag{3.18}$$

$$z(0) = z_0, \qquad \dot{z}(0) = 0,$$

$$p(\tau) = 1 + 2\gamma\tilde{\omega}Bi\int_0^\tau G(\tau-\eta)f(V(1-\dot{z}(\eta)))(1-\dot{z}(\eta))p(\eta)d\eta, \tag{3.19}$$

$$\theta(\tau) = \tilde{\omega}\int_0^\tau g(\tau-\eta)f(V(1-\dot{z}(\eta)))(1-\dot{z}(\eta))p(\eta)d\eta, \tag{3.20}$$

$$u^w(\tau) = k^w\int_0^\tau |1-\dot{z}(\eta)|\, p(\eta)d\eta, \tag{3.21}$$

where

$$z(\tau) = z'(\tau) - z'_{st}, \qquad z_0 = z'_0 - z'_{st}, \qquad z'_{st} = 2\Omega f(V)p_{st} + l' + gt_*/V$$

$$g(\tau) = \sum_{m=1}^{\infty}\frac{\Delta_3(\mu_m)}{\Delta'(\mu_m)}\exp(-\mu_m^2\tilde{\omega}\tau),$$

$$G(\tau) = \sum_{m=1}^{\infty}\frac{\Delta_2(\mu_m)}{\Delta'(\mu_m)}\exp(-\mu_m^2\tilde{\omega}\tau), \tag{3.22}$$

$$\Delta'(\mu_m) = 0.5(2BiS_m + C_m - Bi^2(C_m - S_m)/\mu_m^2),$$

$$\Delta(\mu_m) = 2BiC_m + S_m(Bi^2 - \mu_m^2), \qquad \theta(\tau) = \theta(-L/L_*,\tau) = \theta(L/L_*,\tau),$$

$$\Delta_2(\mu_m) = S_m - BiC_m^0, \qquad \Delta_3(\mu_m) = 1 + C_m + BiS_m,$$

$$C_m = \cosh(\mu_m), \qquad S_m = \frac{\sinh(\mu_m)}{\mu_m}, \qquad C_m^0 = \frac{C_m - 1}{\mu_m},$$

and $s_m = -\mu_m^2$ are the roots of the characteristic equation $\Delta(s) = 0$.

3.1.3 Stationary process subject to kinematic external excitation

Let us first linearize the problem in the vicinity of the stationary solution p_{st}, θ_{st}, z_{st}, assuming that

$$\theta(x,\tau) = \theta_{st}(x) + \theta^*(x,\tau), \qquad p(x) = p_{st} + p^*(\tau), \qquad z(\tau) = z^*(\tau), \tag{3.23}$$

where the stationary solutions are

$$p_{st} = \frac{1}{1-\gamma f(V)}, \qquad \theta_{st} = \frac{1}{Bi}\frac{f(V)}{1-\gamma f(V)}, \qquad z_{st} = 0. \tag{3.24}$$

The characteristic equation of the linearized system has the form

$$\Delta_*(s) = \Omega_2(s)\Delta_1(s) - 2\gamma Bi f(V)\Omega_1(s)\Delta_2(s), \tag{3.25}$$

where

$$\begin{aligned}
&\Omega_1(s) = s^2 - \Omega p_{st} f(V)s + 1, \qquad \Omega_2(s) = s^2 + \Omega p_{st} V f'(V)s + 1,\\
&\Delta_1(s) = 2Bi\ C + S(Bi^2 + s/\tilde{\omega}), \qquad \Delta_2(s) = S + Bi\ C^0, \qquad\qquad (3.26)\\
&\quad S = \sinh\sqrt{s/\tilde{\omega}}/\sqrt{s/\tilde{\omega}}, \qquad C = \cosh\sqrt{s/\tilde{\omega}}, \qquad C^0 = (C-1)/(s/\tilde{\omega}),\\
&\quad \gamma_1 = V_2\frac{f(V)}{V}, \qquad \gamma(V) = V_2 f'(V), \qquad v(V) = \frac{V}{V_1(V)} = \gamma f(V),\\
&V_1(V) = \frac{V}{\gamma f(V)}, \qquad V_2 = \frac{P_0}{\omega_* m}.
\end{aligned}$$

The characteristic function is

$$\Delta_*(s) = \sum_{m=0}^{\infty} d_m s^m, \tag{3.27}$$

where

$$\begin{aligned}
d_0 &= 2Bi(1 + Bi/2)(1 - v(V)),\\
d_m &= d_m^{(1)} + (1-\delta_{m1})\tilde{\omega}^2(d_{m-2}^{(1)} - 2Bi\ v(V)d_{m-2}^{(2)})\\
&\quad - 2Bi\ v(V)d_m^{(2)} + 2p_{st}\tilde{\omega}^2(\gamma d_{m-1}^{(1)} + 2Bi\ v(V)\gamma_1 d_{m-1}^{(2)}),\\
d_m^{(1)} &= \left(\frac{1}{\tilde{\omega}}\right)^m \left[\frac{Bi^2}{(2m+1)!} + \frac{2Bi}{(2m)!} + \frac{1}{(2m-1)!}\right],
\end{aligned}$$

$$d_m^{(2)} = \left(\frac{1}{\tilde{\omega}}\right)^m \left[\frac{1}{(2m+1)!} + \frac{Bi}{(2m+2)!}\right].$$

It is worth noting that the series (3.27) for finite value m provides analytical conditions for stability of a static solution. One may apply the known criteria such as the ones proposed by Routh and/or Hurtwitz [Gantmacher (1959)]. The mentioned criteria are based on the determination of zeros of a polynomial with constant real coefficients with a simultaneous estimation of the right half-plane (RHP) or left half-plane (LHP) of the complex plane. They can be used to derive a first approximation of the critical values of parameters of the considered models.

Let us consider the following limiting cases.

1. In the case when $v(V) = 0$ ($\gamma = 0$) (i.e., thermal expansion of a parallelepiped is not accounted for), a characteristic equation associated with the perturbation equation takes the form $\Omega_2(s) = 0$. Roots of the associated characteristic equation will lie in the RHP if $f'(V) < 0$ ($\gamma < 0$). In this case a stationary solution is unstable, and consequently a corresponding nonstationary solution of the nonlinear problem will move away from the equilibrium state. This kind of motion appears when $0 < V < V_{\min}$. The frequency of vibrations for $\gamma = 0$ is equal to 1.
2. In the case when $\tilde{\omega} = 0$ (i.e., when the spring stiffness approaches an infinite value (unmovable body)), the characteristic equation of the linearized problem is

$$\Delta_1(s) - 2Bi\, v(V)\Delta_2(s) = 0. \tag{3.28}$$

The roots s_m, $m = 1, 2, 3, \ldots$ of Equation (3.28) lie on the real axis ($\mathrm{Im} s_m = 0$). When inequality $v(V) < 1$ holds, the roots are negative ($s_m < 0$) and static solutions are stable. For $v > 1$, root s_1 lies in the RHP; that is, the static solution is unstable.

For $v < 1$, an unstationary solution for $\tau \to \infty$ tends to a stable solution of problem (3.24). For $v > 1$, an unstationary solution increases exponentially $\exp(s_1\tau)$, where s_1 is estimated by the formula

$$s_1 = -\tilde{\omega} d_0/d_1 = (v-1)Bi(2+Bi)/d_1. \tag{3.29}$$

For $v > 1$ the stable solution loses its physical meaning ($p_{st} < 0$, $\theta_{st} < 0$). A small variation of parameter v in the vicinity of 1 gives the exponential divergence of the solution. The latter phenomenon is referred to as the thermoelastic instability [Pyryev et al. (1995)] or heat explosion [Aleksandrov, Annakulova (1990)].

Note that during the thermoelastic instability, the force required to hold up a uniform motion at velocity V increases, and hence energy supplied from the environment increases too. It means physically, that heat generated on the contact surface cannot be received by the surrounding environment and both bodies are heated.

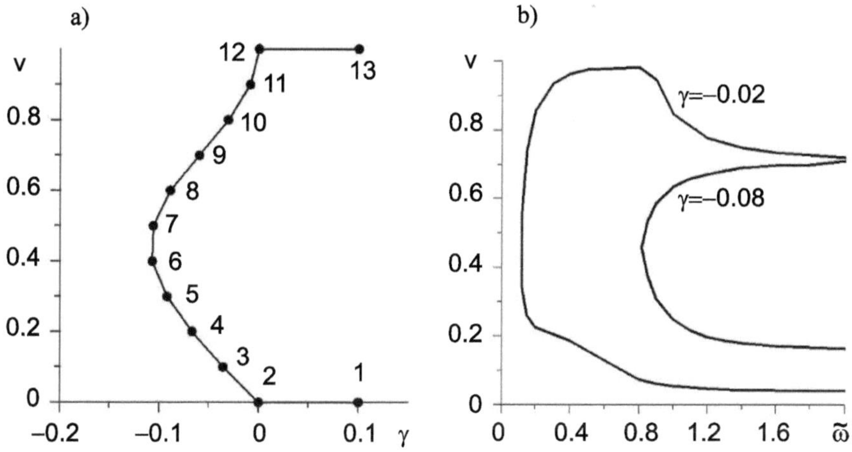

Fig. 3.2: Zones of parameters associated with an unstable stationary solution.

Next, analysis of the zones of parameters for which the stationary solutions are stable (Fig. 3.2) is carried out. During computations the following parameters are fixed: $Bi = 1$, $\gamma_1 = 0.5$.

In Fig. 3.2 zones lying inside the curves correspond to parameters $(\gamma,\ v)$ and $(\tilde{\omega}, v)$. In Fig. 3.2b the curves correspond to the critical values of the parameters $(\tilde{\omega}, v)$ for $\gamma = -0.02$ and $\gamma = -0.08$.

One may conclude through analysis of the figures that during the decrease of the negative value of γ, a stable zone with respect to parameter v decreases (narrowing of the zone appears), whereas the critical values of $\tilde{\omega}$ increase. For $\gamma < -0.107$ a solution for $\tilde{\omega} = 1$, $Bi = 1$, $\gamma_1 = 0.5$ is always unstable. Points 1–13 correspond to oscillations at the frequency $\omega = |\mathrm{Im} s_{1,2}|$. The oscillation frequency increases with increase of v from $\omega = 1$ (heat expansion is not accounted for) to the value of $\omega \approx 4$.

3.1.4 Algorithm and solutions

Recall that the analysed problem is governed by the system of two nonlinear equations, that is, differential and integral ones (3.19). The solutions are sought using the fourth-order Runge–Kutta method and trapezium approximation.

Solution Analysis. A numerical analysis is carried out for a steel parallelepiped ($\alpha = 14 \cdot 10^{-6}\ ^\circ\mathrm{C}^{-1}$, $\lambda_1 = 21\,\mathrm{W\ m}^{-1}\ ^\circ\mathrm{C}^{-1}$, $a_1 = 5.9 \cdot 10^{-6}\,\mathrm{m}^2\,\mathrm{s}^{-1}$, $\nu_1 = 0.3$, $E = 19 \cdot 10^{10}$ Pa) for the following fixed parameters $Bi = 4.76$, $\Omega = 12.8$, $\tilde{\omega} = 0.26 \cdot 10^{-2}$, $V = 1.5 \cdot 10^{-2}\,\mathrm{m\ s}^{-1}$, and for various values of both the wear coefficient k^w and distance z_0. It is assumed that $h_U(\tau) = H(\tau)$. Dashed curves correspond to the case when heat expansion $\gamma = 0$.

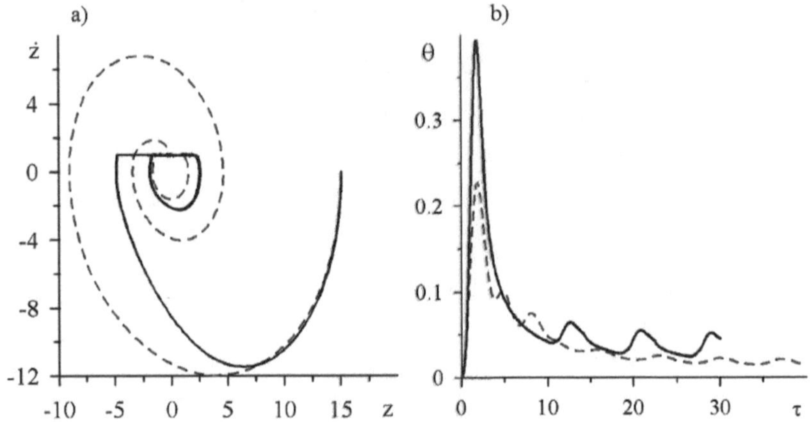

Fig. 3.3: Phase portrait (a) and evolution in time of contact temperature (b) for $z_0 = 15$, $k^w = 0$ (solid curve: $\gamma = 10$; dashed curve: $\gamma = 0$).

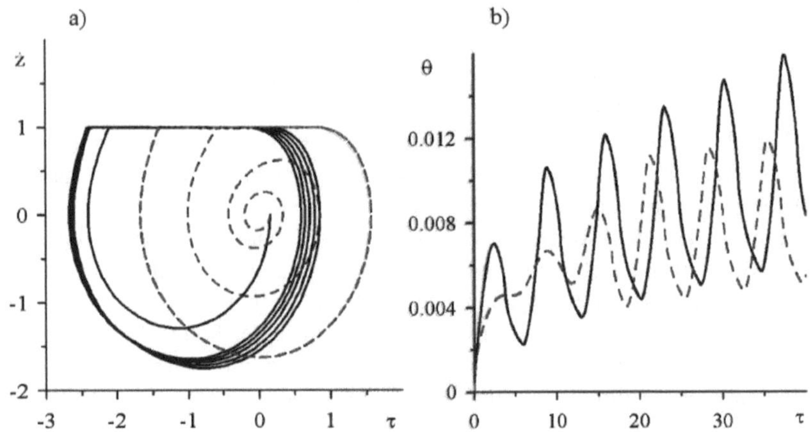

Fig. 3.4: Phase portrait (a) and evolution in time of contact temperature (b) for $z_0 = 0.15$, $k^w = 0$ (solid curve: $\gamma = 10$; dashed curve: $\gamma = 0$).

In Figs. 3.3 and 3.4 the contact characteristics of the system determined numerically without wear ($k^w = 0$) for various initial conditions $z_0 = 15$, $\dot{z}_0 = 0$ (Fig. 3.3), and $z_0 = 0.15$, $\dot{z}_0 = 0$ (Fig. 3.4) are shown. Dashed curves in the figures correspond to the case $\gamma = 0$, whereas solid curves refer to $\gamma = 10$.

Depending on the initial condition (Figs. 3.3 and 3.4), vibrations approach the stick-slip limit cycle either from its inside or outside. The parallelepiped heat expansion yields a decrease of vibration amplitude.

Results of the numerical analysis with reference to wear are shown in Fig. 3.5.

Conclusions. To conclude the results obtained in this section, let us mainly emphasise the study of a new problem of thermoelastic contact of a body with

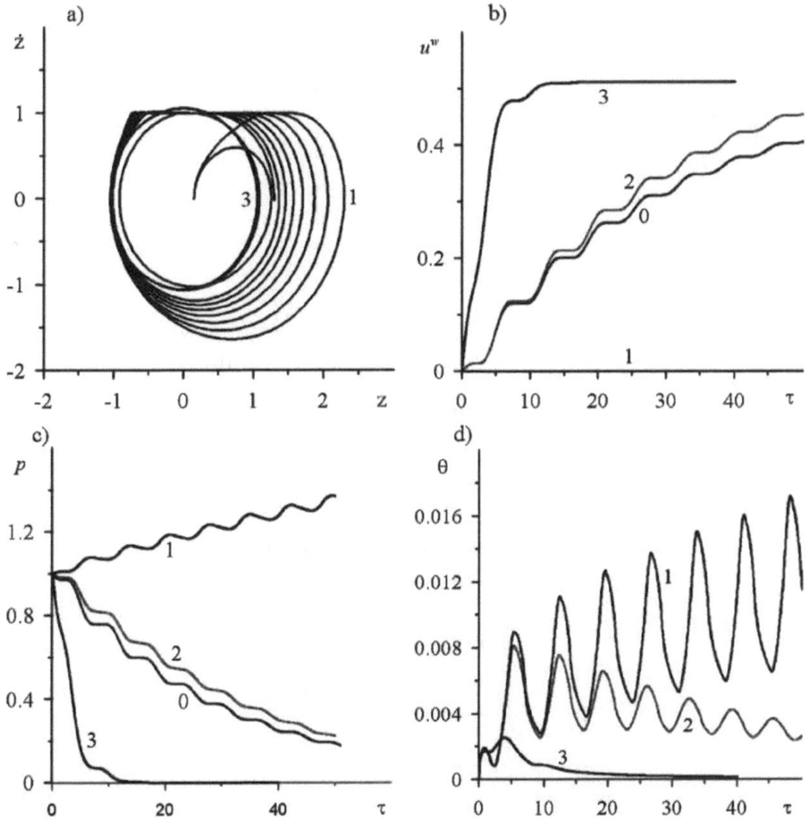

Fig. 3.5: Phase portrait (a), wear evolution with time (b), contact pressure (c), and contact temperature (d) for $z_0 = 0.15$ (curve 0: $\gamma = 0$, $k^w = 0.017$; curve 1: $\gamma = 10$, $k^w = 0$; curve 2: $\gamma = 10$, $k^w = 0.017$; curve 3: $\gamma = 10$, $k^w = 0.17$).

the moving external medium. The main new features of the problem are exhibited through a body inertia and a dependence of the friction coefficient on the relative velocity of contacting bodies. The proposed model can be applied to estimate the contact characteristics of moving tribological systems.

3.2. Rigid plate dynamics subject to temperature perturbation

In the literature there are many examples focused on the analysis of autonomous systems exhibiting regular nonlinear self-excited vibrations (see Figs. 3.6a,b). Vibrations of a mechanical system modelling woodpecker behaviour (Fig. 3.6a) have been analysed by [Pfeiffer (1984), Leine, Campen (2006)] and also studied in the monograph by [Awrejcewicz (1996)]. The ‘mechanical woodpecker’ consists

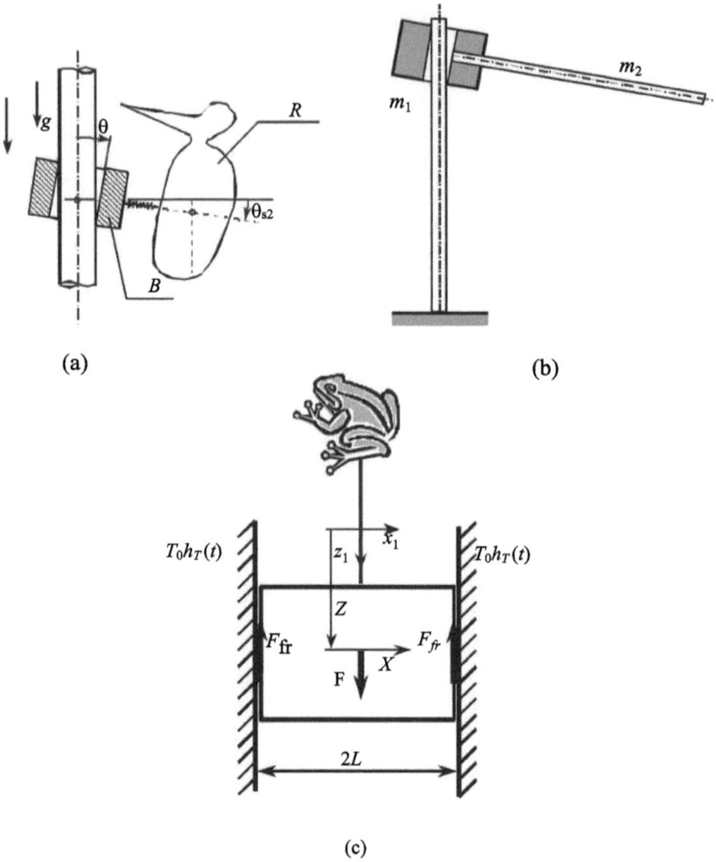

Fig. 3.6: Models of a woodpecker knocking into a cylinder [Pfeiffer (1984)] (a); an ‘Olędzki’s slider’ [Olędzki, Siwicki (1997)] (b); and ‘frog-type’ system (c).

of a stiff body with the moment of inertia B_2 and mass m_2, and is coupled to a bush by a spring with stiffness k. The bush mass inertial moment is denoted by B_1.

Vibrations of the so-called ‘Olędzki’s slider’ have been studied first in the reference [Olędzki, Siwicki (1997)] (see Fig. 3.6b). If the mass centre S_2 of a horizontal rod is located at a distance larger than $b/(2f)$ (b denotes the slider length, f is the friction coefficient), then in static conditions, the rod and slider motions along the vertical rod–guide axis are not possible. In static conditions this distance is

$$l = \left(1 + \frac{m_1}{m_2}\right) \frac{b}{2f} \sqrt{1 + 2\frac{d}{b}\frac{2\delta}{b} + \left(\frac{2\delta}{b}\right)^2},$$

where m_1 is the slider mass, m_2 is the mass of the horizontal rod, d is the cylinder diameter, and δ denotes the small backlash between the cylinder and bush. However, an infinitely small perturbation of the end of the horizontal rod forces the system to move.

Note that the two models of the slider movement discussed are associated with a stick-slip vibration behaviour.

Despite simplicity of the introduced models, important information is obtained. Namely, it is shown that the static self-braking kinematic pairs can initiate movement when vibration appears.

Consider now our new proposed model, which does not have any elastic part, but can exhibit self-excited stick-slip vibrations (Fig. 3.6c). For simplicity, it is further referred to as a 'frog-slider'.

Let us consider a one-dimensional model of the thermoelastic contact of a body with a surrounding medium. Assume that this body is represented by a rectangular plate ($b_1 \times b_2 \times 2L$) (Fig. 3.6c). The plate together with a 'frog' has mass M_1 subject to force $F = F_* h_F(t)$ and moves vertically along the walls in direction z_1 of the rectangular coordinates $0x_1y_1z$. Initially, the body is situated at distance Z_0 and has velocity $\dot{Z}_0$. The distance between the walls is always equal to the initial plate thickness $2L$. The plate moves at nonconstant velocity $\dot{Z}(t)$.

It is assumed that heat transfer between the plate and the walls is ideal and the Newton assumptions hold. Initially, the instant temperature is governed by the formula $T_0 h_T(t)$ ($h_T(t) \to 1$, $t \to \infty$). It causes a parallelepiped heat extension in the direction of $0x_1$, and the body starts to contact the walls. As a result of this process a frictional contact occurs on the parallelepiped sides $X = \pm L$. A simple frictional model is applied further; that is, friction force F_{fr} is the product of normal reaction force $N(t)$ and the friction coefficient. It means that $F_{fr} = f(\dot{Z})N(t)$ is the friction force defining resistance of the movement of bodies. Here, contrary to the assumption made in the reference [Olesiak, Pyryev (2000)], the kinematic friction coefficient $f(\dot{Z})$ depends on the relative velocity of the sliding bodies (Fig. 2.2).

The friction force $\sigma_{XZ}(X,t)$ per unit contact surface $X = -L$, $X = L$, generates heat. According to Ling's assumptions [Ling (1959)], the work of friction forces is transmitted into heat energy. Note that noncontacting plate surfaces are heat-insulated and have the dimensions of $L/b_1 \ll 1$, $L/b_2 \ll 1$, which is in agreement with the assumption of our one-dimensional modelling.

Hence, the problem is reduced to the determination of mass plate centre displacement $Z(t)$; plate velocity $\dot{Z}(t)$; contact pressure $P(t) = N(t)/b_1 b_2 = -\sigma_{XX}(-L,t) = -\sigma_{XX}(L,t)$; plate temperature $T_1(X,t)$; and displacement $U(X,t)$ in the X-axis direction.

3.2.1 Mathematical formulation

In the considered case, the studied problem is governed by the dynamics of the plate mass centre

$$m\ddot{Z}(t) = F_* h_F(t) - 2f(\dot{Z})P(t), \tag{3.30}$$

and equations of the heat stress theory for an isotropic body [Nowacki (1962)] (3.2), (3.3) with the attached mechanical

$$U(-L,t)=0, \qquad U(L,t)=0, \tag{3.31}$$

heat

$$-\lambda_1 \frac{\partial T_1(-L,t)}{\partial X} + \alpha_T(T_1(-L,t) - T_0 h_T(t)) = f(\dot{Z})\dot{Z}(t)P(t), \tag{3.32}$$

$$\lambda_1 \frac{\partial T_1(L,t)}{\partial X} + \alpha_T(T_1(L,t) - T_0 h_T(t)) = f(\dot{Z})\dot{Z}(t)P(t), \tag{3.33}$$

and initial

$$T(X,0)=0, \qquad X\in(-L,L), \qquad Z(0)=Z_0, \qquad \dot{Z}(0)=\dot{Z}_0 \tag{3.34}$$

conditions. Normal stresses occurring in the plate are defined by (3.8).

In the above, the following notation is applied: E_1 is the elasticity modulus; ν_1, λ_1, a_1, α_1, α_T are Poisson's ratio, thermal conductivity, thermal diffusivity, thermal expansion, and heat transfer coefficients, respectively; $m = M_1/b_1b_2$; whereas $P(t) = N(t)/b_1b_2$ denotes the contact pressure.

Integration of Equation (3.2) with an account of (3.8) and boundary conditions (3.31) provides the contact pressure $P(t) = -\sigma_{XX}(-L,t) = -\sigma_{XX}(L,t)$:

$$P(t) = P_0 + \tilde{E}\tilde{\alpha}\frac{1}{L}\int_0^L T_1(\xi,t)d\xi. \tag{3.35}$$

Let us introduce the following similarity coefficients

$$t_* = L^2/a_1 \text{ [s]}, \qquad \mathrm{V}_* = a_1/L \text{ [m s}^{-1}\text{]}, \qquad P_* = T_0E_1\alpha_1/(1-2\nu_1) \text{ [N m}^{-2}\text{]},$$

and the following dimensionless parameters

$$x=\frac{X}{L}, \quad \tau=\frac{t}{t_*}, \quad z=\frac{Z}{L}, \quad p=\frac{P}{P_*}, \quad \theta=\frac{T_1}{T_0}, \quad \varepsilon_1=\frac{2P_*t_*^2}{mL}, \tag{3.36}$$

$$\gamma=\frac{E_1\alpha_1a_1}{(1-2\nu_1)\lambda_1}, \quad Bi=\frac{L\alpha_T}{\lambda_1}, \quad m_0=\frac{F_*}{2P_*}, \quad F(\dot{z})=f(\mathrm{V}_*\dot{z}). \tag{3.37}$$

Our problem is modeled using the following dimensionless equations

$$\frac{\partial^2\theta(x,\tau)}{\partial x^2} = \frac{\partial\theta(x,\tau)}{\partial\tau}, \qquad x\in(-1,1), \qquad \tau\in(0,\infty), \tag{3.38}$$

$$\ddot{z}(\tau) = \varepsilon_1(m_0h_F(\tau) - F(\dot{z})p(\tau)), \tag{3.39}$$

with attached boundary

$$\left[\frac{\partial\theta(x,\tau)}{\partial x} \pm Bi\theta(x,\tau)\right]_{x=\pm 1} = \pm q(\tau), \tag{3.40}$$

and initial conditions

$$\theta(x,0) = 0, \qquad z(0) = z^\circ, \qquad \dot{z}(0) = \dot{z}^\circ, \tag{3.41}$$

where

$$q(\tau) = Bih_T(\tau) + \gamma F(\dot{z})\dot{z}(\tau)p(\tau), \qquad p(\tau) = \frac{1}{2}\int_{-1}^{1} \theta(\xi,\tau)d\xi. \tag{3.42}$$

3.2.2 Application of the Laplace transform

Applying the Laplace transform

$$\{\bar{\theta},\bar{p},\bar{q}\} = \int_0^\infty \{\theta,p,q\}\, e^{-s\tau} d\tau,$$

the following is obtained

$$\frac{d^2\bar{\theta}}{dx^2} = s\bar{\theta}, \tag{3.43}$$

$$\left[\frac{d\bar{\theta}}{dx} - Bi\bar{\theta}\right]_{x=-1} = -\bar{q}, \qquad \left[\frac{d\bar{\theta}}{dx} + Bi\bar{\theta}\right]_{x=1} = \bar{q}, \tag{3.44}$$

$$\bar{p} = \frac{1}{2}\int_{-1}^{1} \bar{\theta}(\xi,s)d\xi. \tag{3.45}$$

A solution to Equation (3.43) is sought in the form

$$\bar{\theta}(x,s) = A(s)\bar{S}(x,s) + B(s)\bar{C}(x,s), \tag{3.46}$$

where

$$\bar{S}(x,s) = \sinh(\sqrt{s}x)/\sqrt{s}, \qquad \bar{C}(x,s) = \cosh(\sqrt{s}x).$$

The quantities $A(s)$ and $B(s)$ are defined by two boundary value problems (3.44). Finally, we get

$$\bar{\theta}(x,s) = s\bar{q}(s)\bar{G}_\theta(x,s), \qquad \bar{p}(s) = s\bar{q}(s)\bar{G}_p(s), \tag{3.47}$$

where

$$\bar{G}_p(x,s) = \frac{\bar{S}(s)}{s\Delta_1(s)}, \qquad \bar{G}_\theta(x,s) = \frac{\bar{C}(x,s)}{s\Delta_1(s)},$$
$$\Delta_1(s) = s\bar{S}(s) + Bi\bar{C}(s),$$
$$\bar{S}(s) = \sinh(\sqrt{s})/\sqrt{s}, \qquad \bar{C}(s) = \cosh(\sqrt{s}). \tag{3.48}$$

Applying an inverse Laplace transformation, the following system of equations is obtained,

$$p(\tau) = Bi\int\limits_0^\tau \dot{h}_T(\xi)G_p(\tau-\xi)d\xi + \gamma\int\limits_0^\tau F(\dot{z})\dot{z}(\xi)p(\xi)\dot{G}_p(\tau-\xi)d\xi, \tag{3.49}$$

$$\dot{z}(\tau) = \varepsilon_1\left[m_0\int\limits_0^\tau h_F(\xi)d\xi - \int\limits_0^\tau F(\dot{z})p(\xi)d\xi\right], \tag{3.50}$$

which gives dimensionless pressure $p(\tau)$ and velocity $\dot{z}(\tau)$. Temperature is defined by the following formula

$$\theta(x,\tau) = Bi\int\limits_0^\tau \dot{h}_T(\xi)G_\theta(x,\tau-\xi)d\xi + \gamma\int\limits_0^\tau F(\dot{z})\dot{z}(\xi)p(\xi)\dot{G}_\theta(x,\tau-\xi)d\xi, \tag{3.51}$$

where

$$\{G_p(\tau), G_\theta(1,\tau)\} = \frac{1}{Bi} - \sum_{m=1}^{\infty}\frac{\{2Bi, 2\mu_m^2\}}{\mu_m^2[Bi(Bi+1)+\mu_m^2]}e^{-\mu_m^2\tau}, \tag{3.52}$$

and μ_m are the roots of the following characteristic equation

$$\mathrm{tg}\mu_m = \frac{Bi}{\mu_m}, \quad m = 1, 2, \ldots. \tag{3.53}$$

Observe that $\mu_m \approx \pi m$, $m \to \infty$.

Note that it is easy to formulate a relation between heat flow and contact pressure velocity variation in the form

$$\frac{dp(\tau)}{d\tau} = \left.\frac{\partial\theta(\tau,x)}{\partial x}\right|_{x=1}, \tag{3.54}$$

which is used further to analyse the contact characteristics.

In order to carry out numerical analysis, a knowledge of the function (3.52) is required. The values of function (3.52) for $\tau \to 0$ and $\tau \to \infty$ are defined using the theorem [Abramowitz, Stegun (1965)] on the limiting values:

$$\{G_p(\tau), G_\theta(1,\tau)\} \approx 1/Bi, \qquad \tau \to \infty, \tag{3.55}$$

$$G_p(\tau) \approx \tau, \quad G_\theta(1,\tau) \approx 2\sqrt{\tau}/\sqrt{\pi}, \qquad \tau \to 0. \tag{3.56}$$

3.2.3 *Stationary process*

A stationary solution to the problem is:

$$p_{st} = \frac{1}{1-\mathrm{v}}, \qquad \theta_{st} = \frac{1}{1-\mathrm{v}}, \qquad \mathrm{v} = F(\mathrm{v}_{st})\frac{\mathrm{v}_{st}\gamma}{Bi}, \tag{3.57}$$

where v_{st} is the solution of the nonlinear equation

$$F(\mathrm{v}_{st}) = \frac{m_0}{1 + \gamma m_0 \mathrm{v}_{st}/Bi}. \tag{3.58}$$

A graphical solution to Equation (3.58) is presented in Fig. 3.7 for various parameters m_0 and Bi. Recall that for steel $\gamma = 1.87$. The right-hand side of Equation (3.58) is represented by a solid curve for different values of the parameters m_0 and Bi. Solid curve 1 is associated with the parameter $m_0 = 0.14$, $Bi = 20$, solid curve 2 with $m_0 = 0.1$, $Bi = 20$, solid curve 3 with $m_0 = 0.1$, $Bi = 5$, and solid curve 4 with $m_0 = 0.14$, $Bi = 5$. The dashed curve is related to the function $F(\mathrm{v}_{st})$.

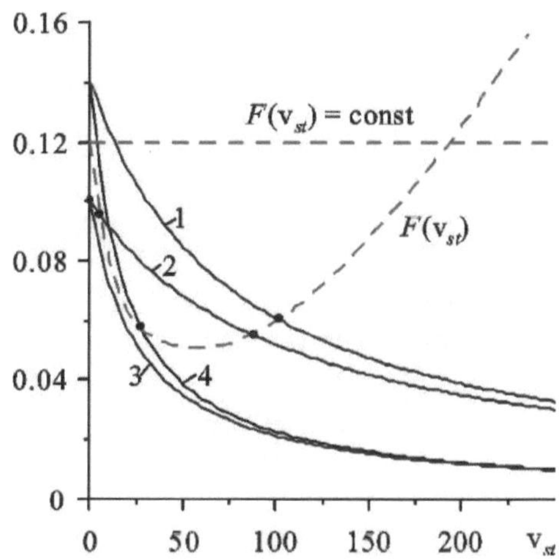

Fig. 3.7: Graphical solution of Equation (3.58) (solid curves: 1, $m_0 = 0.14$, Bi, 20; 2, $m_0 = 0.1$, $Bi = 20$; 3, $m_0 = 0.1$, $Bi = 5$; 4, $m_0 = 0.14$, $Bi = 5$; dashed curve corresponds to $F(\mathrm{v}_{st})$).

At $m_0 = 0.14$, $Bi = 20$ (first case) Equation (3.58) can have one solution v_{st}^3 ($F'(\mathrm{v}_{st}^3) > 0$), at $m_0 = 0.1$, $Bi = 20$ (second case) three solutions v_{st}^1, v_{st}^2, v_{st}^3 ($F'(\mathrm{v}_{st}^1) > 0$, $F'(\mathrm{v}_{st}^1) < 0$, $F'(\mathrm{v}_{st}^3) > 0$), and at $m_0 = 0.1$, $Bi = 5$ (third case) one solution $\mathrm{v}_{st}^1 = 0$ (with approximation (2.58) $\mathrm{v}_{st}^1 \approx \varepsilon_0 m_0/2f_s$, $F'(\mathrm{v}_{st}^1) \approx 2f_s/\varepsilon$). At $m_0 = 0.14$, $Bi = 5$ (fourth case) there is one solution v_{st}^2 ($F'(\mathrm{v}_{st}^2) < 0$).

The case of constant friction represented in Fig. 3.7 by dashed horizontal line $F(\mathrm{v}_{st}) = f_s = \text{const}$ was earlier considered by [Olesiak, Pyryev (2000)], where $\mathrm{v}_{st} = Bi(m_0/f_s - 1)/(m_0\gamma)$.

Let us introduce a perturbation of the stationary solution (3.57) using the following formulas

$$\dot{z} = \mathrm{v}_{st} + \dot{z}^*, \qquad p = p_{st} + p^*, \qquad \theta = \theta_{st} + \theta^*, \qquad h_T = 1 + h_T^*. \quad (3.59)$$

Owing to the linearization of the right-hand sides of (3.39) and taking into account the boundary condition (3.40), the following linear problem is obtained

$$\frac{\partial^2\theta^*(x,\tau)}{\partial\tau^2} = \frac{\partial\theta^*(x,\tau)}{\partial\tau}, \quad (3.60)$$

$$\ddot{z}^*(\tau) = \varepsilon_1[-F(\mathrm{v}_{st})p^*(\tau) - F'(\mathrm{v}_{st})p_{st}\dot{z}^*], \quad (3.61)$$

$$\left[\frac{d\theta^*(x,\tau)}{dx} - Bi\theta^*(x,\tau)\right]_{x=-1} = -q^*(\tau), \quad (3.62)$$

$$\left[\frac{d\theta^*(x,\tau)}{dx} + Bi\theta^*(x,\tau)\right]_{x=1} = q^*(\tau), \quad (3.63)$$

where

$$q^*(\tau) = Bih_T^*(\tau) + \gamma(\mathrm{v}_{st}p_{st}(\beta_1 + \beta_2)\dot{z}^*(\tau) + \mathrm{v}_{st}F(\mathrm{v}_{st})p^*(\tau)),$$

$$p^*(\tau) = \frac{1}{2}\int_{-1}^{1}\theta^*(\xi,\tau)d\xi, \qquad \beta_1 = \frac{F(\mathrm{v}_{st})}{\mathrm{v}_{st}}, \qquad \beta_2 = F'(\mathrm{v}_{st}). \quad (3.64)$$

Furthermore, applying the Laplace transform, a solution of the problem (3.60)–(3.64) in the transform domain is found. For example, the pressure perturbation is

$$p^*(s) = \frac{\bar{S}(s)Bi(\beta_2 p_{st}\varepsilon_1 + s)}{\Delta(s)}h_T^*(s), \quad (3.65)$$

where

$$\Delta(s) = (\varepsilon_1 p_{st}\beta_2 + s)\Delta_1(s) + Bi\mathrm{v}(\varepsilon_1 p_{st}\beta_1 - s)\bar{S}(s) = 0 \quad (3.66)$$

is the characteristic equation of the linearized problem. The roots s_m ($\mathrm{Re}s_1 > \mathrm{Re}s_2 > \cdots > \mathrm{Re}s_m > \cdots$, $m = 1, 2, 3, \ldots$) of the characteristic Equation (3.66) lie either on the left-hand side of the complex plane Re$s < 0$ (stationary solution is stable) or on the right-hand side of the complex plane Re$s > 0$ (stationary solution is unstable) of the complex variable s.

The characteristic function $\Delta(s)$, has the form of infinite-order polynomial

$$\Delta(s) = \sum_{m=0}^{\infty} s^m b_m, \tag{3.67}$$

where

$$b_0 = \varepsilon_1 p_{st} c_0, \qquad b_m = \varepsilon_1 p_{st} c_m + d_{m-1}, \qquad m = 1, 2, \ldots$$

$$c_0 = Bi(\beta_2 + \mathrm{v}\beta_1), \qquad d_m = d_m^{(1)} - Bi\mathrm{v}d_m^{(2)}, \qquad c_m = \beta_2 d_m^{(1)} + Bi\mathrm{v}\beta_1 d_m^{(2)},$$

$$d_m^{(1)} = \frac{2m + Bi}{(2m)!}, \qquad d_m^{(2)} = \frac{1}{(2m+1)!}.$$

Observe that during analysis of the roots of the characteristic equation (3.66), the parameter v_{st} represents a solution of the nonlinear Equation (3.58). Furthermore, note that in accordance with (3.57) $0 < \mathrm{v} < 1$. It is easy to prove that $d_m > 0$ $(m = 0, 1, 2, \ldots)$ and $b_m > 0$, if $\beta_2 \geq 0$.

Assuming that the body moves at constant velocity $\mathrm{v}_{st} = \mathrm{const}$, the so-called frictional thermoelastic instability occurs (Re$s_1 > 0$) for $\mathrm{v} > 1$. The latter is characterized by an exponential increase of the contact characteristics, and the moving body is overheated.

3.2.4 Constant friction coefficient and nonstationary process

Analysis of the announced problem is carried out for $F(\dot{z}) = f_s = \mathrm{const}$ using the Runge–Kutta and quadrature methods, and taking into account the asymptotic estimation (3.55), (3.56).

For a steel plate (parameter $\gamma = 1.87$) of the parallelepiped shape ($\alpha_1 = 14 \cdot 10^{-6}$ °C^{-1}, $\lambda_1 = 21$ W m^{-1} ·° C^{-1}), $\nu_1 = 0.3$, $a_1 = 5.9 \cdot 10^{-6}$ m^{2}s^{-1}, $E_1 = 19 \cdot 10^{10}$ Pa), and for $L = 0.01$ m, $Bi = 10$, $T_0 = 5$°C, $f_s = 0.12$ the computational results are shown in Fig. 3.8 through 3.10 for $\varepsilon_1 = 1000;\ 500;\ 100$. Figures 3.8a to 3.10a correspond to body braking ($\dot{z}(0) = \dot{z}^\circ = 169$, $m_0 = 0$), whereas Figs. 3.8b–3.10b refer to body acceleration generated by the applied force $F_* = 9.31 \cdot 10^6$ N m^{-2} ($m_0 = 0.14$), $z^\circ = \dot{z}^\circ = 0$. In the latter case $t_* = 16.95$ s, $\mathrm{V}_* = 0.59\dot{1}0^{-3}$ m s^{-1}, $P_* = 3.3 \cdot 10^7$ Pa.

Braking process. Dimensionless body velocity $\dot{z}(\tau)$ versus dimensionless time τ is shown in Fig. 3.8a. Curve 1 corresponds to the dimensionless parameter $\varepsilon_1 = 1000$ ($m = 1.9 \cdot 10^9$ kg m^{-2}), curve 2 corresponds to parameter $\varepsilon_1 = 500$ ($m = 2.8 \cdot 10^9$ kg m^{-2}), and curve 3 refers to parameter $\varepsilon_1 = 100$

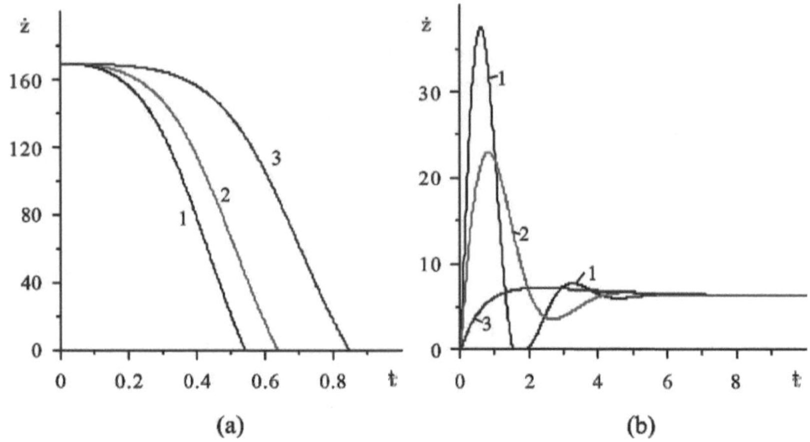

Fig. 3.8: Time histories of the body velocity during braking (a) and acceleration (b) for various values of parameter ε_1 (curves 1: $\varepsilon_1 = 1000$; curves 2: $\varepsilon_1 = 500$; curves 3: $\varepsilon_1 = 100$).

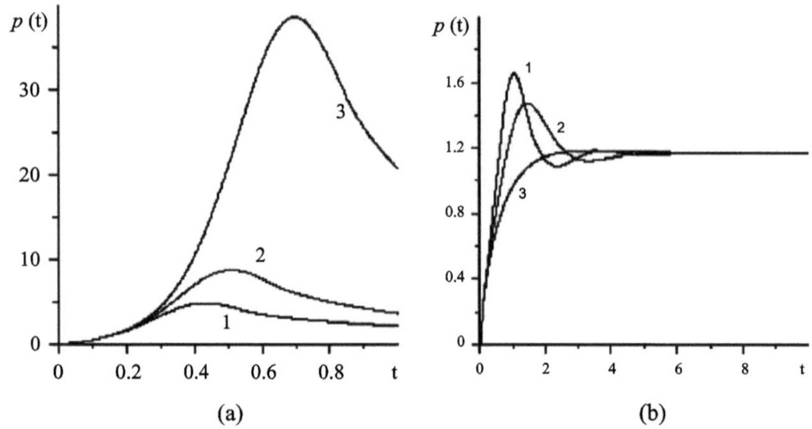

Fig. 3.9: Time history of contact pressure during braking (a) and acceleration (b) for various values of ε_1 (curves 1: $\varepsilon_1 = 1000$; curves 2: $\varepsilon_1 = 500$; curves 3: $\varepsilon_1 = 100$).

($m = 1.9 \cdot 10^{10}$ kg m^{-2}). Dimensionless contact pressure $p(\tau)$ and dimensionless contact temperature $\theta(-1, \tau) = \theta(1, \tau)$ are reported in Figs. 3.9a and 3.10a (curve 1: $\varepsilon_1 = 1000$; curve 2: $\varepsilon_1 = 500$; curve 3: $\varepsilon_1 = 100$).

A body with the initial velocity $\dot{z}^\circ = 169$ ($\dot{Z}_0 = 0.1$ m s^{-1}) is stopped by the action of friction force if the following inequality holds: $m_0 < f_s$ ($F_* < 2f_s P_*$). Then the body is cooled, and after reaching wall temperature T_0, it starts to move again. A reason is that the applied force is smaller than the friction force.

An increase of the body inertia (decrease of parameter ε_1) causes an increase of both braking duration (see Fig. 3.8a) and contact temperature (Fig. 3.10a), and the contact pressure (Fig. 3.9a) is increased. Contrary to the braking process

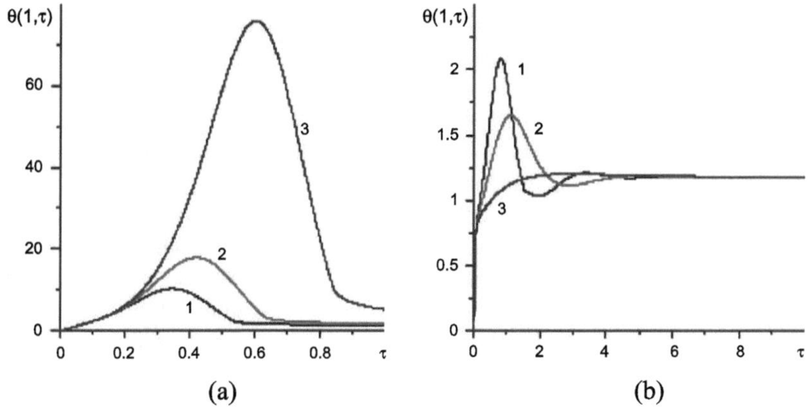

Fig. 3.10: Time histories of contact temperature during braking (a) and acceleration (b) for different ε_1 (curves 1: $\varepsilon_1 = 1000$; curves 2: $\varepsilon_1 = 500$; curves 3: $\varepsilon_1 = 100$).

associated with the constant friction [Chichinadze et al. (1979)], the braking time is not proportional to the mass body. Both contact temperature and pressure achieve their maximal values just before the body stops, and then the body temperature approaches surrounding medium temperature.

Acceleration process. As Figs. 3.8b to 3.10b show, the contact characteristics during acceleration are represented by damped oscillations, which tend to steady states. According to (3.57), one gets: $\mathrm{v}_{st} = 6.37$, $p_{st} = 1.17$, and $\theta_{st} = 1.17$. The roots of characteristic equation (3.67) with the maximal real parts ($\varepsilon_1 = 1000$) are $s_{1,2} = -1.1 \pm i2.3$ for the first case (expected 'period' of damped oscillations $T = 2\pi/\mathrm{Im}s_{1,2} = 2.7$), for the second case ($\varepsilon_1 = 500$) $s_{1,2} = -0.99 \pm i1.5$ (expected 'period' of damped oscillations $T_p = 4.1$), and for the third case ($\varepsilon_1 = 100$) one gets real root values $s_1 = -0.49$, $s_2 = -1.33$. It means that the contact characteristics should change in an aperiodic manner.

The numerical analysis fully confirms our theoretical prediction. In the first case the period of damped oscillations of the contact characteristic is $T_p = 2.7$, whereas in the second case $T_p = 4.1$. Note that before reaching a stationary state, the body may exhibit stick-slip dynamics (curves 1 in Fig. 3.8b). It is worth noting that during the acceleration process, owing to an increase of the body inertia (decrease of parameter ε_1) both contact pressure (see Fig. 3.9b) and temperature (see Fig. 3.10b) decrease, whereas a 'period' of damped oscillations increases (see Fig. 3.8b).

3.2.5 *Variable friction coefficient and nonstationary process*

In this case the steel parallelepiped plate ($\alpha_1 = 14 \cdot 10^{-6}{}^\circ\mathrm{C}^{-1}$, $\lambda_1 = 21$ W m$^{-1} \cdot {}^\circ\mathrm{C}^{-1}$), $\nu_1 = 0.3$, $a_1 = 5.9 \cdot 10^{-6}$ m^2 s^{-1}, $E_1 = 19 \cdot 10^{10}$Pa) with $L = 0.01$ m,

$T_0 = 5°$C, $z^\circ = \dot{z}^\circ = 0$ and with nonconstant friction coefficient is studied. One gets $t_* = 16.95$ s, $V_* = 0.59 \cdot 10^{-3}$m s^{-1}, $P_* = 3.3 \cdot 10^7$ Pa. The function $F(\dot{z}) = f(V_*\dot{z})$ is defined by formula (2.57).

In the first case (for $m_0 = 0.14$, $Bi = 20$) one solution $v^3_{st} = 101.1$, $p^3_{st} = 2.32$, $\theta^3_{st} = 2.32$, ($\beta_1 = 0.6 \cdot 10^{-3}$, $\beta_2 = 0.4 \cdot 10^{-3}$, $v = 0.57$) is found. It is always stable (for instance, for $\varepsilon_1 = 800$ the roots of Equation (3.67) $s_{1,2} = -1.02 \pm 1.57i$, whereas for $\varepsilon_1 = 75.45$ the roots of Equation (3.67) $s_{1,2} = -0.57$ lie on the left-hand side of the complex variable s). In this case after a transitional process, the body starts to move at constant velocity $v^3_{st} = 101.1$. The contact characteristics for $\varepsilon_1 = 800$ achieve their limiting values through damped oscillation process with the expected 'period' $T_p = 4$, whereas for $\varepsilon_1 < 75.45$ the contact characteristics are overdamped.

In the second case ($m_0 = 0.1$, $Bi = 20$) three solutions appear. The solution $v^3_{st} = 87.1$, $p^3_{st} = \theta^3_{st} = 1.8$, $\theta^3_{st} = 2.25$, ($\beta_1 = 0.6 \cdot 10^{-3}$, $\beta_2 = 0.3 \cdot 10^{-3}$, $v = 0.45$) is stable (e.g., for $\varepsilon_1 = 800$ the roots of Equation (3.66) $s_{1,2} = -0.93 \pm 1.1i$ lie on the left-hand side of the complex plane s). The solution $v^2_{st} = 5.26$ ($\beta_1 = 0.2 \cdot 10^{-1}$, $\beta_2 = -0.4 \cdot 10^{-2}$, $v = 0.05$) can be unstable (e.g., for $\varepsilon_1 = 800$ the root of Equation (3.67) $s_1 = 2.84$). The solution described through approximation (2.58) $v^1_{st} = 0.4 \cdot 10^{-4}$ ($\beta_1 = \beta_2 = 2.4 \cdot 10^3$, $v = 4 \cdot 10^{-7}$) corresponds to an equilibrium state (roots of Equation (3.67) $s_1 = -2.2$, $s_2 = -20$ lie on the left-hand side of the complex plane). In the second case, the contact characteristics for $\varepsilon_1 = 800$ tend, depending on initial conditions, to one of two stable solutions.

In the third case ($m_0 = 0.1$, $Bi = 5$) one solution appears. Approximation of solution (2.58) $v^1_{st} = 0.4 \cdot 10^{-4} \approx 0$, $p^1_{st} = 1.0$, $\theta^1_{st} = 1.0$, ($\beta_1 = \beta_2 = 2.4 \cdot 10^3$, $v = 1.5 \cdot 10^{-6}$) corresponds to the equilibrium state (for $\varepsilon_1 = 800$ roots of Equation (3.67) $s_1 = -1.7$, $s_2 = -16.3$ lie on the left-hand side of the complex plane). Note that in this case the braking process always occurs (the applied external force is smaller than the friction force).

In the fourth case ($m_0 = 0.14$, $Bi = 5$) we have one solution of the form: $v^2_{st} = 27.8$, $p^2_{st} = \theta^2_{st} = 2.45$, ($\beta_1 = 0.2 \cdot 10^{-2}$, $\beta_2 = -0.58 \cdot 10^{-3}$, $v = 0.59$). It is unstable if parameter ε_1 is larger than its critical value ($\varepsilon_1 \geq \tilde{\varepsilon}$). One may use the characteristic function (3.67) for $m = 3$ to estimate a stability zone of the stationary solution of (3.57). Recall that one of the conditions for cubic characteristic equation roots to lie on the right-hand side of the complex plane is exhibited by the inequality $b_1 b_2 - b_0 b_3 < 0$. It results in our case in the following instability condition

$$\varepsilon_1 > \tilde{\varepsilon}, \quad \tilde{\varepsilon} = (1 - v)(-B - \sqrt{B^2 - 4AC})/(2A), \tag{3.68}$$

where $A = c_1 c_2 - c_0 c_3$, $B = c_1 d_1 + c_2 d_0 - c_0 d_2$, $C = d_0 d_1$. In Fig. 3.11 the dashed curve represents dependence of function $\tilde{\varepsilon}$ on parameter Bi. The stability loss curve derived through analysis of the characteristic Equation (3.66) is denoted by the solid line. Observe that for considered parameters good agreement of unstable zone estimation by Equations (3.66) and (3.67) for $m = 3$ is achieved. Furthermore,

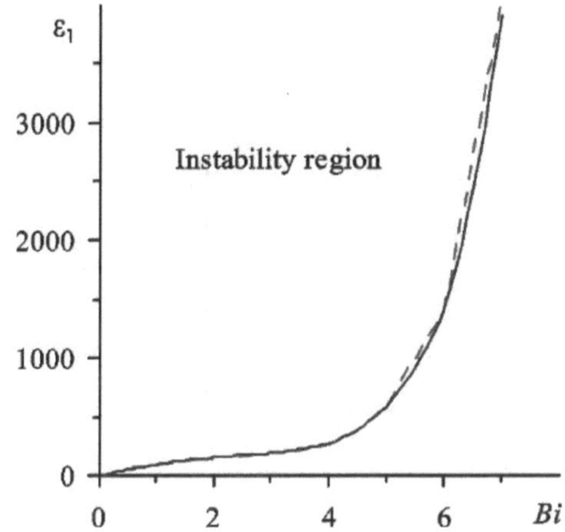

Fig. 3.11: Critical parameter ε_1 versus Bi.

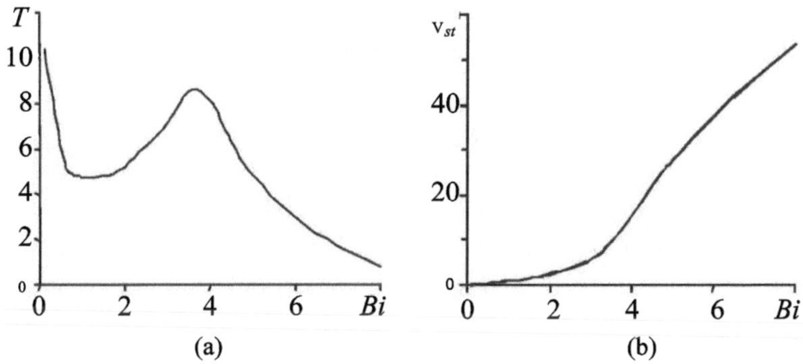

Fig. 3.12: Dimensionless period T_p versus Bi (a), and dimensional velocity v_{st} versus Bi.

an associated analytical formula is given. An increase of heat taken up (increase of the dimensionless parameter Bi) causes an increase of critical parameter $\tilde{\varepsilon}$ (stable solution zone is increased). Furthermore, for any fixed material body and for its loading parameters there is a parameter Bi such that a stationary solution will be always stable (in the considered case $Bi > 8$). The period of oscillation versus Bi is shown in Fig. 3.12a. An increase of Bi causes variation of the period of oscillations (first it decreases, then increases, and then decreases again). Figure 3.12b shows dependence of a solution to Equation (3.58) on parameter Bi. Note that a physical sense of dimensionless stationary velocity v_{st} exists for $\varepsilon_1 < \tilde{\varepsilon}$.

For example, in the fourth case for $\varepsilon_1 = 400$ ($\varepsilon_1 < \tilde{\varepsilon}$) the roots of Equation (3.67) $s_{1,2} = -0.12 \pm 1.06i$ lie on the left-hand side of the complex plane s. A solution is exhibited by 'periodic' damped oscillation (expected period $T_p = 2\pi/\mathrm{Im}s_1 = 5.94$). For the critical value $\varepsilon_1 \approx \tilde{\varepsilon} = 587$, the roots

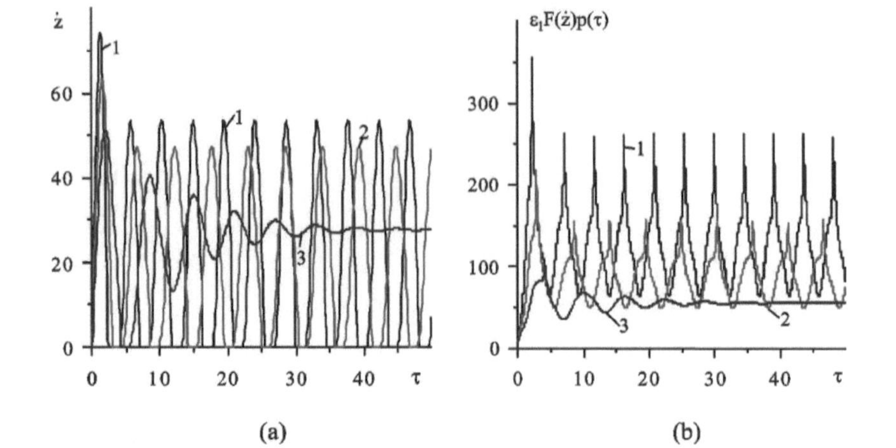

Fig. 3.13: Time history of dimensionless body velocity (a) and friction force (b) for various values of ε_1 (curve 1: $\varepsilon_1 = 800$; curve 2: $\varepsilon_1 = 586.5$; curve 3: $\varepsilon_1 = 400$).

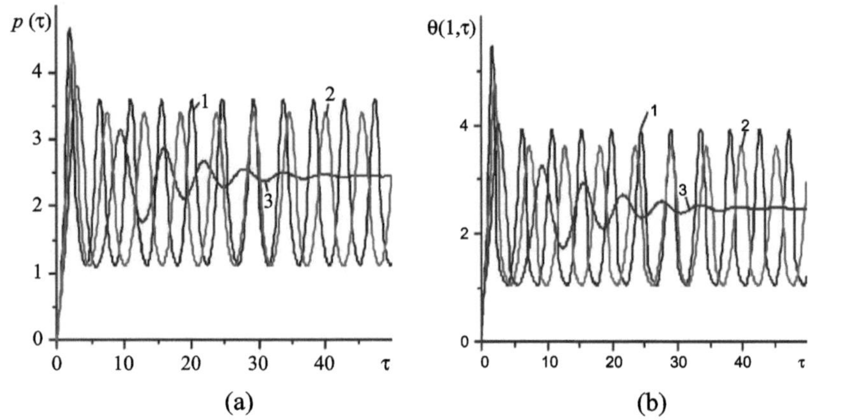

Fig. 3.14: Time history of dimensionless contact pressure (a) and temperature (b) for various values of ε_1 (curve 1: $\varepsilon_1 = 800$; curve 2: $\varepsilon_1 = 586.5$; curve 3: $\varepsilon_1 = 400$).

$s_{1,2} = \pm 1.29i$ lie on the imaginary axis. In this case, the period $T_p = 4.87$ is expected. For $\varepsilon_1 = 800$, the roots $s_{1,2} = 0.14 \pm 1.5i$ lie on the right-hand side of the complex plane s. A stationary solution is unstable, and a limiting stick-slip cycle appears with the expected period $T_p = 4.18$.

In order to confirm the given conclusions, a numerical analysis is carried out for the fourth case for $Bi = 5$ (now $\varepsilon_1 \approx \tilde{\varepsilon} = 586.5$), and the computational results are shown in Figs. 3.13–3.14 for a few values of the parameter $\varepsilon_1 = 400; 586.5; 800$. In Fig. 3.13a, the dependence of dimensionless body velocity $\dot{z}(\tau)$ on dimensionless time τ is shown. Curve 1 corresponds to the case when $\varepsilon_1 = 800$ ($m = 2.4 \cdot 2.4 \cdot 10^9$kg m^{-2}2), curve 2 corresponds to $\varepsilon_1 = 586.5$ ($m = 3.25 \cdot 10^9$kg m^{-2}), and

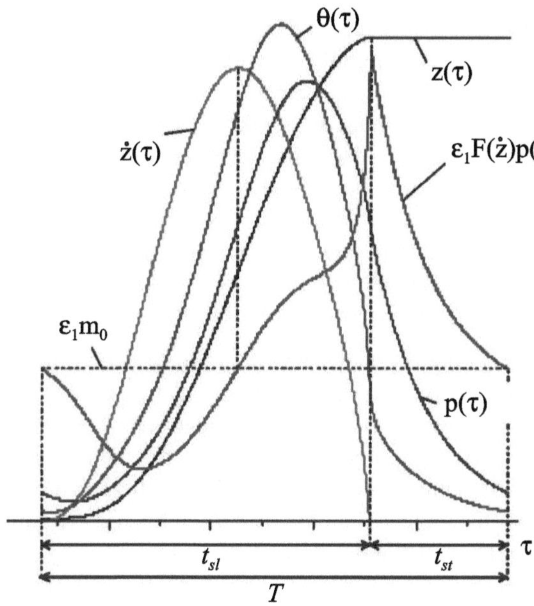

Fig. 3.15: T_p: periodic oscillations of dimensionless contact characteristics (body displacement $z(\tau)$, body velocity $\dot{z}(\tau)$, contact pressure $p(\tau)$, contact temperature $\theta(\tau)$, and friction force $\varepsilon_1 F(\dot{z})p(\tau)$) for $\varepsilon_1 = 800$.

curve 3 corresponds to $\varepsilon_1 = 400$ ($m = 4.8 \cdot 10^{10}$kg m^{-2}). Figure 3.13b shows the time history of the dimensionless friction force $\varepsilon_1 F(\dot{z})p(\tau)$ occurring on the right-hand side of Equation (3.39). Time evolution of both dimensionless contact pressure $p(\tau)$ and temperature $\theta(-1,\tau) = \theta(1,\tau)$ is illustrated in Figs. 3.14a and 3.14b (curve 1: $\varepsilon_1 = 800$; curve 2: $\varepsilon_1 = 586.5$; curve 3: $\varepsilon_1 = 400$).

As Figs. 3.13 and 3.14 show, the contact characteristics either have damped oscillatory shape (curves 3) approaching stationary states or they are periodic (curves 2) or stick-slip periodic (curves 1). The numerical analysis confirms the theoretical prediction. It is found that: $\mathrm{v}_{st}^2 = 27.8$, $p_{st}^2 = \theta_{st}^2 = 2.45$. In the case associated with curves 1, the period of stick-slip contacts $T_p = 2.7$. In the critical case $T_p = 4.1$, for $\varepsilon_1 = 400$ (curve 3) the period of damped oscillations $T_p = 2.7$.

To facilitate analysis of the contact stick-slip characteristics ($\varepsilon_1 = 800$), the associated periodic orbits are shown in Fig. 3.15.

Note that the period of oscillations T includes slip phase t_{sl} and stick phase t_{st} ($T = t_{sl} + t_{st}$). The slip phase consists of acceleration ($\ddot{z} > 0$) and braking. It begins at the time instant when friction force $\varepsilon_1 F(\dot{z})p(\tau)$ is smaller than force $\varepsilon_1 m_0$ applied to the body. Beginning from this time instant, the body velocity increases and contact temperature increases, but $\dot{p}(\tau) < 0$ (according to formula (3.54) the body is cooled). In the time instant corresponding to $\dot{p}(\tau) = 0$, the heat stream equals zero and the body starts to be heated. Because heat expansion is increased, the contact pressure also increases. The kinetic friction coefficient

$F(\dot{z})$ goes down, the friction force $\varepsilon_1 F(\dot{z})p(\tau)$ also decreases initially, because the contact pressure increases slightly. Further pressure increase causes friction increase. When the friction force achieves the value of external force $\varepsilon_1 m_0$, the body velocity achieves its maximum and braking begins ($\ddot{z} < 0$) and the kinematic friction coefficient is decreased. Also, after some slight delay, the contact pressure achieves its maximum. Then the body begins to cool down. When $\dot{z} = 0$, the slip phase is finished and the friction force achieves its maximal value. A stick phase begins. The contact temperature, contact pressure, and friction force decrease. The latter process stops when the frictional force achieves a value of the applied external force. Then the stick-slip process is repeated.

Conclusions. In this section a novel problem of the so-called 'frog-slider' mechanical system exhibiting frictional thermoelastic contact of a moving body subject to both constant and nonconstant friction coefficients has been presented and discussed. It has been shown, among others, that in the case of a constant friction coefficient, the contact characteristics can achieve their stationary stable value oscillating processes (the roots of the associated characteristic equation are imaginary and they lie on the left-hand side of the complex plane). It is worth noting that in the case of a nonconstant friction coefficient, the self-excited vibration can appear in our system without an elastic part (stiffness). The last phenomenon is caused by heating of the body while accelerating, the friction increase, and then the braking and cooling of the system. The characteristic changes of both displacement (Fig. 3.15) and velocity of the analysed system inspired us to use the expression: a 'frog-slider' system.

3.3. Dynamics of a two degrees-of-freedom system with friction and heat generation

A novel thermomechanical model of frictional self-excited stick-slip vibrations is proposed. A mechanical system consisting of two masses that are coupled by an elastic spring and moving vertically between two walls is considered. It is assumed that between masses and walls a Coulomb friction occurs, and stick-slip motion of the system is studied. The applied friction force depends on a relative velocity of the sliding bodies. Stability of stationary solutions is considered. A computation of contact parameters during heating of the bodies is performed. The possibility of existence of frictional autovibrations is illustrated and discussed.

Stick-slip motion is intimately related to the nature of frictional phenomena and is often attributed to the difference between the static and kinematic friction coefficients. Even though the topic of friction is a relatively old one and plays an important role in many practical and engineering applications, surprisingly it is not as well understood as might be expected.

Research reported in this section extends the authors' earlier results, where both regular and chaotic vibrations in a cylinder–bush system have been analysed

[Awrejcewicz (1996), Awrejcewicz, Pyryev (2002), (2003c), (2004a), (2004b)]. In addition, the two degrees-of-freedom system, where a friction force depends on a distance between two masses, has been studied in reference [Wikieł, Hill (2000)].

It is worth noting that the main conditions for occurrence of self-excited vibrations in the models discussed earlier are associated with a difference between static and kinematic frictions, and with existence of an elastic coupling in a tribomechanical system.

Analysis of various references [Andronov et al. (1966), Awrejcewicz (1996), Awrejcewicz, Pyryev (2002), Kragelsky, Gitis (1987), Martins et al. (1990), Olędzki, Siwicki (1997)] leads to a conclusion that velocity of one of the contacting bodies is always given. The system in the condition of self-excited vibrations takes energy from a body moving at constant velocity. The self-excited vibrations do not appear when inertia of the contacting bodies is taken into account. The latter case is considered in this section. It has been shown that owing to heat extension, a body can be periodically heated, braked, cooled, and accelerated. In some conditions, stick-slip self-excited vibrations may also appear.

3.3.1 Statement of the problem

We consider two masses M_1 (body 1) and M_2 (body 2) which are coupled by an elastic spring as indicated in Fig. 3.16. We assume that the initial length of the spring is l_0 and that the spring has stiffness k_{12}, which represents the overall elastic properties of the system. We also assume that the masses are constrained by walls to move only in the vertical direction, and that Z_1 and Z_2 denote positions of masses M_1 and M_2, respectively, as indicated in Fig. 3.16. Let us consider a one-dimensional model of the thermoelastic contact of body 1 with a surrounding

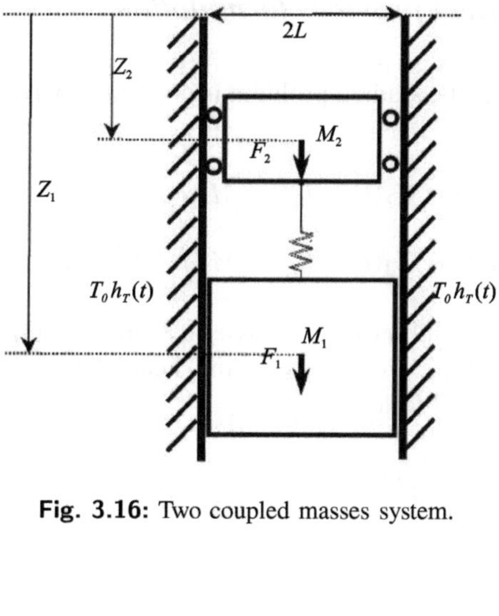

Fig. 3.16: Two coupled masses system.

medium. Assume that this body 1 is represented by a rectangular plate $(l_1 \times l_2 \times 2L)$ (Fig. 3.16). Both bodies are subjected to an action of the forces $F_n = F_*^n h_F(t)$, $n = 1, 2$, $(h_F(t) \to 1,\ t \to \infty)$. At the initial time instant, body 1 (2) is situated at distance Z_1° (Z_2°) and its velocity reads $\dot{Z}_1^\circ$ $(\dot{Z}_2^\circ)$. The distance between walls is always equal to the initial plate thickness $2L$.

It is assumed that heat conduction between the bodies and the walls obeys Newton's law. At an initial instant the temperature is governed by the formula $T_0 h_T(t)$ $(h_T(t) \to 1,\quad t \to \infty)$. It causes heat extension of the parallelepiped in the direction of $0X$, and body 1 starts to contact the walls. As a result of this process, a frictional contact on the parallelepiped sides $X = \pm L$ occurs. A simple frictional model is applied in the further considerations; that is, friction force F_{fr} is a product of normal reaction force $N(t)$ and a friction coefficient. That means that $F_{fr} = f(\dot{Z}_1)\,N(t)$ is the friction force defining resistance of the movement of two sliding bodies. Here, owing to the assumption made in references [Andronov et al. (1966), Awrejcewicz, Pyryev (2002), Kragelsky, Gitis (1987)], the kinematic friction coefficient $f(\dot{Z}_1)$ depends on the relative velocity $V_r = \dot{Z}_1$ of the sliding bodies (Fig. 2.2).

The friction force $\sigma_{xz}(X, t)$ per unit contact surface $X = -L$, $X = L$ generates heat. According to Ling's assumptions (cf. [Ling (1959)]), the work of the friction forces is transmitted into heat energy. Note, that the noncontacting plate surfaces are heat-isolated and have the dimensions of $L/l_1 \ll 1$, $L/l_2 \ll 1$, which is in agreement with the assumption of our one-dimensional modelling for body 1. Quantities M_n, F_*^n, k_{12} are related to a unit contacting surface.

Below, the problem is reduced to determination of the mass plate (body 2) center displacement $Z_1(t)$ $(Z_2(t))$, plate (body 2) velocity $\dot{Z}_1(t)$ $(\dot{Z}_2(t))$, contact pressure $P(t) = N(t)/l_1 l_2 = -\sigma_{XX}(-L, t) = -\sigma_{XX}(L, t)$, plate temperature $T_1(X, t)$, and displacement $U_1(X, t)$ in the direction of the X-axis.

3.3.2 *Mathematical problem formulation*

In the considered case, the studied problem is governed by two equations of motion in the form

$$M_1 \ddot{Z}_1(t) + k_{12}(Z_1(t) - Z_2(t) - l_0) = M_1 g + F_*^1 h_F(t) - 2f(\dot{Z}_1)P(t), \quad (3.69)$$

$$M_2 \ddot{Z}_2(t) - k_{12}(Z_1(t) - Z_2(t) - l_0) = M_2 g + F_*^2 h_F(t), \quad (3.70)$$

where Z_1, Z_2 denote position of both masses as shown in Fig. 3.16; $\dot{Z}_1$, $\dot{Z}_2$ denote their respective velocities. Equations of the heat stress theory for an isotropic body 1 [Nowacki (1962)] follow

$$\frac{\partial}{\partial X}\left[\frac{\partial}{\partial X} U_1(X, t) - \alpha_1 \frac{1 + \nu_1}{1 - \nu_1} T_1(X, t)\right] = 0, \quad (3.71)$$

$$\frac{\partial^2}{\partial X^2} T_1(X,t) = \frac{1}{a_1} \frac{\partial}{\partial t} T_1(X,t), \qquad X \in (-L, L), \tag{3.72}$$

and mechanical

$$U_1(-L,t) = 0, \qquad U_1(L,t) = 0, \tag{3.73}$$

heat

$$-\lambda_1 \frac{\partial T_1(-L,t)}{\partial X} + \alpha_T (T_1(-L,t) - T_0 h_T(t)) = f(\dot{Z}_1) \dot{Z}_1(t) P(t), \tag{3.74}$$

$$\lambda_1 \frac{\partial T_1(L,t)}{\partial X} + \alpha_T (T_1(L,t) - T_0 h_T(t)) = f(\dot{Z}_1) \dot{Z}_1(t) P(t), \tag{3.75}$$

and initial conditions

$$T_1(X,0) = 0, \qquad X \in (-L, L),$$

$$Z_1(0) = Z_1^\circ, \qquad Z_2(0) = Z_2^\circ, \qquad \dot{Z}_1(0) = \dot{Z}_1^\circ, \qquad \dot{Z}_2(0) = \dot{Z}_2^\circ \tag{3.76}$$

are attached. Normal stresses that occur in the plate are defined via the relation

$$\sigma_{XX} = \frac{E_1}{1 - 2\nu_1} \left[\frac{1 - \nu_1}{1 + \nu_1} \frac{\partial U_1}{\partial X} - \alpha_1 T_1 \right]. \tag{3.77}$$

In the above, the following notation is applied. E_1 is the elasticity modulus; and ν_1, λ_1, a_1, α_1, α_T are Poisson's ratio, thermal conductivity, thermal diffusivity, thermal expansion and heat transfer coefficients, respectively.

Integration of Equation (3.71), owing to (3.77) and boundary conditions (3.73), yields the contact pressure $P(t) = -\sigma_{XX}(-L,t) = -\sigma_{XX}(L,t)$ cast in the form

$$P(t) = P_0 + \tilde{E}\tilde{\alpha} \frac{1}{L} \int_0^L T_1(\xi, t) d\xi. \tag{3.78}$$

Let us introduce the following similarity coefficients

$$t_* = L^2/a_1[\mathrm{s}], \qquad \mathrm{v}_* = a_1/L[\mathrm{m/s}], \qquad P_* = T_0 E_1 \alpha_1/(1 - 2\nu_1)[\mathrm{N/m^2}], \tag{3.79}$$

and the following nondimensional parameters

$$x = \frac{X}{L}, \quad \tau = \frac{t}{t_*}, \quad z_n = \frac{Z_n}{L}, \quad p = \frac{P}{P_*}, \quad \theta = \frac{T_1}{T_0}, \quad z_n^\circ = \frac{Z_n^\circ}{L}, \quad \dot{z}_n^\circ = \frac{\dot{Z}_n^\circ}{\mathrm{v}_*},$$

$$\tau_{nD} = t_*/t_{nD}, \quad m_{n0} = (M_n g + F_n)/2P_*, \quad \varepsilon_n = 2P_* t_*^2/M_n L, \quad n = 1, 2,$$

$$l = \frac{l_0}{L}, \quad \gamma = \frac{E_1 \alpha_1 a_1}{(1-2\nu_1)\lambda_1}, \quad Bi = \frac{L\alpha_T}{\lambda_1}, \quad F(\dot{z}) = f(\mathrm{v}_* \dot{z}), \tag{3.80}$$

where $t_{nD} = \sqrt{M_n/k_{12}}$, $n = 1, 2$.

The examined problem is governed by the following nondimensional equations

$$\frac{\partial^2 \theta(x,\tau)}{\partial x^2} = \frac{\partial \theta(x,\tau)}{\partial \tau}, \qquad x \in (-1,1), \qquad \tau \in (0,\infty), \tag{3.81}$$

$$\ddot{z}_1(\tau) + (z_1(\tau) - z_2(\tau) - l)\tau_{1D}^2 = \varepsilon_1(m_{10} - F(\dot{z}_1)p(\tau)), \tag{3.82}$$

$$\ddot{z}_2(\tau) - (z_1(\tau) - z_2(\tau) - l)\tau_{2D}^2 = \varepsilon_2 m_{20}, \tag{3.83}$$

with both boundaries

$$\left[\frac{\partial \theta(x,\tau)}{\partial x} \pm Bi\theta(x,\tau)\right]_{x=\pm 1} = \pm q(\tau), \tag{3.84}$$

and initial conditions,

$$\theta(x,0) = 0, \quad z_1(0) = z_1^\circ, \quad z_2(0) = z_2^\circ, \quad \dot{z}_1(0) = \dot{z}_1^\circ, \quad \dot{z}_2(0) = \dot{z}_2^\circ \tag{3.85}$$

where

$$q(\tau) = Bih_T(\tau) + \gamma F(\dot{z}_1)\dot{z}_1(\tau)p(\tau), \qquad p(\tau) = \frac{1}{2}\int_{-1}^{1} \theta(\xi,\tau)d\xi. \tag{3.86}$$

3.3.3 *Solution of the problem*

Applying an inverse Laplace transform [Carslaw, Jaeger (1959)], our nonlinear problem governed by Equations (3.81), (3.84), and (3.85) is reduced to the following integral equation

$$p(\tau) = Bi \int_0^\tau \dot{h}_T(\xi) G_p(\tau - \xi) d\xi + \gamma \int_0^\tau F(\dot{z}_1)\dot{z}_1(\xi)p(\xi)\dot{G}_p(\tau - \xi)d\xi, \tag{3.87}$$

which yields both nondimensional pressure $p(\tau)$ and velocity $\dot{z}_1(\tau)$. The temperature is defined by the following formula

$$\theta(x,\tau) = Bi \int_0^\tau \dot{h}_T(\xi) G_\theta(x, \tau - \xi) d\xi + \gamma \int_0^\tau F(\dot{z}_1)\dot{z}_1(\xi)p(\xi)\dot{G}_\theta(x, \tau - \xi)d\xi, \tag{3.88}$$

where

$$\{G_p(\tau), \qquad G_\theta(1,\tau)\} = \frac{1}{Bi} - \sum_{m=1}^{\infty} \frac{\{2Bi, \qquad 2\mu_m^2\}}{\mu_m^2[Bi(Bi+1)+\mu_m^2]} e^{-\mu_m^2 \tau}, \tag{3.89}$$

and μ_m are the roots of the following characteristic equation

$$\text{tg}\mu_m = \frac{Bi}{\mu_m}, \qquad m = 1, 2, \ldots. \tag{3.90}$$

3.3.4 Steady-state solution analysis

A stationary solution to the problem reads

$$p_{st} = \theta_{st} = \frac{1}{1 - \text{v}}, \qquad \text{v} = F(\text{v}_{st}) \frac{\text{v}_{st}\gamma}{Bi}, \tag{3.91}$$

where v_{st} is the solution to the nonlinear equation

$$F(\text{v}_{st}) = \frac{m_0}{1 + \gamma\, m_0 \text{v}_{st}/Bi}, \qquad m_0 = m_{10}^{st} + m_{20}^{st}, \qquad m_{n0}^{st} = \frac{M_n g + F_*^n}{2P_*}. \tag{3.92}$$

A graphical solution of Equation (3.92) is presented in Fig. 3.17 for various parameters m_0 and Bi. Recall that for steel $\gamma = 1.87$.

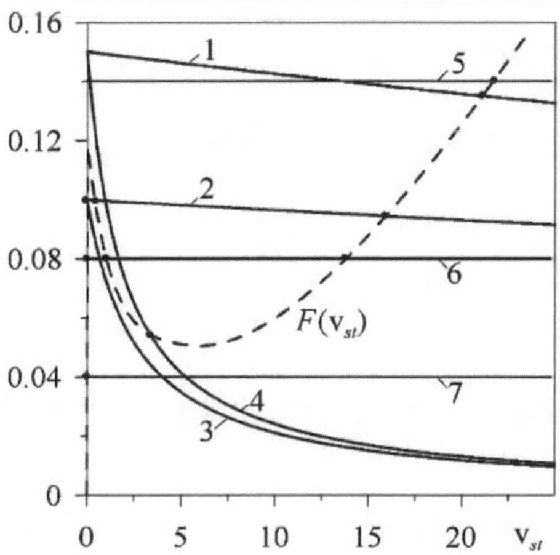

Fig. 3.17: Graphical solution of Equation (3.92) (solid curves: 1, $m_0 = 0.15$, $Bi = 50$, $\gamma = 1.87$; 2, $m_0 = 0.1$, $Bi = 50$, $\gamma = 1.87$; 3, $m_0 = 0.1$, $Bi = 0.5$, $\gamma = 1.87$; 4, $m_0 = 0.15$, $Bi = 0.5$, $\gamma = 1.87$; 5, $m_0 = 0.14$, $\gamma = 0$; 6, $m_0 = 0.8$, $\gamma = 0$; 7, $m_0 = 0.04$, $\gamma = 0$; dashed curve corresponds to $F(\text{v}_{st})$).

The right-hand side of Equation (3.92) is represented by solid curves for different values of the parameters m_0 and Bi. Solid curve 1 is associated with parameters $m_0 = 0.15$, $Bi = 50$; solid curve 2 with $m_0 = 0.1$, $Bi = 50$; solid curve 3 with $m_0 = 0.1$, $Bi = 0.5$; and solid curve 4 with $m_0 = 0.15$, $Bi = 0.5$. The dashed curve displays the function $F(\mathrm{v}_{st})$.

For different values of the parameters, Equation (3.92) can have a different number of solutions: for $m_0 = 0.15$, $Bi = 50$ (first case) it has one solution $\mathrm{v}_{st}^{(3)}$ ($F'(\mathrm{v}_{st}^{(3)}) > 0$); for $m_0 = 0.1$, $Bi = 50$ (second case) it has three solutions $\mathrm{v}_{st}^{(1)}$, $\mathrm{v}_{st}^{(2)}$, $\mathrm{v}_{st}^{(3)}$ ($F'(\mathrm{v}_{st}^{(1)}) > 0$, $F'(\mathrm{v}_{st}^{(2)}) < 0$, $F'(\mathrm{v}_{st}^{(3)}) > 0$); and for $m_0 = 0.1$, $Bi = 0.5$ (third case) one solution $\mathrm{v}_{st}^{(1)}$ ($F'(\mathrm{v}_{st}^{(1)}) > 0$). Owing to approximation (2.58) we have $\mathrm{v}_{st}^{(1)} \approx \varepsilon_0 m_0/2f_s$ and $F'(\mathrm{v}_{st}^{(1)}) \approx 2f_s/\varepsilon_0$. For $m_0 = 0.15$, $Bi = 0.5$ (fourth case), again one solution exists $\mathrm{v}_{st}^{(2)}$ ($F'(\mathrm{v}_{st}^{(2)}) < 0$).

Let us introduce a perturbation of the stationary solution (3.91) by means of the following formulas

$$z_n = \mathrm{v}_{st}\tau + z_n^*, \quad \dot{z}_n = \mathrm{v}_{st} + \dot{z}_n^*, \quad n = 1, 2, \quad p = p_{st} + p^*, \quad \theta = \theta_{st} + \theta^*,$$

$$h_T = 1 + h_T^*, \quad h_F = 1 + h_F^*, \quad |h_F^*| \ll 1, \quad |h_T^*| \ll 1. \tag{3.93}$$

Owing to linearization of the right-hand sides of (3.82) and with boundary condition (3.84) the following linear problem is obtained

$$\frac{\partial^2\theta^*(x,\tau)}{\partial x^2} = \frac{\partial\theta^*(x,\tau)}{\partial\tau}, \tag{3.94}$$

$$\ddot{z}_1^*(\tau) + (z_1^* - z_2^*)\tau_{1D}^2 = \varepsilon_1[m_{10}^*(\tau) - F(\mathrm{v}_{st})p^*(\tau) - F'(\mathrm{v}_{st})p_{st}\dot{z}_1^*], \tag{3.95}$$

$$\ddot{z}_2^*(\tau) - (z_1^* - z_2^*)\tau_{2D}^2 = \varepsilon_2 m_{20}^*(\tau), \tag{3.96}$$

$$\left[\frac{d\theta^*(x,\tau)}{dx} - Bi\theta^*(x,\tau)\right]_{x=-1} = -q^*(\tau),$$

$$\left[\frac{d\theta^*(x,\tau)}{dx} + Bi\theta^*(x,\tau)\right]_{x=1} = q^*(\tau), \tag{3.97}$$

where

$$q^*(\tau) = Bih_T^*(\tau) + \gamma(\mathrm{v}_{st}p_{st}(\beta_1 + \beta_2)\dot{z}_1^*(\tau) + \mathrm{v}_{st}F(\mathrm{v}_{st})p^*(\tau)),$$

$$m_{n0}^*(\tau) = F_*^n h_F^*(\tau)/(2P_*), \qquad p^*(\tau) = \frac{1}{2}\int_{-1}^{1}\theta^*(\xi,\tau)d\xi,$$

$$\beta_1 = \frac{F(\mathrm{v}_{st})}{\mathrm{v}_{st}}, \qquad \beta_2 = F'(\mathrm{v}_{st}). \tag{3.98}$$

Furthermore, applying the Laplace transform, a solution of problem (3.94)–(3.97) in the transform domain is found. For example, the Laplace transform of the velocity perturbation of body 1 reads

$$s\bar{z}_1^*(s) = -\Delta^{-1}(s)\{\varepsilon_1\beta_1 \mathrm{v}_{st} Bi(s^2+\tau_{2D}^2)\bar{S}(s)\bar{h}_T^*(s)$$
$$+(Bi\,\mathrm{v}\,\bar{S}(s)-\Delta_1(s))\,[\varepsilon_1(s^2+\tau_{2D}^2)\bar{m}_{10}^*(s)+\varepsilon_2\tau_{1D}^2\bar{m}_{20}^*(s)]\}, \quad (3.99)$$

where

$$\{\bar{z}_1^*(s),\bar{h}_T^*(s),\bar{h}_F^*(s),\bar{m}_{n0}^*(s)\} = \int_0^\infty \{z_1^*(\tau),h_T^*(\tau),h_F^*(\tau),m_{n0}^*(\tau)\}\,e^{-s\tau}d\tau.$$

The characteristic equation of the linearized problem reads

$$\Delta(s) = (\varepsilon_1 p_{st}\beta_2(s^2+\tau_{2D}^2)+s(s^2+\tau_{1D}^2+\tau_{2D}^2))\Delta_1(s)$$
$$+\,Bi\,\mathrm{v}(\varepsilon_1 p_{st}\beta_1(s^2+\tau_{2D}^2)-s(s^2+\tau_{1D}^2+\tau_{2D}^2))\bar{S}(s)=0, \quad (3.100)$$

where $\Delta_1(s)=s\bar{S}(s)+Bi\bar{C}(s)$, $\bar{S}(s)=\sinh(\sqrt{s})/\sqrt{s}$, $\bar{C}(s)=\cosh(\sqrt{s})$.

The characteristic function $\Delta(s)$, in the form of an infinite-order polynomial takes the form

$$\Delta(s)=\sum_{m=0}^{\infty} s^m a_m, \quad (3.101)$$

where

$$a_0=\tau_{2D}^2 b_0,\qquad a_1=\tau_{2D}^2 b_1+\tau_{1D}^2(d_0^{(1)}-Bi\,\mathrm{v}),$$

$$a_m=b_{m-2}+\tau_{2D}^2 b_m+\tau_{1D}^2(d_{m-1}^{(1)}-Bi\,\mathrm{v}\,d_{m-1}^{(2)}),\qquad m=2,3,\ldots,$$

$$b_0=\varepsilon_1 p_{st}c_0,\qquad b_m=\varepsilon_1 p_{st}c_m+d_{m-1},\qquad m=1,2,\ldots,$$

$$c_0=Bi(\beta_2+\mathrm{v}\beta_1),\qquad d_m=d_m^{(1)}-Bi\,\mathrm{v}d_m^{(2)},\qquad c_m=\beta_2 d_m^{(1)}+Bi\,\mathrm{v}\beta_1 d_m^{(2)},$$

$$d_m^{(1)}=\frac{2m+Bi}{(2m)!},\qquad d_m^{(2)}=\frac{1}{(2m+1)!}\qquad m=0,1,2,\ldots.$$

Observe that owing to analysis of the roots of characteristic Equation (3.101), the parameter v_{st} represents a solution to nonlinear Equation (3.92).

If the frictional heat generation is not taken into account ($\gamma=0$), the characteristic equations are governed by the following cubic equation: $s^3+\varepsilon_1\beta_2 s^2+(\tau_{1D}^2+\tau_{2D}^2)s+\varepsilon_1\beta_2\tau_{2D}^2=0$. Its roots lie in the right-hand part of the complex plane if $\beta_2<0$.

In the case of a perfectly stiff spring ($k\to\infty$), we have $\tau_{1D}\to\infty$, $\tau_{2D}\to\infty$. The Laplace transformation of the velocity perturbation of body 1 reads

$$s\bar{z}_1^*(s) = -\Delta^{-1}(s)\left\{\varepsilon_1\beta_1 \mathrm{v}_{st} Bi\bar{S}(s)\bar{h}_T^*(s)\right.$$
$$\left.+(Bi\,\mathrm{v}\,\bar{S}(s) - \Delta_1(s))(\varepsilon_1\bar{m}_{10}^*(s) + \varepsilon_2\bar{m}_{20}^*(s))\right\}, \tag{3.102}$$

where

$$\Delta(s) = (\varepsilon_1 p_{st}\beta_2 + s)\Delta_1(s) + Bi\,\mathrm{v}(\varepsilon_1 p_{st}\beta_1 - s)\bar{S}(s) = 0 \tag{3.103}$$

is the characteristic equation of the linearized problem, and velocity v_{st} is a solution to Equation (3.92). Note that a detailed analysis of roots of Equation (3.103) for $\beta_2 = 0$ has been carried out in reference [Olesiak, Pyryev (2000)].

In the case when $M_2 \to 0$, we have $\tau_{2D} \to \infty$, and for $F_*^2 = 0$ ($\bar{m}_{20}^*(s) = 0$) the Laplace transform of the velocity perturbation of body 1 is defined by Equation (3.102), whereas the characteristic equation is given by (3.103).

In this case, the examination concerns the steel parallelepiped plate ($\alpha_1 = 14 \cdot 10^{-6}{}^\circ\mathrm{C}^{-1}$, $\lambda_1 = 21\,W/(\mathrm{m}\cdot{}^\circ\mathrm{C}^{-1})$, $\nu_1 = 0.3$, $a_1 = 5.9 \cdot 10^{-6}\mathrm{m}^2/s$, $E_1 = 19 \cdot 10^{10}\mathrm{Pa}$) with $\varepsilon_1 = 100$, $\varepsilon_2 = 900$, $\tau_{1D} = 2$, $\tau_{2D} = 6$ and with a nonconstant friction coefficient. The function $F(\dot{z}) = f(\mathrm{v}_*\dot{z})$ is defined by the formula (2.57).

In the first case (for $m_0 = 0.15$, $Bi = 50$), one solution $\mathrm{v}_{st}^{(3)} = 21$, $p_{st}^{(3)} = \theta_{st}^{(3)} = 1.12$ ($\mathrm{v} = 0.105$) is found. It is always stable (the roots of Equation (3.100) $s_{1,2} = -0.05 \pm 6.3\,i$ lie in the left-hand side of the complex variable s). The contact characteristics achieve their limiting values through a damped oscillation process with the expected ‘period’ $T = 0.99$.

In the second case ($m_0 = 0.1$, $Bi = 50$), three solutions appear. The solution $\mathrm{v}_{st}^{(3)} = 15.94$, $p_{st}^{(3)} = \theta_{st}^{(3)} = 1.06$, ($\mathrm{v} = 0.056$) is stable (the roots of Equation (3.100) $s_{1,2} = -0.04 \pm 6.32\,i$ lie in the left-hand side of the complex plane s). The solution $\mathrm{v}_{st}^{(2)} = 0.41$ ($\mathrm{v} = 0.0015$) is unstable ($s_1 = 3.8$, $s_{2,3} = 0.15 \pm 6.23\,i$). The solution $\mathrm{v}_{st}^{(1)} \approx 0$ corresponds to an equilibrium state. In the considered case, the contact characteristics, depending on initial conditions, tend to one of the two stable solutions.

In the third case ($m_0 = 0.1$, $Bi = 0.5$), there is only one solution, which is stable. Note that in this case, a braking process always occurs (the applied external force is smaller than the friction force).

In the fourth case ($m_0 = 0.15$, $Bi = 0.5$), the only solution that exists is of the form: $\mathrm{v}_{st}^{(2)} = 3.01$, $p_{st}^{(2)} = \theta_{st}^{(2)} = 2.69$, ($\mathrm{v} = 0.63$) and it is unstable ($s_{1,2} = 0.5 \pm 0.68\,i$, $s_{3,4} = 0.06 \pm 6.32\,i$). If a solution is unstable, then in solving a nonstationary problem this solution may approach a stable limit cycle or it can be expressed via oscillations increasing in time (its behaviour depends on other nonlinear terms).

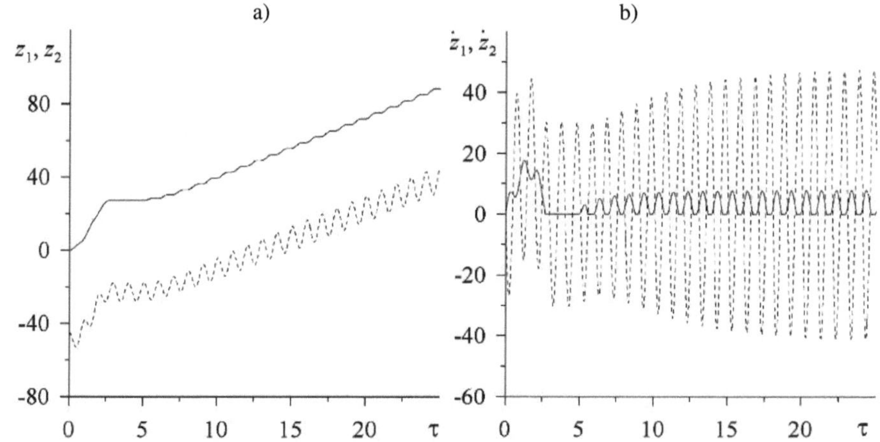

Fig. 3.18: Time histories of nondimensional body displacement (a) and velocity (b). Solid curves correspond to body 1, dashed curves correspond to body 2.

3.3.5 Numerical analysis of transient solution

Let us consider the fourth case as an example. Figure 3.18a shows the dependence of displacement of nondimensional body 1 (body 2) $z_1(\tau)$ ($z_2(\tau)$) versus nondimensional time τ, whereas Fig. 3.18b displays a dependence of velocity of nondimensional body 1 (body 2) $\dot{z}_1(\tau)$ ($\dot{z}_2(\tau)$) versus nondimensional time τ. Solid curves correspond to body 1; dashed curves correspond to body 2. Note that body 1 is in a stick-slip state. Zones with stick ($\dot{z}_1 = 0$) are substituted by zones of slips.

Evolution of nondimensional contact pressure in time (curve 1) and a temperature on the contact surface (curve 2) is shown in Fig. 3.19.

In this section a new physical and mathematical model of a two degrees-of-freedom system with an account of friction and heating processes is studied. It is assumed that the friction coefficient depends on sliding velocity.

It has been shown that when a heat transfer is not taken into account ($\gamma = 0$), the system cannot exhibit a stick-slip motion. This potential behaviour of the studied system is displayed by a solid curve in Fig. 3.17. For $m_0 > f_s$ (solid curve 5) equation $F(v_{st}) = m_0$ has one stable static solution, which attracts a nonstationary one. For $f_{\min} < m_0 < f_s$ (solid curve 6) the mentioned equation has three static solutions, and one of them is unstable. In this case, a nonstationary solution will be attracted by one of two static stable solutions. In the cases when $m_0 < f_{\min}$ (solid curve 7) the discussed equation has only one stable solution.

In order to realize a stick-slip motion the parameter γ should be positive (see case 4). In this case, one deals with only one solution, which is unstable and a nonstationary solution can be attracted by a limiting cycle. The numerical analysis is in agreement with theoretical prediction of the occurrence of stick-slip dynamics of our investigated system with friction and heat generation.

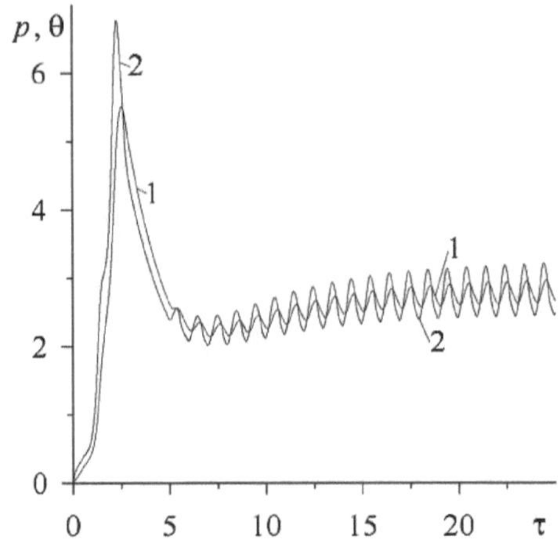

Fig. 3.19: Dimensional contact pressure p (curve 1) and dimensionless contact temperature θ (curve 2) versus dimensionless time τ.

3.4. Tribological dynamical damper of vibrations with thermoelastic contact

Nonlinear oscillations of mechanical systems are described extensively in a series of classical monographs [Den Hartog (1952), Kauderer (1958), Hayashi (1968), Andronov et al. (1966), Bogoliubov et al. (1961), Awrejcewicz (1996), Arnold et al. (1997)], and some asymptotic methods for solution of equations of nonlinear oscillations are presented in well-known books [Andronov et al. (1966), Bogoliubov et al. (1961), Awrejcewicz (1996), Arnold et al. (1997)]. They mainly address classical approaches to study vibrations exhibited by various engineering systems.

In many cases in engineering, harmful effects of vibrations are suppressed by inclusion of the so-called dynamic dampers of vibration [Den Hartog (1952)]. They have rather a wide spectrum of application and can be used to damp various longitudinal, torsional, and transversal vibrations of both machines and civil engineering constructions [Giergiel (1990)]. Now, it is well known in engineering that in order to avoid harmful effects of resonance, the majority of the externally driven mechanical systems should be damped.

A drawback of the currently designed vibration dampers is associated with heat transfer to the contacting bodies induced by frictional processes. It causes extension of the contacting bodies, and a change of contacting pressure and friction often resulting in the harmful damper wedging effects.

In this section a one degree-of-freedom system driven by either a force or kinematic excitation is studied. An additional mass is added to the mentioned mechanical system via a special pressing device initiating dry friction occurrence on the contacting surfaces. Our proposed mathematical model of the preliminary described system includes thermal effects that appear on the contacting bodies [Pyryev (1994), Awrejcewicz, Pyryev (2002)]. Note that the damper geometrical properties, heat transfer between the bodies, and a surrounding medium yield a change of friction on the contacting surface. We focus on a solution to the nonlinear problem of thermal stresses and strongly nonlinear equations governing the dynamics of the investigated system. Based on the analysis, the directions for proper construction of the mechanical vibration dampers are given.

3.4.1 Mathematical modelling

Below, we consider a damper of torsion vibrations of the so-called Lanchester system shown in Fig. 3.20a. In this figure the following notation is introduced: 1 denotes a bush for mounting the damper on a shaft; 2 denotes two coupled discs serving as flywheels and freely rotating on bush 3 mounted on shaft 4. The bush is coupled with friction washer 5, where discs are pressed by screw 6. A relatively weak power tight screw generates small friction forces and weak energy damping. On the other hand, strong power tight screws may eliminate sliding and hence energy dissipation does not occur.

Figure 3.20b shows a dynamic model of the Lanchester system with dry friction. Body with mass m_1 models a fundamental part of the system (shaft 4, bushes 1

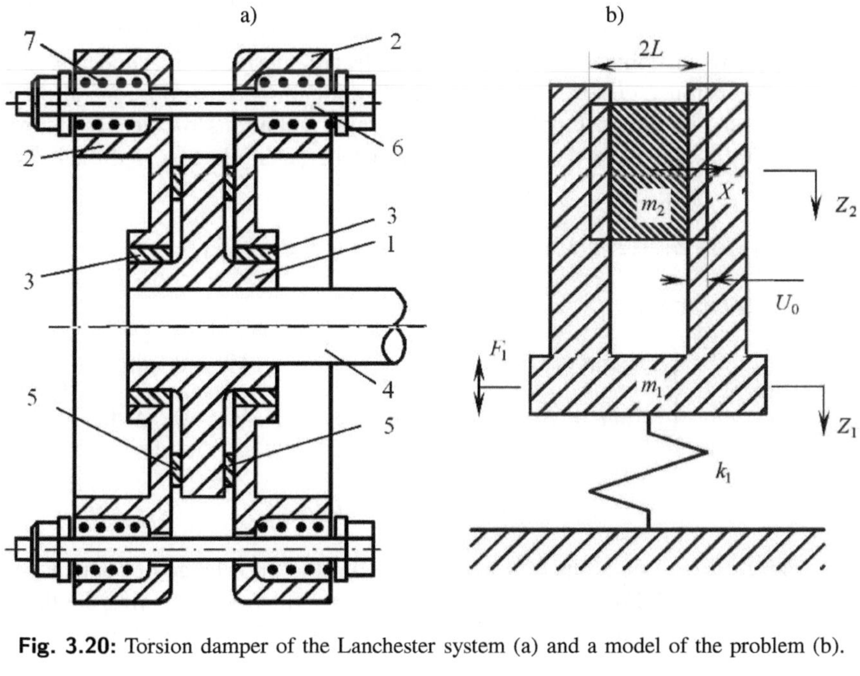

Fig. 3.20: Torsion damper of the Lanchester system (a) and a model of the problem (b).

and 3). System vibrations are generated by harmonic force $F_1 = F_0 \sin \omega_0' t$. The following notation is applied. m_1 is mass; k_1 is the elasticity coefficient of the main system constraints; F_0, ω_0' are the amplitude and frequency of the driving force, respectively. The damper (two coupled discs 2 and screws 6 shown in Fig. 3.20a) of mass m_2 is added. It is coupled with mass m_1 by a pressing system and hence dry friction force F_{fr} occurs. Assume that the tested damper has the shape of parallelepiped $(2L \times L_1 \times L_2)$ moving in direction Z_2 along the walls of the main system. The initial value of the distance between the walls, that is, between discs 2 and the bush 1 in Fig. 3.20a is equal to the plate thickness $2L$ (thickness of friction washers 5). Then, this distance is decreased according to the formula $2U_0 h_U(t)$, which is realized via power tight of screws 6, where U_0 is a constant larger than zero, and $h_U(t)$ is a known dimensionless function of time ($h_U(t) \to 1$, $t \to \infty$). Then, this distance is decreased according to the formula $2U_0 h_U(t)$, which is realized via power tight of screws 6. As a result of this process, dry friction occurs on the parallelepiped surfaces$X = \pm L$. It is defined by the function $F_{fr}(V_r)$, where V_r is the relative velocity of the plate and walls; that is, $V_r = \dot{Z}_1 - \dot{Z}_2$ $(dZ_i/dt \equiv \dot{Z}_i,\ i = 1, 2)$.

According to Amonton's assumption, the friction force $F_{fr} = 2f(V_r)P$ is equal to the product of the normal reaction component and friction coefficient, $f(V_r)$ denotes the kinetic friction coefficient ($f(-V_r) = -f(V_r)$), and we take $f(V_r) = f_s \operatorname{sgn}(V_r)$. The action of friction on the contact surface $X = \pm L$ generates heat. We follow the generally accepted assumption [Ling (1959)] that friction work is transformed into heat energy. Furthermore, we assume that walls ideally transform heat and that between the plate and walls heat transfer is governed by Newton's law, and a surrounding medium temperature is equal to zero. Plate surfaces not being in contact with movable walls are thermally isolated and they have the dimension $L/L_1 \ll 1$, $L/L_2 \ll 1$. The mentioned assumptions allow us to introduce a one-dimensional model. Governing equations of the system presented in Fig. 3.20b have the following form (see [Den Hartog (1952), Pyryev (1994)])

$$m_1 \ddot{Z}_1 + k_1 Z_1 + F_{fr}(V_r, Z_1) = F_1, m_2 \ddot{Z}_2 - F_{fr}(V_r, Z_1) = 0, \tag{3.104}$$

with the friction model

$$F_{fr}(V_r, Z_1) = \begin{cases} 2f_s \operatorname{sgn}(V_r) P(t), & V_r \neq 0, \text{ slip}, \\ \min(F_{st}, 2f_s P(t)) \operatorname{sgn}(F_1 - k_1 Z_1), & V_r = 0, \text{ stick}, \end{cases} \tag{3.105}$$

where

$$F_{st} = \frac{m_2}{m_1 + m_2} \left| F_1 - k_1 Z_1 \right| .$$

In order to solve Equations (3.104) and (3.105) the knowledge of contact pressure $P(t)$ is required. For this purpose the following equation governing the theory of thermal stresses for an isotropic body [Nowacki (1962)] is solved first,

$$\frac{\partial}{\partial X}\left[\frac{\partial U_2(X,t)}{\partial X} - \alpha_2\frac{1+\nu_2}{1-\nu_2}T_2(X,t)\right] = 0, \tag{3.106}$$

$$\frac{\partial^2 T_2(X,t)}{\partial X^2} = \frac{1}{a_2}\frac{\partial T_2(X,t)}{\partial t}, \qquad X \in (-L, L), \tag{3.107}$$

with the attached mechanical

$$U_2(\mp L, t) = \pm U_0 h_U(t), \tag{3.108}$$

and heat boundary conditions

$$\mp\lambda_2\frac{\partial T_2(\mp L,t)}{\partial X} + \alpha^T T_2(\mp L,t) = f(V_r)V_r P(t), \tag{3.109}$$

as well as zero initial conditions.

Normal stress occurring in the plate is defined by the following formula

$$\sigma_{XX}(X,t) = \frac{E_2}{1-2\nu_2}\left[\frac{1-\nu_2}{1+\nu_2}\frac{\partial U_2(X,t)}{\partial X} - \alpha_2 T_2(X,t)\right]. \tag{3.110}$$

In the above, the following notation is taken: E_2 is the Young's modulus of the plate; ν_2, λ_2, a_2, α_2, α^T are Poisson's ratio of the plate, thermal conductivity of the plate, thermal diffusivity, coefficient of thermal expansion of the plate, and heat transfer coefficient (from the wall to plate), respectively; and $P(t) = -\sigma_{XX}(\pm L, t)$ denotes contact pressure. Quantities m_1, m_2, k_1, P, F_1 are measured per unit of the contact surface $S = L_1 \times L_2$ of the moving rigid plate and the wall. Note that when the damper is neglected, the considered system is reduced to that with 1-dof with its natural frequency $\omega_{01} = \sqrt{k_1/m_1}$.

Integration of Equation (3.106) with Equation (3.110) and boundary conditions (3.108) gives the contact pressure $P(t) = -\sigma_{XX}(\pm L, t)$ of the form

$$P(t) = \frac{E_2(1-\nu_2)U_0 h_U(t)}{(1+\nu_2)(1-2\nu_2)L} + \frac{E_2\alpha_2}{(1-2\nu_2)L}\frac{1}{2}\int_{-L}^{L} T_2(\xi,t)d\xi. \tag{3.111}$$

Motion of the investigated system depends on both ratios F_{fr}/F_0 and ω_0'/ω_{01}. For various values of the F_{fr}/F_0 ratio the system moves with one or more sticks within half of the period of motion. It should be emphasized that the exact solution of this problem without tribologic processes and for the case with one stop and without stops has been already reported by Den Hartog [Den Hartog (1952)].

Dimensionless differential and integral equations. Let us introduce the following similarity coefficients

$$t_* = 1/\omega_{01}[\mathrm{s}], \qquad L_* = \frac{F_0}{k_1}[\mathrm{m}],$$

$$N_* = \frac{E_2(1-\nu_2)U_0 S}{(1+\nu_2)(1-2\nu_2)L}[\mathrm{N}], \qquad T_* = \frac{(1-\nu_2)U_0}{\alpha_2(1+\nu_2)L}[^0\mathrm{C}], \tag{3.112}$$

and the following dimensionless parameters

$$x = \frac{X}{L}, \tau = \frac{t}{t_*}, \theta = \frac{T_2}{T_*}, \quad p = \frac{PS}{N_*}, \quad z_n = \frac{Z_n}{L_*}, \quad n = 1,2, \quad \varepsilon = \frac{2N_* f_s}{F_0 S},$$

$$Bi = \frac{\alpha^T L}{\lambda_2}, \quad \tau_T = \frac{t_*}{t_T}, \quad l = \frac{L_*}{L}, \quad \gamma = \frac{E_2 \alpha_2 a_2}{(1-2\nu_2)\lambda_2}, \quad \Omega = \frac{\Omega_1}{\tau_T}, \quad \Omega_1 = \gamma l,$$

$$\omega_0 = \omega_0'/\omega_{01}, \quad \kappa = \frac{C}{2m_2\omega_{01}}, \quad \mu = \frac{m_2}{m_1}, \quad b_1 = B_1/L_*, \quad b_2 = B_2/L_*,$$

$$f(L_* t_*^{-1} \mathrm{v}_r) = f_s F(\mathrm{v}_r), \quad \mathrm{v}_r = \dot{z}_1 - \dot{z}_2, \quad dz_n/d\tau \equiv \dot{z}_n, \quad n = 1,2, \tag{3.113}$$

where $F(\mathrm{v}_r) = \mathrm{sgn}(\mathrm{v}_r)$; $t_T = L^2/a$ is the characteristic time of thermal inertia; Bi is the Biot number. Dimensionless parameter ε represents the friction force, γ governs the body heat extension, and parameter Ω is responsible for heat generation on the surface contact.

In the dimensionless form the considered mathematical model reads

$$\ddot{z}_1 + z_1 + f_{fr}(\mathrm{v}_r, z_1) = \sin(\omega_0 \tau), \qquad \mu \ddot{z}_2 - f_{fr}(\mathrm{v}_r, z_1) = 0, \tag{3.114}$$

$$\frac{\partial^2 \theta(x,\tau)}{\partial x^2} = \frac{1}{\tau_T}\frac{\partial \theta}{\partial \tau}, \qquad x \in (-1,1), \qquad \tau \in (0,\infty), \tag{3.115}$$

with the friction model

$$f_{fr}(\mathrm{v}_r, z_1) = \begin{cases} \varepsilon F(\mathrm{v}_r) p(\tau), & \mathrm{v}_r \neq 0, \text{ slip}, \\ \min(f_{st}, \varepsilon p(\tau)) \mathrm{sgn}(\sin(\omega_0\tau) - z_1), & \mathrm{v}_r = 0, \text{ stick}, \end{cases} \tag{3.116}$$

where

$$f_{st} = \frac{\mu}{1+\mu}\left|\sin(\omega_0\tau) - z_1\right|,$$

with the following boundary

$$\left[\frac{\partial \theta(x,\tau)}{\partial x} \mp Bi\,\theta(x,\tau)\right]_{x=\mp 1} = \mp q(\tau), \tag{3.117}$$

and initial conditions

$$\theta(x,0) = 0, \quad z_1(0) = 0, \quad \dot{z}_1(0) = 0, \quad z_2(0) = 0, \quad \dot{z}_2(0) = 0, \tag{3.118}$$

where

$$q(\tau) = \Omega F(\dot{z}_1 - \dot{z}_2) p(\tau)(\dot{z}_1 - \dot{z}_2), \tag{3.119}$$

$$p(\tau) = h_U(\tau) + \frac{1}{2}\int_{-1}^{1} \theta(\xi,\tau)d\xi. \tag{3.120}$$

In order to solve problems (3.115) and (3.117), the Laplace transform is applied with respect to time τ. The theorem on 'convolution' is used [Abramowitz, Stegun (1965)] to find an inverse transform. Finally, we get

$$p(\tau) = h_U(\tau) + \Omega_1 \int_0^{\tau} F(\dot{z}_1 - \dot{z}_2)p(\xi)(\dot{z}_1 - \dot{z}_2)\dot{G}_p(\tau-\xi)d\xi, \tag{3.121}$$

$$\theta(x,\tau) = \Omega_1 \int_0^{\tau} F(\dot{z}_1 - \dot{z}_2)p(\xi)(\dot{z}_1 - \dot{z}_2)\dot{G}_\theta(x,\tau-\xi)d\xi, \tag{3.122}$$

where

$$\{G_p(\tau), G_\theta(\pm1,\tau)\} = \frac{1}{\tau_T Bi} - \sum_{m=1}^{\infty} \frac{\{2Bi,\, 2\mu_m^2\}\exp(-\tau_T\mu_m^2\tau)}{\tau_T\mu_m^2(Bi(Bi+1)+\mu_m^2)}. \tag{3.123}$$

In the above $\mu_m (m = 1, 2, 3, \ldots)$ are the roots of the characteristic equation $\tan(\mu) = Bi/\mu$. The functions $G_p(\tau), G_\theta(\pm1,\tau)$ have the following asymptotic estimations

$$G_p(\tau) \approx \tau, \qquad G_\theta(\pm1,\tau) \approx 2\sqrt{\tau/\tau_T\pi}, \qquad \tau \to 0, \tag{3.124}$$

$$\{G_p(\tau), G_\theta(\pm1,\tau)\} \approx 1/(\tau_T Bi), \qquad \tau \to \infty. \tag{3.125}$$

Observe that the considered problem is reduced to the system of nonlinear differential equations (3.114) and (3.116), and the integral Equation (3.121) describing dimensionless velocities $\dot{z}_1$ and $\dot{z}_2$, and the dimensionless contact pressure $p(\tau)$. Dimensionless temperature $\theta(x,\tau)$ is governed by Equation (3.122).

Solution with a lack of heating. We take $\Omega_1 = 0$. Because $P(t) = N_*/S = P_*$, one gets

$$\ddot{z}_1 + z_1 + \varepsilon F(\dot{z}_1 - \dot{z}_2) = \sin(\omega_0\tau), \tag{3.126}$$

$$\ddot{z}_2 - \varepsilon\mu^{-1}F(\dot{z}_1 - \dot{z}_2) = 0. \tag{3.127}$$

Assuming that the system motion is close to harmonic, one may apply linearisation and use an equivalent viscous damping instead of dry friction. In other words, the equivalent damping is found comparing energy loss in a real A_T and in an equivalent A_C viscous system during the period $T_p = 2\pi/\omega_0'$.

Assuming the harmonic motion

$$z_1(\tau) = b_{11}\sin(\omega_0\tau) + b_{12}\cos(\omega_0\tau) = b_1\sin(\omega_0\tau + \varphi_1),$$

$$z_2(\tau) = b_{21}\sin(\omega_0\tau) + b_{22}\cos(\omega_0\tau) = b_2\sin(\omega_0\tau + \varphi_2), \tag{3.128}$$

the relative displacement is

$$z_1 - z_2 = d\sin(\omega_0\tau + \psi),$$

where

$$d = \sqrt{(b_{11} - b_{21})^2 + (b_{12} - b_{22})^2}. \tag{3.129}$$

Comparing works done by the two dampers one gets

$$A_T = \int_{t_1}^{t_1+T_0} 2f(\dot{Z}_1 - \dot{Z}_2)P_*(\dot{Z}_1 - \dot{Z}_2)dt,$$

$$A_C = \int_{t_1}^{t_1+T_0} C(\dot{Z}_1 - \dot{Z}_2)^2 dt, \tag{3.130}$$

whereas comparing two works $A_T = 8P_* f_s L_* d$ and $A_C = C\pi\omega_0' L_*^2 d^2$ one gets the following equivalent dimensional

$$C = \frac{8P_* f_s}{\pi\omega_0' L_* d} \tag{3.131}$$

or nondimensional damping coefficient of the form

$$\kappa = \frac{2\varepsilon}{\pi\mu\omega_0 d}. \tag{3.132}$$

Taking into account the equivalent damping coefficient, the governing Equations (3.126) and (3.127) assume the form

$$\ddot{z}_1 + z_1 + 2\mu\kappa(\dot{z}_1 - \dot{z}_2) = \sin(\omega_0\tau), \tag{3.133}$$

$$\ddot{z}_2 - 2\kappa(\dot{z}_1 - \dot{z}_2) = 0. \tag{3.134}$$

A solution to Equations (3.133) and (3.134) is given by (3.128), where the corresponding dimensionless amplitudes are

$$b_{11} = \frac{(1 - \omega_0^2(1+\mu))4\kappa^2 + \omega_0^2(1 - \omega_0^2)}{(1 - \omega_0^2(1+\mu))^2 4\kappa^2 + \omega_0^2(1 - \omega_0^2)^2}, \tag{3.135}$$

$$b_{12} = -\frac{2\mu\kappa\omega_0^3}{(1-\omega_0^2(1+\mu))^2 4\kappa^2 + \omega_0^2(1-\omega_0^2)^2}, \tag{3.136}$$

$$b_{21} = \frac{(1-\omega_0^2(1+\mu))4\kappa^2}{(1-\omega_0^2(1+\mu))^2 4\kappa^2 + \omega_0^2(1-\omega_0^2)^2}, \tag{3.137}$$

$$b_{22} = -\frac{2\omega_0\kappa(1-\omega_0^2)}{(1-\omega_0^2(1+\mu))^2 4\kappa^2 + \omega_0^2(1-\omega_0^2)^2}, \tag{3.138}$$

$$b_1 = \sqrt{\frac{\omega_0^2 + 4\kappa^2}{(1-\omega_0^2(1+\mu))^2 4\kappa^2 + \omega_0^2(1-\omega_0^2)^2}}, \tag{3.139}$$

$$b_2 = \sqrt{\frac{4\kappa^2}{(1-\omega_0^2(1+\mu))^2 4\kappa^2 + \omega_0^2(1-\omega_0^2)^2}}. \tag{3.140}$$

According to (3.129) the dimensionless amplitude d is

$$d = \sqrt{\frac{\omega_0^2}{(1-\omega_0^2(1+\mu))^2 4\kappa^2 + \omega_0^2(1-\omega_0^2)^2}}. \tag{3.141}$$

According to formula (3.141), d depends on κ, and the equivalent damping depends on d (see (3.132)). Solving (3.132), (3.141) one gets

$$\kappa = \frac{\omega_0|1-\omega_0^2|}{2}\sqrt{\frac{\omega_+\omega_-}{(\omega_0^2-\omega_+)(\omega_- - \omega_0^2)}}. \tag{3.142}$$

The amplitudes are

$$b_1 = \frac{1}{\omega_0|1-\omega_0^2|}\sqrt{\omega_0^2\left(1-\left(1+\frac{2}{\mu}\right)\left(\frac{4\varepsilon}{\pi}\right)^2\right)+\frac{2}{\mu}\left(\frac{4\varepsilon}{\pi}\right)^2},$$

$$b_2 = \frac{4\varepsilon}{\pi\mu\omega_0^2} \tag{3.143}$$

where

$$\omega_\pm = \frac{\varepsilon}{(1+\mu)(\varepsilon \pm \varepsilon_0)}, \qquad \epsilon_0 = \frac{\pi}{4}\left(1-\frac{1}{1+\mu}\right). \tag{3.144}$$

For $\varepsilon < \epsilon_0$ and $\sqrt{\omega_+} < \omega_0$, dry friction does not bound resonance amplitude, and for $\omega_0 \to 1$, $b_1 \to \infty$. In the case $\epsilon_0 < \varepsilon < \pi/4$ and for $\sqrt{\omega_+} < \omega_0 < \sqrt{\omega_-}$ dry friction does not bound resonance amplitude either and for $\omega_0 \to 1$, $b_1 \to \infty$. When $\varepsilon > \pi/4$, we get the frequency interval $\sqrt{\omega_+} < \omega_0 < \sqrt{\omega_-}$, where the resonance frequency $\omega_0 = 1$ does not appear.

For $\mu \to \infty$ ($m_2 \to \infty$, $\epsilon_0 = \pi/4$) we obtain the case of 1-dof vibration with friction considered by Den Hartog [Den Hartog (1952)], where

$$b_1 = \frac{1}{|1-\omega_0^2|}\sqrt{1-\left(\frac{4\varepsilon}{\pi}\right)^2}. \tag{3.145}$$

Observe that for $\varepsilon < \pi/4$ (b_1 is a real number) one gets: $\omega_0 \to 1$, $b_1 \to \infty$.

Note that the velocity of body m_2 is governed by

$$\dot{z}_2(\tau) = \frac{\varepsilon}{\mu}\tau + C_+, \qquad \text{if } \dot{z}_1 - \dot{z}_2 > 0, \tag{3.146}$$

$$\dot{z}_2(\tau) = -\frac{\varepsilon}{\mu}\tau + C_-, \qquad \text{if } \dot{z}_1 - \dot{z}_2 < 0. \tag{3.147}$$

Constants C_+, C_- are different for all intervals of motion. Assuming that only sliding occurs, the work done within one period by the damping force $2f_sP_*\text{sgn}(\dot{Z}_1 - \dot{Z}_2)$ is [Den Hartog (1952)]:

$$A_T = 4F_0L_*b_1\varepsilon\sqrt{1-\frac{\pi^2}{4}\left(\frac{\varepsilon}{\mu\omega_0^2 b_1}\right)^2}. \tag{3.148}$$

An optimal value of friction is obtained from the following equation $\partial A_T/\partial\varepsilon = 0$, if $(\partial^2 A_T/\partial\varepsilon^2\big|_{\varepsilon=\varepsilon_{opt}} < 0)$.

The optimal dimensionless friction is

$$\varepsilon_{opt} = \frac{\sqrt{2}}{\pi}\mu\omega_0^2 b_1, \tag{3.149}$$

and

$$A_{T\,opt} = \frac{4}{\pi}F_0L_*b_1^2\mu\omega_0^2. \tag{3.150}$$

The obtained value corresponds to the maximally damped fundamental system. However, in this case the amplitude b_1 should be known (say, from an experiment).

3.4.2 *On the heat transfer influence on dynamical damper of self-vibrations*

Numerical analysis of the considered problem is carried out using both numerical and analytical computations. The latter ones include asymptotic estimation (3.124). The function $\text{sgn}(\text{v}_r)$ has been approximated in the following way (2.58).

If heat is not generated by friction ($\gamma = 0$), then the contact pressure $p(\tau) = h_U(\tau)$. Let us assume that $h_U(\tau) = H(\tau)$, where $H(\cdot)$ is the Heaviside step function ($H(\tau) = 1, \quad \tau > 0, H(\tau) = 0, \quad \tau < 0$). The previous analysis (part 3.4.1) indicates that resonance occurs in the system. Taking $\mu = 0.5$, $\varepsilon = 0.5$,

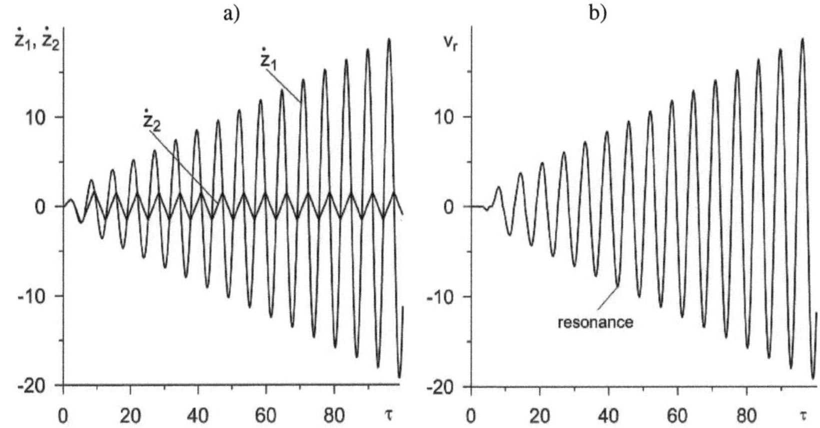

Fig. 3.21: The dimensionless velocity $\dot{z}_1(\tau)$ of the fundamental body and velocity $\dot{z}_2(\tau)$ of the damper (a) versus dimensionless time τ during resonance ($\omega_0 = 1$); dimensionless dependence of sliding velocity $v_r(\tau) = \dot{z}_1(\tau) - \dot{z}_2(\tau)$ versus time τ (b) ($\Omega_1 = 0$, $\mu = 0.5$, $\varepsilon = 0.5$, $\omega_0 = 1$).

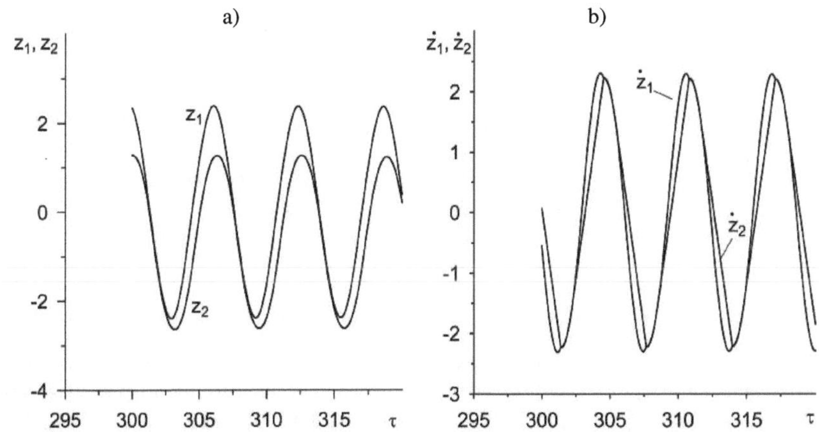

Fig. 3.22: The dependence of the dimensionless displacements z_1 of the fundamental body and the displacements z_2 of the damper (a) the dimensionless velocity $\dot{z}_1(\tau)$ of the fundamental body and the velocity $\dot{z}_2(\tau)$ of the damper (b) versus dimensionless time τ for $\tau \in (300, 320)$, taking into account heat generation ($\Omega_1 = 0.1$, $\mu = 0.5$, $\varepsilon = 0.5$, $Bi = 1$, $\tau_T = 0.1$, $\omega_0 = 1$).

and using (3.144) one finds $\epsilon_0 = 0.26$, $\omega_+ = 0.44$, $\omega_- = 1.4$. We have $\varepsilon > \epsilon_0$, $\sqrt{\omega_+} < \omega_0 < \sqrt{\omega_-}$ and for $\omega_0 = 1$ the system is in resonance. In Fig. 3.21a for a lack of heat extension ($\Omega_1 = 0$), time histories of both dimensionless velocities $\dot{z}_1$ and $\dot{z}_2$, and dimensionless relative velocity (Fig. 3.21b) $v_r = \dot{z}_1 - \dot{z}_2$ are reported.

For a general case, numerical analysis of the considered problem (differential Equations (3.114), (3.115) and integral Equation (3.121)) is carried out using the Runge–Kutta method and the method of quadrature with estimations (3.124). Temperature on the contact surface is given in formula (3.122).

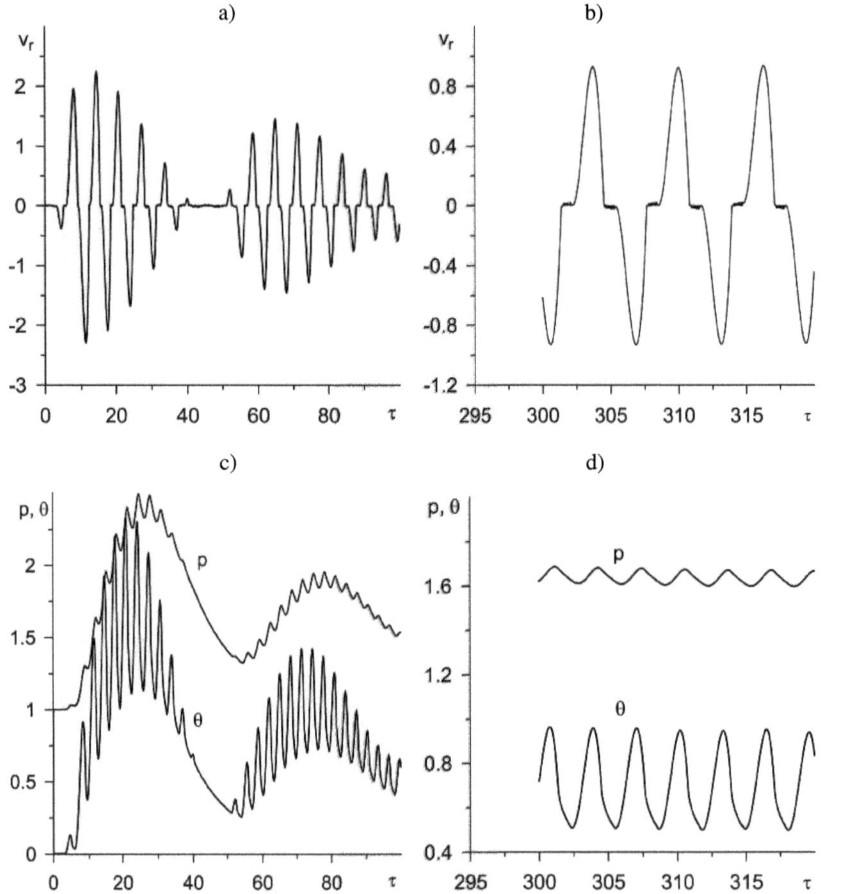

Fig. 3.23: The dimensionless sliding velocity $v_r(\tau) = \dot{z}_1(\tau) - \dot{z}_2(\tau)$ versus time τ for $\tau \in (0, 100)$(a) and for $\tau \in (300, 320)$(b) taking into account heat generation. Time histories of dimensionless contact pressure $p(\tau)$ and contact surface temperature $\theta(\tau)$ for $\tau \in (0, 100)$(c) and for $\tau \in (300, 320)$(d), taking into account heat generation ($\Omega_1 = 0.1$, $\mu = 0.5$, $\varepsilon = 0.5$, $Bi = 1$, $\tau_T = 0.1$, $\omega_0 = 1$).

Numerical computations are carried out for various values of the parameters μ, Ω_1. Figures 3.22 to 3.24 show results of the numerical analysis for $\Omega_1 = 0.1$ and $Bi = 1$, $\tau_T = 0.1$. In Fig. 3.22a dimensionless displacements z_1 of the body with mass m_1 and z_2 of the damper with mass m_2 versus time τ are shown (the same is done for the velocities in Fig. 3.22b). Figure 3.23 illustrates the dimensionless relative velocities v_r of two bodies versus dimensionless time τ. The system oscillations are out of resonance and they reach periodic attractor (Figs. 3.22, 3.23b) with the period of $T_p = 2\pi/\omega_0 = 2\pi$, whereas the damper undergoes the stick-slip oscillations (Fig. 3.23b). Evolutions of the dimensionless contact pressure $p(\tau)$ and the temperature $\theta(\tau)$ on the contacting surface are shown in Fig. 3.23.

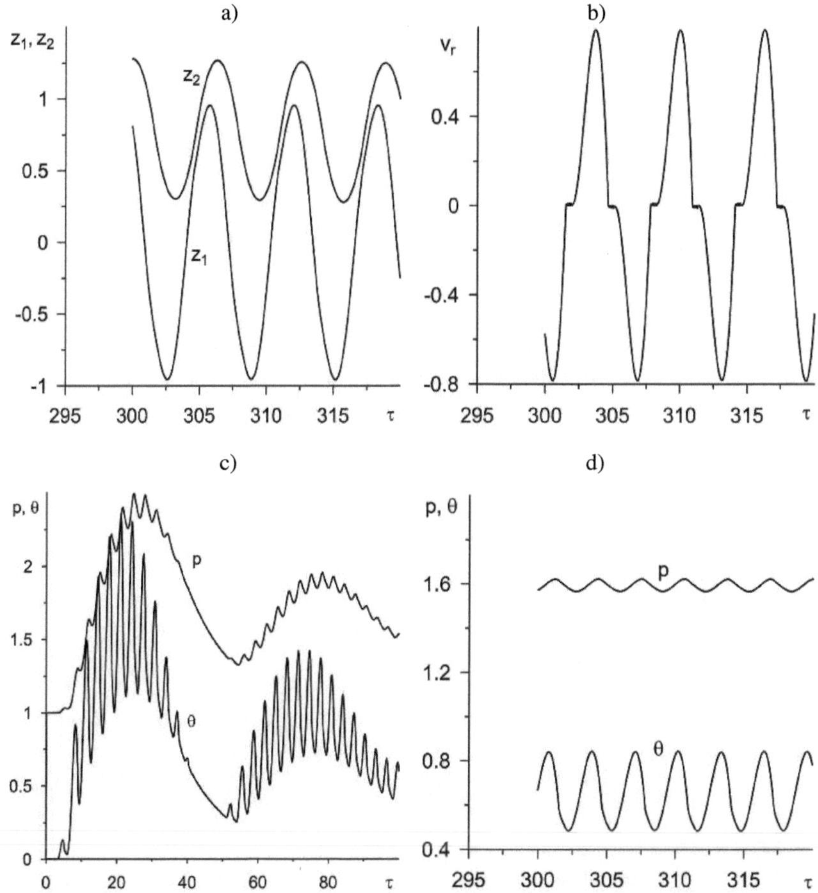

Fig. 3.24: The dependence of the dimensionless displacements z_1 of the fundamental body and the displacements z_2 of the damper (a) the dimensionless dependence of the sliding velocity $\mathrm{v}_r(\tau) = \dot{z}_1(\tau) - \dot{z}_2(\tau)$ (b) versus dimensionless time τ for $\tau \in (300, 320)$, taking into account heat generation. Time histories of dimensionless contact pressure $p(\tau)$ and contact surface temperature $\theta(\tau)$ for $\tau \in (0, 100)$(c) and for $\tau \in (300, 320)$(d), taking into account heat generation ($\Omega_1 = 0.1$, $\mu = 2$, $\varepsilon = 0.5$, $Bi = 1$, $\tau_T = 0.1$, $\omega_0 = 1$).

Both mentioned characteristics change periodically with the dimensionless period of $T_p = \pi$. The same period has the variable $|\mathrm{v}_r|$ occurring in (3.119), which governs heat generation on the sliding surface.

In order to investigate how the damper mass influences the system motion we also investigated our system for $\mu = 2$ ($m_2 = 2m_1$). In this case one gets $\epsilon_0 = 0.52$, $\omega_+ = 0.16$. We have $\varepsilon < \epsilon_0$, $\sqrt{\omega_+} < \omega_0$ and for $\omega_0 = 1$ the system is in resonance assuming that the heat extension is omitted. In Fig. 3.24 time evolutions of the contact characteristics are reported. In Fig. 3.24a the dependence of dimensionless displacement z_1 and z_2 versus dimensionless time is shown, whereas

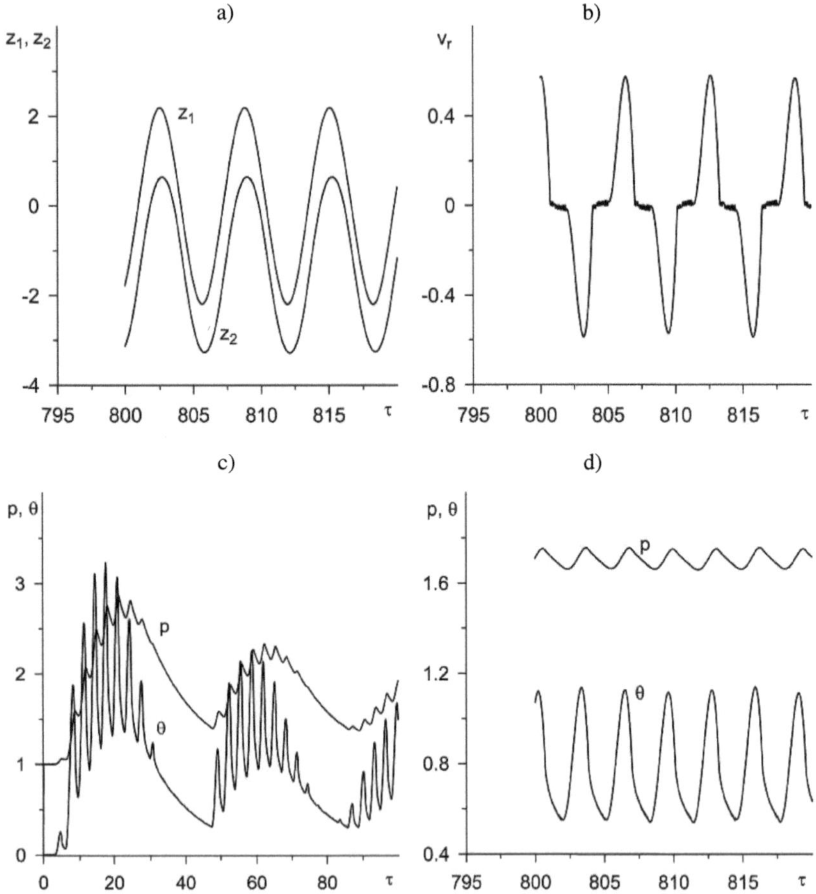

Fig. 3.25: The dependence of the dimensionless displacements z_1 of the fundamental body and the displacements z_2 of the damper (a) the dimensionless dependence of the sliding velocity $v_r(\tau) = \dot{z}_1(\tau) - \dot{z}_2(\tau)$ (b) versus dimensionless time τ for $\tau \in (800, 820)$, taking into account heat generation. Time histories of the dimensionless contact pressure $p(\tau)$ and the contact surface temperature $\theta(\tau)$ for $\tau \in (0, 100)$(c) and for $\tau \in (800, 820)$(d), taking into account heat generation ($\Omega_1 = 0.2$, $\mu = 0.5$, $\varepsilon = 0.5$, $Bi = 1$, $\tau_T = 0.1$, $\omega_0 = 1$).

Fig. 3.24b illustrates the corresponding relative velocity. Figures 3.24 show dimensionless contact pressure and temperature versus dimensionless time. Note that an increase of parameter μ causes a decrease of the contact time (see Figs. 3.23b and 3.24b) and vibration amplitude decreases (see Figs. 3.22a and 3.24a), but the temperature amplitude in the periodic state increases (see Figs. 3.23d and 3.24d).

A numerical analysis of results for $\Omega_1 = 0.2$, $\varepsilon = 0.5$, $Bi = 1$, and $\tau_T = 0.1$ is illustrated in Fig. 3.25. An increase of coefficient Ω_1 causes an increase of time τ_r (the so-called time of passive regulation), when the trajectory achieves periodic motion. For $\Omega_1 = 0.1$ time $\tau_r \approx 250$, whereas for $\Omega_1 = 0.2$ time

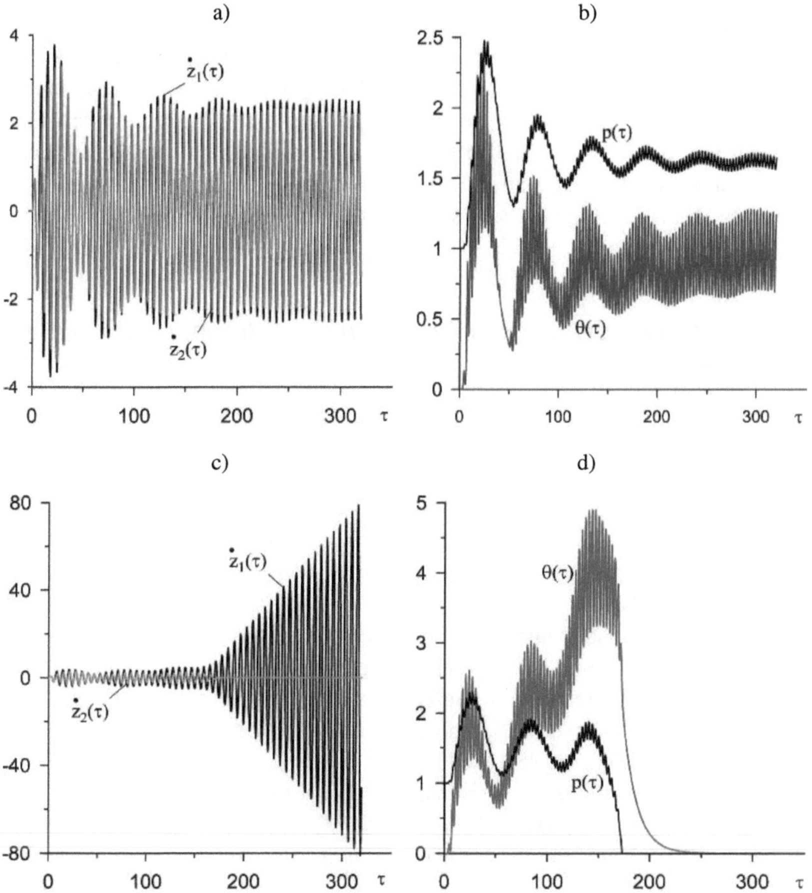

Fig. 3.26: Time evolution of nondimensional velocity $\dot{z}_1(\tau)$ i $\dot{z}_2(\tau)$ (a); nondimensional contact pressure $p(\tau)$ and contact surface temperature $\theta(\tau)$ (b) taking into account heat generation and wear ($k^w = 0.001$). Time evolution of nondimensional velocity $\dot{z}_1(\tau)$ i $\dot{z}_2(\tau)$ (c); nondimensional contact pressure $p(\tau)$ and contact surface temperature $\theta(\tau)$ (d) taking into account wear and heat generation ($k^w = 0.01$).

$\tau_r \approx 500$. In addition, the time of both bodies being in contact also increases (see Figs. 3.23b and 3.25b). An increase of Ω_1 causes a decrease of the relative velocity amplitude (see Figs. 3.23b and 3.25b).

A dynamic 2-dof damper with dry friction and heat generation has been modelled mathematically. The proposed method of solution may also be applied in modelling of any other nonlinear problem of dynamics of thermoelastic contacting bodies. A series of practical results regarding kinetics of the main system and of the dynamic damper are formulated as a result of the analysis of various contact characteristics (contact pressure, temperature on the contacting surface).

It should be emphasized that the dynamic damper with dry friction may not achieve the expected properties. As we have shown, heat generation on the contacting surface between the damper and the oscillating body as well as heat expansion eliminate, for certain parameters, resonance phenomena. The real system, in certain conditions, behaves as a self-regulating one; that is, it controls achievement of an optimal contacting pressure. The thermoelastic parallelepiped extends itself according to the conditions of both sliding velocity and heat transfer.

3.4.3 Nonlinear dynamics of a dynamical damper with wear processes

In this work analysis of influence of the wear coefficient k^w on the system (with dynamic damper and friction) motion is carried out. The following nondimensional parameters are introduced: $\mu = m_2/m_1 = 0.5$, and $\varepsilon = 0.5$. For dimensionless frequency $\omega_0 = \omega_0'/\omega_{01} = 1$ ($\omega_{01} = \sqrt{k_1/m_1}$) a resonance occurs [Den Hartog (1952)]. However, taking into account heat plate extension ($\Omega_1 = \gamma l = 0.1$, $t_T = L^2/a_2 = 0.1$, $Bi = \alpha^T L/\lambda_2 = 1$, where $\gamma = E_2\alpha_2 a_2 \lambda_2^{-1}/(1-2\nu_2)$, $l = L_*/L$) yields disappearance of the resonance. The computational results are shown in Figs. 3.26–3.27 for various wear parameters k^w. In Fig. 3.26a the dependence of nondimensional velocity $\dot{z}_1(\tau)$ i $\dot{z}_2(\tau)$ ($z_n(\tau) = Z_n/L_*$) versus nondimensional time $\tau = t/t_*$ for nondimensional wear coefficient $k^w = 0.001$ ($k_w = K^w N_* L/(U_0 S)$) are reported.

The dependencies of the nondimensional contact pressure $p(\tau)$ ($p = PS/N_*$) and contact surface temperature $\theta(\tau)$ ($\theta = T/T_*$) versus nondimensional time τ are shown in Figs. 3.26b for the wear coefficient $k^w = 0.001$.

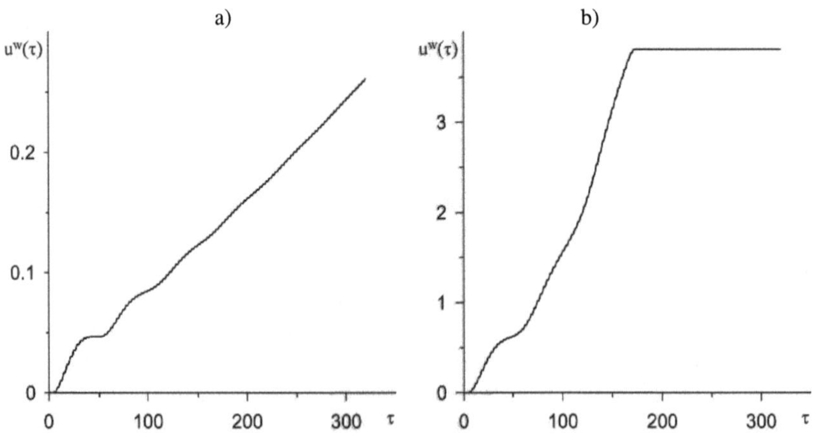

Fig. 3.27: Time evolution of nondimensional wear $u^w(\tau)$ taking into account the tribologic processes: (a) – $k^w = 0.001$, (b) – $k^w = 0.01$.

In addition, for $k^w = 0.001$ in Fig. 3.27a the function of nondimensional wear $u^w(\tau)$ ($u^w = U^w/U_0$) against nondimensional time τ is exhibited.

For wear coefficient $k^w = 0.01$ the appropriate contact parameter dependencies are shown in Figs. 3.26c and 3.27b

Observe that if wear is less than plate heat extension the time of body contact is infinite (Fig. 3.26b), wear increases almost linearly (Fig. 3.27a), temperature increases, and compensates wear (Fig. 3.26b). However, if plate wear is larger than its heat extension the time of body contact is limited (Fig. 3.26d), and the next resonance phenomenon occurs (Fig. 3.26c).

Chapter 4
Contact Characteristics During Braking Process

A braking process belongs to nonstationary frictional behaviour, which is associated with frictional heat generation. Note that usually temperatures on the rubbing surfaces achieve large values and play a key role while formulating their wear-resistance criterions.

In brake systems metallic–ceramic and mineral–ceramic materials are widely used. The mentioned materials are wear- and heat-resistant and they counteract both chemical bond (frictional contact) and seizing. A frictional cover strap consists of a thin metallic–ceramic layer put on a copper or steel base. During a braking process, the frictional cover strap is pressed to a body surface (a brake drum, a friction disk, or a wheel rim). The action of friction forces causes conversion of kinematic energy of the moving bodies into heat energy. The brake elements are overheated and the ability of working conditions of the cover strap are decreased; that is, the wear increases and the value of the friction coefficient is decreased. Owing to the described phenomenon, the braking process is extended and hence a failure may occur. On the basis of the above considerations one may conclude that heating constraint estimation belongs to one of the most important problems referring to the computation of strength and working time of brakes.

Owing to recent computations of brakes with respect to heating, the contact characteristics can be defined with a priori given tolerance (braking time, temperature, and wear). The mentioned computational techniques are based on the solution of the system of equations governing Heat Frictional Dynamics (HFD) [Chichinadze et al. (1979), Chichinadze (1995)]. An analytical solution of HFD for two uniform homogeneous half-spaces transferring heat is proposed by [Olesiak, Pyryev et al. (1997)], whereas a numerical one is given in references [Yevtushenko, Pyryev (1997), (1998)]. In reference [Pyryev (2004)] a solution of the HFD equations for a three-layered tribomechanical system consisting of three different layers is proposed. The proposed model takes into account the thickness of frictional elements. It also provides knowledge of the associated heating and wear processes of the tribomechanical system composed of a metallic–ceramic cover strap and metallic friction disc.

J. Awrejcewicz and Yu. Pyryev, *Nonsmooth Dynamics of Contacting Thermoelastic Bodies*, DOI: 10.1007/978-0-387-09653-7_4, © Springer Science+Business Media, LLC 2009

A brake temperature model is built [Chichinadze et al. (1979), Chichinadze (1995)], in which a maximum contact temperature is the sum of the average temperature of the real frictional surface and the 'flash' temperature. The average temperature is obtained through analysis of a solution to the one-dimensional heat transfer problem for two contacting bodies assuming that temperature time variation is proportional to an increase of the heat volume. The flash temperature is derived via a solution of the problem dealing with a sliding rod (arbor) along a half-space surface assuming that heat transfer intensity is constant in the contact zone. Note that the application of one-dimensional models to solve the heat frictional problem is reasonable if in typical heat transfer conditions [Chichinadze et al. (1979)], the braking time t_s is less than $10b_l/a_l$ ($l = 1, 2$), where b_l, a_l are the characteristic dimensions of the friction pair and the temperature wear coefficient, respectively.

In [Olesiak et al. (1997)] a solution of the heat transfer problem for two semi-infinite bodies without any limitations on heat stream density is proposed. The cases of constant friction coefficients [Pyryev (2004)] as well as their linear dependence on temperature are accounted for [Olesiak et al. (1997)]. [Yevtushenko, Pyryev (1999b)] found a solution of the heat problem related to nonlinear dependence between both friction and temperature coefficients.

4.1. Contact characteristics of three-layer brake models

Let us now consider the problem of friction of a lateral disk brake surface rubbing on a friction surface of a two-layer cover strap subject to external load $P(t)$ (Fig. 4.1). The top layer of thickness H_1 (body 1) slides at constant velocity $V(t)$ on the layer of thickness H_2 (body 2) along axis Y of the Cartesian coordinates XYZ

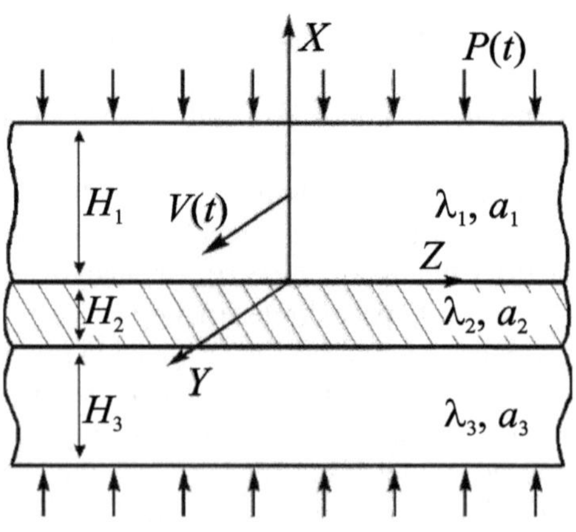

Fig. 4.1: Scheme of a three-layer brake system.

with the origin situated on the friction surface. Body 2 is attached to nonmovable layer of thickness H_3 (body 3).

Owing to the action of frictional surface on the contact surface $X = 0$, body 1 is braked with a simultaneous generation of heat and wear. The sum of heat stream directions into the first and second layers is equal to the power density of friction forces Q. It is assumed that between the second and third bodies an ideal heat contact occurs. The external surface of the layer system keeps zero value temperature of the external environment.

Our attention if focused on the derivation of velocity V, temperature T_n, $n = 1, 2, 3$, and the wear amount U^w in an arbitrary time instant $0 \leq t \leq t_s$, where t_s is the braking time.

4.1.1 Mathematical modelling

A decrease of velocity $V(t)$ from the initial value $V(0) = V_0$ to zero during braking is defined by the equation

$$m\frac{d}{dt}V(t) = -F_{fr}, \qquad V(0) = V_0, \qquad t \in (0, t_s), \tag{4.1}$$

where M is the mass per contact surface unit, whereas F_{fr} denotes friction force per surface unit.

Owing to Amonton's law $F_{fr} = f(T)P(t)$, where $f(T) = f_0 f^*(T)$ is the friction coefficient depending on the temperature. It is worth noting that functions $f^*(T)$ can be (i) decreasing for metallic–ceramic joints; (ii) increased ones with a minimum or maximum for critical material; or (iii) constant for frictional metallo–plastic materials. In a general case, a dependence of friction coefficient on temperature $f^*(T)$ has the form

$$f^*(T) = f_1 + \frac{f_2}{(f_3(T - T_f))^2 + 1}. \tag{4.2}$$

In the case of rubbing of the metallic–ceramic pair and anti-friction cast iron, the following function is used [Chichinadze et al. (1979)],

$$f^*(T) = \exp(-\lambda_f T), \tag{4.3}$$

where f_0, λ_f, T_f, f_j $(j = 1, 2, 3)$ are the coefficients defined experimentally [Chichinadze et al. (1979)]. The curves approximating the experimental results are shown in Fig. 4.2a.

Curve 1 corresponds to the case $f^*(T) = 1$ (metal–artificial material); curve 2 corresponds to the values: $f_1 = -0.552$, $f_2 = 1.83$, $f_3 = 2.05 \cdot 10^{-3\circ}\mathrm{C}^{-1}$, $T_f = 200°\mathrm{C}$ (plastic); and curve 3 corresponds to $f_1 = 0.1$, $f_2 = 0.845$, $f_3 = 3.02 \cdot 10^{-3\circ}\mathrm{C}^{-1}$, $f_3 = 3.02 \cdot 10^{-3\circ}\mathrm{C}^{-1}$, $T_f = -100°\mathrm{C}$ (ceramics).

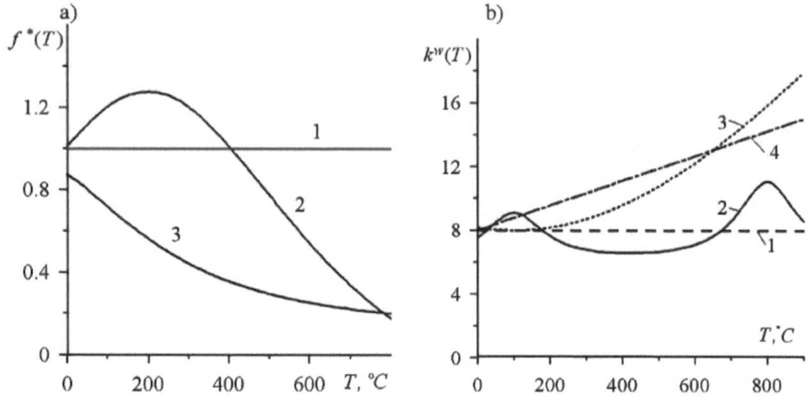

Fig. 4.2: Dependence of friction coefficient $f^*(T)$ (a) and wear $k^w(T)$ (b) on the contact temperature T.

The following pressure variation in time is assumed

$$P(t) = P_0[1 - \exp(-t/t_m)][1 + B_1 \sin(B_2 t/t_m)], \tag{4.4}$$

where t_m is the parameter responsible for a loading from 0 to P_0; B_1 is the amplitude of periodically driven vibrations; B_2 denotes the dimensionless period of vibrations.

The temperature T_n $(n = 1, 2, 3)$ is found using the following nonstationary heat transfer equation of the form

$$\frac{\partial^2}{\partial X^2} T_n(X,t) = \frac{1}{a_n} \frac{\partial}{\partial t} T_n(X,t), \qquad 0 \le t \le t_s, \tag{4.5}$$

$$(0 < X < H_1 \ \text{for} \ n = 1; \quad -H_2 < X < 0 \ \text{for} \ n = 2;$$

$$- H_2 - H_3 < X < -X_2 \ \text{for} \ n = 3),$$

with the following boundary conditions

$$\lambda_2 \frac{\partial}{\partial X} T_2(0,t) - \lambda_1 \frac{\partial}{\partial X} T_1(0,t) = Q(t), \qquad T_1(0,t) = T_2(0,t),$$

$$T_1(H_1,t) = 0, \qquad T_2(-H_2,t) = T_3(-H_2,t),$$

$$\lambda_2 \frac{\partial}{\partial X} T_2(-H_2,t) = \lambda_3 \frac{\partial}{\partial X} T_3(-H_2,t),$$

$$T_3(-H_2 - H_3, t) = 0, \qquad 0 \le t \le t_s, \tag{4.6}$$

and initial conditions

$$T_n(X,0) = 0, \qquad n = 1,2,3,$$

where λ_n are the heat transfer coefficients, and a_n are the temperature compensating coefficients.

The heat stream density Q generated by friction forces on the surface between different materials is equal to friction work [Aleksandrov, Annakulova (1990)]

$$Q(t) = (1-\eta)f(T)V(t)P(t), \qquad 0 \le t \le t_s, \tag{4.7}$$

where η is the part of friction work devoted to wear.

The following friction wear law is applied [Archard (1959), Goryacheva (1988)],

$$U^w(t) = \int_0^t f(T)V(t^*)P(t^*)K^w(T)dt^*, \tag{4.8}$$

where $U^w(t) = U_1^w(t) + U_2^w(t)$; U_n^w is the displacement of the surface working along the X-axis; $K^w = K_1^w + K_2^w$ is the wear coefficient; and $T(t) = T_1(0,t) = T_2(0,t)$ denotes the contact temperature.

Owing to the work of [Chichinadze et al. (1979)], in general one may take $K^w(T) = K_0^w k^w(T)$, where

$$k^w(T) = d_0 + d_1T(t) + \frac{d_2}{[d_3(T(t)-T_1^w)]^2+1} + \frac{d_4}{[d_5(T(t)-T_2^w)]^2+1} \tag{4.9}$$

and T_n^w, $n = 1,2$, d_j, $j = 0,1,\dots,5$ are the coefficients found experimentally. The curves approximating experimental results are shown in Fig. 4.2b. Curve 1 corresponds to the case $k^w(T) = 0$; curve 2 corresponds to the values: $d_0 = 6$, $d_1 = 0$, $d_2 = 3$, $d_3 = 10^{-2\circ}\mathrm{C}^{-1}$, $d_4 = 5$, $d_5 = 10^{-2\circ}\mathrm{C}^{-1}$, $T_1^w = 100°\mathrm{C}$, $T_2^w = 800°\mathrm{C}$; curve 3 to $d_0 = 80$, $d_1 = 0$, $d_2 = -72$, $d_3 = 5\cdot 10^{-4\circ}\mathrm{C}^{-1}$, $d_4 = 0$, $T_1^w = 100°\mathrm{C}$; curve 4 to $d_0 = 8$, $d_1 = 0.78\cdot 10^{-2\circ}\mathrm{C}^{-1}$, $d_2 = 0$, $d_4 = 0$.

Let us introduce the following dimensionless parameters

$$\tau = t/t_m, \qquad \tau_s = t_s/t_m, \qquad \delta = t_0/t_m, \qquad q = Q/Q_0, \qquad q_0 = Q_0R/T_0,$$

$$r_n = R_n/R, \qquad \eta_n = r_n\sqrt{Fo_n}, \qquad Fo_n = a_nt_m/H_n^2, \qquad n = 1,2,3, \tag{4.10}$$

and the following dimensionless sought function

$$\mathrm{v} = V/V_0, \qquad p = P/P_0, \qquad \theta = T/T_0,$$

$$u^w = U^w/U_0, \qquad \theta_n = T_n/T_0, \qquad n = 1,2,3, \tag{4.11}$$

where $R = R_1 + R_2 + R_3$, $R_n = H_n/\lambda_n$, $n = 1, 2, 3$,

$$Q_0 = (1-\eta) f_0 V_0 P_0, \qquad t_0 = mV_0/(f_0 P_0) = 2W/Q_0,$$

$$U_0 = f_0 t_m V_0 P_0 K_0^w, \qquad T_0 = Q_0 \sqrt{a_1 a_2 t_m/\pi}/(\lambda_2\sqrt{a_1} + \lambda_1\sqrt{a_2}). \quad (4.12)$$

In the above t_0 denotes braking time in the case of constant load action P_0, and $W = mV_0^2/2$ denotes the initial kinetic energy per unit contact surface, whereas R_n is the heat resistance of the layers.

Applying the Laplace transform with respect to time t_0 equations of motion (4.1) and boundary conditions (4.5) and (4.6), the following system of HDF equations is obtained.

$$\mathrm{v}(\tau) = 1 - \delta^{-1}\int_0^{\tau} f^*(T_0\theta) p(\tau^*) d\tau^*, \quad (4.13)$$

$$p(\tau) = [1 - \exp(-\tau)][1 + B_1 \sin(B_2\tau)], \quad \theta(\tau) = \theta_n(0,\tau), \quad n = 1, 2, \quad (4.14)$$

$$\theta_n(X,\tau) = \int_0^{\tau} G_n(X, \tau - \tau^*) f^*(T_0\theta) \mathrm{v}(\tau^*) p(\tau^*) d\tau^*, \quad (4.15)$$

$$u^w(\tau) = \int_0^{\tau} f^*(T_0\theta) \mathrm{v}(\tau^*) p(\tau^*) k^w(T_0\theta) d\tau^*, \quad (4.16)$$

$$G_n(0,\tau) = G(\tau) = 2q_0 \sum_{m=1}^{\infty} \frac{\eta_1 \mathrm{tg}\xi_{1m}(\eta_2 \mathrm{tg}\xi_{2m} + \eta_3 \mathrm{tg}\xi_{3m})}{D(\mu_m)} e^{-\mu_m^2 \tau} \quad (4.17)$$

$$\begin{aligned} D(\mu_m) = {} & tg\xi_{1m} tg\xi_{3m}(\eta_3/\sqrt{Fo_1} + \eta_1/\sqrt{Fo_3} + \eta_1\eta_3/(\eta_2\sqrt{Fo_2})) \\ & + tg\xi_{2m} tg\xi_{3m}(\eta_2/\sqrt{Fo_3} + \eta_3/\sqrt{Fo_2} + \eta_1\eta_3/(\eta_2\sqrt{Fo_1})) \\ & + tg\xi_{1m} tg\xi_{2m}(\eta_1/\sqrt{Fo_2} + \eta_2/\sqrt{Fo_1} + \eta_1\eta_3/(\eta_2\sqrt{Fo_3})) - 1, \end{aligned}$$

where $\xi_{nn} = \mu_m/\sqrt{Fo_n}$, where μ_m are the roots of the following characteristic equation

$$\begin{aligned} & \eta_3 \cos\xi_1 \cos\xi_2 \sin\xi_3 + \eta_1 \cos\xi_2 \cos\xi_3 \sin\xi_1 \\ & \quad + \eta_2 \cos\xi_1 \cos\xi_3 \sin\xi_2 - \eta_1\eta_3/\eta_2 \sin\xi_1 \sin\xi_2 \sin\xi_3 = 0, \end{aligned} \quad (4.18)$$

$$\xi_n = \mu/\sqrt{Fo_n}, \qquad n = 1, 2, 3.$$

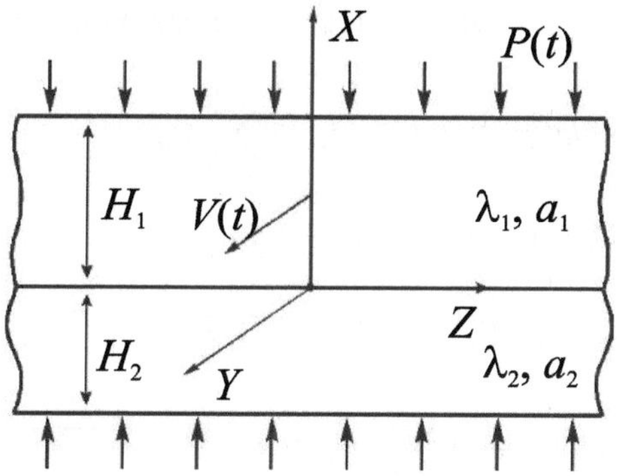

Fig. 4.3: Scheme of two-layer braking system.

For the tribological system consisting of two layers (Fig. 4.3), the system of Equations (4.13)–(4.16) has a similar form; that is, the difference is only manifested through another function

$$G_n(X,\tau) = 2\sum_{m=1}^{\infty} \frac{\cos(\xi^x_{nm}) - \text{cotan}(\xi_{nm})\sin(\xi^x_{nm})}{D(\mu_m)} e^{-\mu_m^2\tau}, \tag{4.19}$$

where

$$D(\mu_m) = \frac{\lambda^{(1)}}{Fo_1^2} + \frac{\lambda^{(2)}}{Fo_2^2} - (\lambda^{(1)} + \lambda^{(2)})\frac{\text{cotan}(\xi_{1\text{m}})\text{cotan}(\xi_{2\text{m}})}{\mu_m^2}, \tag{4.20}$$

$$\xi^x_{nm} = (\mu_m X)/(\sqrt{Fo_n}H_n), \qquad \xi_{nm} = \mu_m/\sqrt{Fo_n},$$

$$\lambda^{(n)} = \lambda_n T_0/(H_n(1-\eta)Q_0) \qquad n = 1, 2.$$

In the above, μ_m denotes the roots of the characteristic equation associated with two layers of the form

$$\lambda^{(1)}\text{cotan}(\xi_1) + \lambda^{(2)}\text{cotan}(\xi_2) = 0. \tag{4.21}$$

For small values of time τ or for large values of thickness of layer 1 and 2, the following value of two contacting half-spaces is obtained

$$G_n(X,\tau) = \tau^{-1/2}\exp(-X^2/(4\tau t_m a_n)). \tag{4.22}$$

A solution to the system of nonlinear integral equations (4.13)–(4.16) is found numerically using iterations with an account of trapezoidal approximation [Abramowitz and Stegun (1965)].

In time instants $\tau = \tau_i = ih$, $i = 1, 2, \ldots$ (h, step) one gets

$$\mathrm{v}_i^{(j)} = 1 - \delta^{-1}\left[0.5h\Phi_i^{(j-1)} + (1-\delta_{i1})h\sum_{m=1}^{i-1}\Phi_m^{(j-1)}\right], \tag{4.23}$$

$$\begin{aligned}\theta_i^{(j)} &= (2/3)\sqrt{h}[2\Phi_i^{(j-1)}\mathrm{v}_i^{(j-1)} + \Phi_{i-1}^{(j-1)}\mathrm{v}_{i-1}^{(j-1)}]\\ &+ 0.5h(1-\delta_{i1})\Phi_{i-1}^{(j-1)}\mathrm{v}_{i-1}^{(j-1)}/\sqrt{\tau_i - \tau_1}\\ &+ h(1-\delta_{i1})(1-\delta_{i2})\sum_{m=1}^{i-2}\Phi_m^{(j-1)}\mathrm{v}_m^{(j-1)}/\sqrt{\tau_i - \tau_m}, \qquad j = 1, 2, \ldots,\end{aligned} \tag{4.24}$$

$$\theta_i^{(1)} = \theta_{i-1}^{(j_{i-1})}, \quad \mathrm{v}_i^{(1)} = \mathrm{v}_{i-1}^{(j_{i-1})}, \qquad \theta_0 = 0, \qquad v_0 = 1,$$

where

$$\Phi_i^{(j-1)} = f^* F(T_0\theta_i^{(j-1)})p_i, \qquad \delta_{ij} = 1, \qquad i = j, \qquad \delta_{ij} = 0, \qquad i \neq j,$$

$$\theta_i = \theta_m(0, \tau_i), \qquad \mathrm{v}_i = \mathrm{v}(\tau_t), \qquad p_i = p(\tau_i).$$

The iterational process is finished, when for $j = j_i$ the following conditions hold

$$|(\theta_i^{(j_i)} - \theta_i^{(j_i-1)})/\theta_i^{(j_i)}| < \varepsilon_b, \qquad |(\mathrm{v}_i^{(j_i)} - \mathrm{v}_i^{(j_i-1)})/\mathrm{v}_i^{(j_i)}| < \varepsilon_b,$$

where ε_b is the relative error. The amount of wear is estimated by the relation

$$u_i^w = 0.5h\Phi_i\mathrm{v}_i k^w(T_0\theta_i) + (1-\delta_{i1})h\sum_{m=1}^{i-1}\Phi_m\mathrm{v}_m k^w(T_0\theta_m). \tag{4.25}$$

4.1.2 *Contact characteristics of the metallic–ceramic frictional strap and the metal disk during braking*

The numerical analysis is carried out using Equations (4.13)–(4.16) for the frictional pair of alloy cast iron (disc, body 1), metallic–ceramic (frictional strap element, body 2), and steel 30ChGSA (strap foundation, body 3) for the following fixed parameters [Chichinadze et al. (1979)]: $P_0 = 0.98\,\mathrm{MPa}$, $t_m = 0.2\,\mathrm{s}$, $V_0 = 30\,\mathrm{m\,s^{-1}}$, $W = 35.4\,\mathrm{MN\,m^{-1}}$, $f_0 = 0.8$, $f_1 = 0.1$, $f_2 = 0.845$, $f_3 = 3.02\cdot 10^{-3\circ}\mathrm{C}^{-1}$, $T_f = -100°\mathrm{C}$ (curve 3 in the Fig. 4.2a), $d_0 = 6$, $d_1 = 0$, $d_2 = 3$, $d_3 = 10^{-2\circ}\mathrm{C}^{-1}$, $d_4 = 5$, $d_5 = 10^{-2\circ}\mathrm{C}^{-1}$, $T_1^w = 100°\mathrm{C}$, $T_2^w = 800°\mathrm{C}$ (curve 2 in Fig. 4.2b), $K_0^w = 1\,\mathrm{m^2\,N^{-1}}$, $\lambda_1 = 51\,\mathrm{W\,m^{-1\circ}C^{-1}}$, $a_1 = 14\,\mathrm{mm^2\,s^{-1}}$,

Table 4.1: Definitions of curves for various thickness of layers

Curve Number	Curves Solid $H_2 = 1\,\mathrm{mm}$ $H_3 = 15\,\mathrm{mm}$	Dashed $H_1 = 15\,\mathrm{mm}$ $H_2 = 1\,\mathrm{mm}$	Dotted $H_1 = 5\,\mathrm{mm}$ $H_3 = 5\,\mathrm{mm}$
	H_1, mm	H_3, mm	H_2, mm
1	15	15	15
2	10	10	10
3	5	5	5
4	1	1	1

$\lambda_2 = 34.3\,\mathrm{W\,m^{-1}{}^\circ C^{-1}}$, $a_2 = 15.2\,\mathrm{mm^2s^{-1}}$, $\lambda_3 = 37.2\,\mathrm{W\,m^{-1}{}^\circ C^{-1}}$, $a_3 = 10.3\,\mathrm{mm^2s^{-1}}$, $B_1 = 0$.

Owing to (4.10)–(4.12) one gets $t_0 = 3.01\,\mathrm{s}$, $\delta = 15.05$, $T_0 = 265°\mathrm{C}$.

Kinetics of the contact temperature T, dimensionless sliding velocity v, and wear u^w during brake are illustrated in Fig. 4.4. The curves characterize various values of body thickness H_n (Table 4.1).

The highest contact temperature is achieved for the minimal thickness values of the metallic–ceramic strap, the steel strap, and the disc made from alloy cast iron (Fig. 4.4a). In this case one deals with the longest braking time (Fig. 4.4b) and the smallest wear (Fig. 4.4c). Increase of the contact temperature is accompanied by a decrease of wear of the rubbing elements. In this aspect, temperature plays a positive role for the considered tribological system.

For fixed thickness of base and disc layers ($H_1 = H_3 = 5\,\mathrm{mm}$, dotted curves), an increase of thickness of the metallic–ceramic layer causes an increase of contact temperature, whereas an increase of braking time yields the wear decrease. Hence, minimization of wear of the contacting bodies can be achieved through an increase of the thickness of the frictional layer of the metallic–ceramic strap.

As the effective measure of element thickness occurred in the considered tribological system, the thickness of a hypothetical case is taken, that is, when the contact temperature differs by 5% in comparison to the case for infinite elements. The computations have shown that the effective thickness of the frictional strap is $H_{2,ef} = 2\,\mathrm{mm}$ for $H_1 = H_3 = 5\,\mathrm{mm}$; $H_{2,ef} = 3\,\mathrm{mm}$ for $H_1 = 5\,\mathrm{mm}$, $H_3 = 3\,\mathrm{mm}$. Hence, one may conclude that the decrease of thickness of the metallic base causes an increase of thickness of the effective metallic–ceramic layer. For $H_n > H_{n,ef}$ $n = 1, 2, 3$ one may consider the half-infinite layers and the contact characteristics may be found through consideration of rubbing of two half-spaces [Olesiak et al. (1997), Pyryev (2004)].

Below, the exponential dependence of the frictional coefficient on the temperature (4.3) is analysed for the following fixed parameters: $P_0 = 0.98\,\mathrm{MPa}$, $f_0 = 0.7$, $\lambda_f = 1.5 \cdot 10^{-3}$, $V_0 = 30\,\mathrm{m\,s^{-1}}$, $t_m = 0.2\,\mathrm{s}$, $W = 35.4\,\mathrm{MN\,m^{-1}}$, $\lambda_1 = 51\,\mathrm{W\,m^{-1}\,{}^\circ C^{-1}}$, $a_1 = 14\,\mathrm{mm^2\,s^{-1}}$, $\lambda_2 = 34.3\,\mathrm{W\,m^{-1}{}^\circ C^{-1}}$, $a_2 = 15.2\,\mathrm{mm^2\,s^{-1}}$, $\lambda_3 = 37.2\,\mathrm{W\,m^{-1}{}^\circ C^{-1}}$, $a_3 = 10.3\,\mathrm{mm^2\,s^{-1}}$, $d_0 = 6$, $d_1 = 0$, $d_3 = d_5 = 0.01°\mathrm{C}^{-1}$, $d_4 = 5$, $T_1^2 = 100°\mathrm{C}$, $T_2^w = 800°\mathrm{C}$ (curve 2 in Fig. 4.2b), $B_1 = 0$.

(iii) For the constant thickness of the frictional strap an increase of the disc thickness (braking drum) causes the temperature increase and wear decrease at least up to a certain extent. For thickness greater than the critical values, the maximal values of temperature and wear do not change.

4.2. Computation of the contact characteristics of the two-layer brake model

Equations (4.13)–(4.16) and (4.19) served for numerical computations of the contact brake characteristics for the frictional pair and the metallic–ceramic material (frictional strap element, body 2) for the parameters: $P_0 = 0.98\,\mathrm{MPa}$, $t_m = 0.2\,\mathrm{s}$, $V_0 = 30\,\mathrm{m\,s^{-1}}$, $W = 35.4\,\mathrm{MN\,m^{-1}}$, $\eta = 0.5$, $B_1 = 0$, $f_0 = 0.8$, $f_1 = 0.1$, $f_2 = 0.845$, $f_3 = 3.02 \cdot 10^{-3}{}^\circ\mathrm{C}^{-1}$, $T_f = -100^\circ\mathrm{C}$ (curve 3 in Fig. 4.2a), $d_0 = 6$, $d_1 = 0$, $d_2 = 3$, $d_3 = 10^{-2}{}^\circ\mathrm{C}^{-1}$, $d_4 = 5$, $d_5 = 10^{-2}{}^\circ\mathrm{C}^{-1}$, $T_1^w = 100^\circ\mathrm{C}$, $T_2^w = 800^\circ\mathrm{C}$ (curve 2 in Fig. 4.2b), $K_0^w = 1\,\mathrm{m^2N^{-1}}$, $\lambda_1 = 51\,\mathrm{W\,m^{-1}{}^\circ C^{-1}}$), $a_1 = 14\,\mathrm{mm^2\,s^{-1}}$, $\lambda_2 = 34.3\,\mathrm{W\,m^{-1}{}^\circ C^{-1}}$), $a_2 = 15.2\,\mathrm{mm^2\,s^{-1}}$.

From the computations we got $t_0 = 3.01\,\mathrm{s}$, $\delta = 15.05$, and $T_0 = 132^\circ\mathrm{C}$.

Figure 4.6a shows the dependencies of the contact temperature on time τ for various layer thicknesses H_2. The layer thickness is $H_1 = 15\,\mathrm{mm}$. It is assumed that the friction coefficient depends on the contact temperature in agreement with curve 2 in Fig. 4.2a. The corresponding computational results of the sliding velocity and wear are reported in Figs. 4.6b and c.

The dependence of maximal contact temperature $T_{\max}$ on thickness H_2 for the constant thickness $H_1 = 15\,\mathrm{mm}$ for different types of functions $f^*(T)$ (see Fig. 4.2a) is shown in Fig. 4.7a. The corresponding computational results of the dimensionless time τ_s and time $\tau_{\max}$ are shown in Fig. 4.7b. Numbers of curves in Fig. 4.7 correspond to the numbers of curves in Fig. 4.2a.

Note that an important brake characteristic is the effective body thickness H_n^*. The latter is defined as a minimum of the rubbing element thickness, which does not interact with the maximum of the contact temperature. In other words, $T_{\max}$ computed for H_n^* with the accuracy of 5% in agreement with the computational result of $T_{\max}$ for $H_n = \infty$. The value of H_n^* is obtained from the formula $H_n^* = v_e\sqrt{a_n t_s}$. The parameter $v_e = 1.75$ is due to Fazekas [Fazekas (1953)], $v_e = 1.73$ is due to Chichinadze [Chichinadze et al. (1979)], whereas $v_e = 1.29$ is due to Hasselgruber [Hasselgruber (1963)].

The computational results shown in Fig. 4.7 provide an estimation of $v_e = 0.8$. The largest values of τ_s and $\tau_{\max}$ for the fixed thickness H_2 during the braking process are obtained for the function $f^*(T)$ depicted by curve 3 in Fig. 4.2a. From numerical analysis it follows that a decrease of the friction coefficient with a simultaneous temperature increase causes an increase of both t_s and $t_{\max}$. On the other hand, the contact temperature and wear decrease.

Table 4.1: Definitions of curves for various thickness of layers

Curve Number	Curves Solid $H_2 = 1\,\text{mm}$ $H_3 = 15\,\text{mm}$	Dashed $H_1 = 15\,\text{mm}$ $H_2 = 1\,\text{mm}$	Dotted $H_1 = 5\,\text{mm}$ $H_3 = 5\,\text{mm}$
	H_1, mm	H_3, mm	H_2, mm
1	15	15	15
2	10	10	10
3	5	5	5
4	1	1	1

$\lambda_2 = 34.3\,\text{W m}^{-1}{}^\circ\text{C}^{-1}$, $a_2 = 15.2\,\text{mm}^2\text{s}^{-1}$, $\lambda_3 = 37.2\,\text{W m}^{-1}{}^\circ\text{C}^{-1}$, $a_3 = 10.3\,\text{mm}^2\text{s}^{-1}$, $B_1 = 0$.

Owing to (4.10)–(4.12) one gets $t_0 = 3.01\,\text{s}$, $\delta = 15.05$, $T_0 = 265°\text{C}$.

Kinetics of the contact temperature T, dimensionless sliding velocity v, and wear u^w during brake are illustrated in Fig. 4.4. The curves characterize various values of body thickness H_n (Table 4.1).

The highest contact temperature is achieved for the minimal thickness values of the metallic–ceramic strap, the steel strap, and the disc made from alloy cast iron (Fig. 4.4a). In this case one deals with the longest braking time (Fig. 4.4b) and the smallest wear (Fig. 4.4c). Increase of the contact temperature is accompanied by a decrease of wear of the rubbing elements. In this aspect, temperature plays a positive role for the considered tribological system.

For fixed thickness of base and disc layers ($H_1 = H_3 = 5\,\text{mm}$, dotted curves), an increase of thickness of the metallic–ceramic layer causes an increase of contact temperature, whereas an increase of braking time yields the wear decrease. Hence, minimization of wear of the contacting bodies can be achieved through an increase of the thickness of the frictional layer of the metallic–ceramic strap.

As the effective measure of element thickness occurred in the considered tribological system, the thickness of a hypothetical case is taken, that is, when the contact temperature differs by 5% in comparison to the case for infinite elements. The computations have shown that the effective thickness of the frictional strap is $H_{2,ef} = 2\,\text{mm}$ for $H_1 = H_3 = 5\,\text{mm}$; $H_{2,ef} = 3\,\text{mm}$ for $H_1 = 5\,\text{mm}$, $H_3 = 3\,\text{mm}$. Hence, one may conclude that the decrease of thickness of the metallic base causes an increase of thickness of the effective metallic–ceramic layer. For $H_n > H_{n,ef}$ $n = 1, 2, 3$ one may consider the half-infinite layers and the contact characteristics may be found through consideration of rubbing of two half-spaces [Olesiak et al. (1997), Pyryev (2004)].

Below, the exponential dependence of the frictional coefficient on the temperature (4.3) is analysed for the following fixed parameters: $P_0 = 0.98\,\text{MPa}$, $f_0 = 0.7$, $\lambda_f = 1.5 \cdot 10^{-3}$, $V_0 = 30\,\text{m s}^{-1}$, $t_m = 0.2\,\text{s}$, $W = 35.4\,\text{MN m}^{-1}$, $\lambda_1 = 51\,\text{W m}^{-1}\,{}^\circ\text{C}^{-1}$, $a_1 = 14\,\text{mm}^2\,\text{s}^{-1}$, $\lambda_2 = 34.3\,\text{W m}^{-1}{}^\circ\text{C}^{-1}$, $a_2 = 15.2\,\text{mm}^2\,\text{s}^{-1}$, $\lambda_3 = 37.2\,\text{W m}^{-1}{}^\circ\text{C}^{-1}$, $a_3 = 10.3\,\text{mm}^2\,\text{s}^{-1}$, $d_0 = 6$, $d_1 = 0$, $d_3 = d_5 = 0.01°\text{C}^{-1}$, $d_4 = 5$, $T_1^2 = 100°\text{C}$, $T_2^w = 800°\text{C}$ (curve 2 in Fig. 4.2b), $B_1 = 0$.

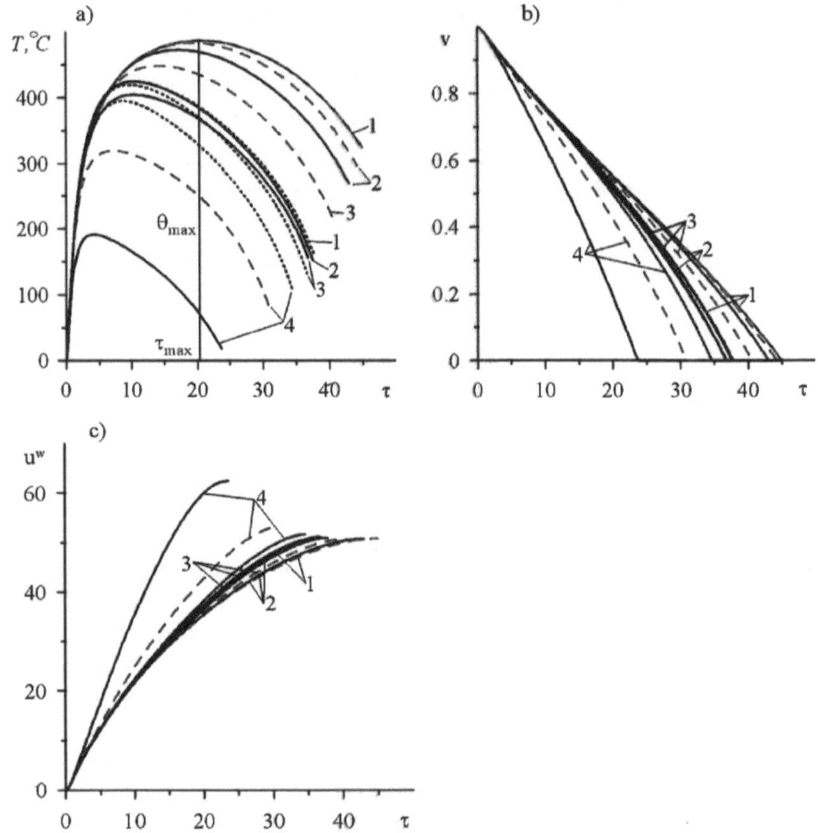

Fig. 4.4: Time history of the contact temperature $T(\tau)$ (a), dimensionless sliding velocity $\mathrm{v}(\tau)$ (b), and wear $u^w(\tau)$ (c) during braking for various layer thicknesses.

The computational results give $t_0 = 3.44\,\mathrm{s}$, $\delta = 17.2$, and $T_0 = 232°\mathrm{C}$.

Numerical results are shown in Fig. 4.5. For constant thickness of friction strap ($H_2 = H_3 = 5\,\mathrm{mm}$), an increase of disc thickness from 1 mm to 5 mm caused an increase of braking time τ_s, maximum contact temperature $\theta_{\max}$, and time to achieve it $\tau_{\max}$ (Fig. 4.5a). Simultaneously, wear $u = u(\tau_s)$ quickly decreased. The latter results from the fact that the friction coefficient decreases with an increase of temperature. Further increase of $H_1 > 5\,\mathrm{mm}$ does not influence the variation of the earlier mentioned contact characteristics.

Thickness H_2 of the metallic–ceramic layer essentially influences the contact characteristics only in the case when thickness H_3 of the steel strap base is not greater than 2 mm (broken curves in Fig. 4.5b). For $H_3 = 5\,\mathrm{mm}$, the contact temperature and wear do not change with the increase of metallic–ceramic layer thickness (solid curves in Fig. 4.5b). Hence, one may conclude that a way of obtaining resistance on the heat action (i.e., stability of the contact characteristics

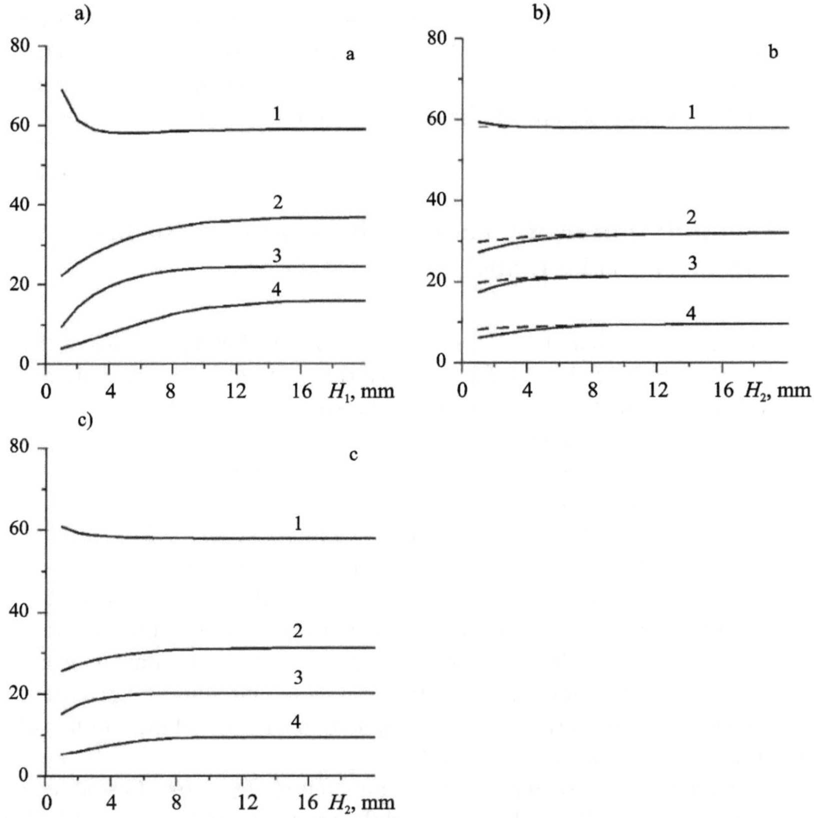

Fig. 4.5: Dependencies of dimensionless wear u^w (curves 1), braking time τ_s (2), maximal contact temperature $10 \cdot \theta_{\max}$ (3), and time associated with maximal temperature $\tau_{\max}$ (4) on (a): disc thickness H_1 for $H_2 = 5\,\text{mm}$, $H_3 = 5\,\text{mm}$; (b): frictional strap thickness H_2 (solid curves, $H_1 = 5\,\text{mm}$, $H_3 = 2\,\text{mm}$; dashed curves, $H_1 = 5\,\text{mm}$, $H_3 = 5\,\text{mm}$); (c) strap base thickness H_3 for $H_1 = 5\,\text{mm}$, $H_2 = 1\,\text{mm}$.

in a wide temperature range) is to cover the steel base of thickness greater than 5 mm (Fig. 4.5) by the metallic–ceramic layer ($\approx$1 mm).

Conclusions. During the frictional braking process of the metallic–ceramic layer lying on the steel base (steel 30HGSA) of the frictional pad on the transversal disc surface (cast iron) the following observations have been made.

(i) For the constant thickness of the strap and disc, an increase of thickness of the metallic–ceramic layer causes an increase of the contact temperature and wear decrease.

(ii) There are thickness values of the base of the strap and disc such that a variation of the metallic–ceramic layer thickness does not influence the contact characteristics; in this case, for computational purposes one may use the results of HFD for two semi-infinite bodies [Yevtushenko and Pyryev(1998)].

(iii) For the constant thickness of the frictional strap an increase of the disc thickness (braking drum) causes the temperature increase and wear decrease at least up to a certain extent. For thickness greater than the critical values, the maximal values of temperature and wear do not change.

4.2. Computation of the contact characteristics of the two-layer brake model

Equations (4.13)–(4.16) and (4.19) served for numerical computations of the contact brake characteristics for the frictional pair and the metallic–ceramic material (frictional strap element, body 2) for the parameters: $P_0 = 0.98\,\mathrm{MPa}$, $t_m = 0.2\,\mathrm{s}$, $V_0 = 30\,\mathrm{m\,s^{-1}}$, $W = 35.4\,\mathrm{MN\,m^{-1}}$, $\eta = 0.5$, $B_1 = 0$, $f_0 = 0.8$, $f_1 = 0.1$, $f_2 = 0.845$, $f_3 = 3.02 \cdot 10^{-3}{}^\circ\mathrm{C}^{-1}$, $T_f = -100^\circ\mathrm{C}$ (curve 3 in Fig. 4.2a), $d_0 = 6$, $d_1 = 0$, $d_2 = 3$, $d_3 = 10^{-2}{}^\circ\mathrm{C}^{-1}$, $d_4 = 5$, $d_5 = 10^{-2}{}^\circ\mathrm{C}^{-1}$, $T_1^w = 100^\circ\mathrm{C}$, $T_2^w = 800^\circ\mathrm{C}$ (curve 2 in Fig. 4.2b), $K_0^w = 1\,\mathrm{m^2N^{-1}}$, $\lambda_1 = 51\,\mathrm{W\,m^{-1}{}^\circ C^{-1}}$), $a_1 = 14\,\mathrm{mm^2\,s^{-1}}$, $\lambda_2 = 34.3\,\mathrm{W\,m^{-1}{}^\circ C^{-1}}$), $a_2 = 15.2\,\mathrm{mm^2\,s^{-1}}$.

From the computations we got $t_0 = 3.01$ s, $\delta = 15.05$, and $T_0 = 132^\circ$C.

Figure 4.6a shows the dependencies of the contact temperature on time τ for various layer thicknesses H_2. The layer thickness is $H_1 = 15$ mm. It is assumed that the friction coefficient depends on the contact temperature in agreement with curve 2 in Fig. 4.2a. The corresponding computational results of the sliding velocity and wear are reported in Figs. 4.6b and c.

The dependence of maximal contact temperature $T_{\max}$ on thickness H_2 for the constant thickness $H_1 = 15$ mm for different types of functions $f^*(T)$ (see Fig. 4.2a) is shown in Fig. 4.7a. The corresponding computational results of the dimensionless time τ_s and time $\tau_{\max}$ are shown in Fig. 4.7b. Numbers of curves in Fig. 4.7 correspond to the numbers of curves in Fig. 4.2a.

Note that an important brake characteristic is the effective body thickness H_n^*. The latter is defined as a minimum of the rubbing element thickness, which does not interact with the maximum of the contact temperature. In other words, $T_{\max}$ computed for H_n^* with the accuracy of 5% in agreement with the computational result of $T_{\max}$ for $H_n = \infty$. The value of H_n^* is obtained from the formula $H_n^* = v_e\sqrt{a_n t_s}$. The parameter $v_e = 1.75$ is due to Fazekas [Fazekas (1953)], $v_e = 1.73$ is due to Chichinadze [Chichinadze et al. (1979)], whereas $v_e = 1.29$ is due to Hasselgruber [Hasselgruber (1963)].

The computational results shown in Fig. 4.7 provide an estimation of $v_e = 0.8$. The largest values of τ_s and $\tau_{\max}$ for the fixed thickness H_2 during the braking process are obtained for the function $f^*(T)$ depicted by curve 3 in Fig. 4.2a. From numerical analysis it follows that a decrease of the friction coefficient with a simultaneous temperature increase causes an increase of both t_s and $t_{\max}$. On the other hand, the contact temperature and wear decrease.

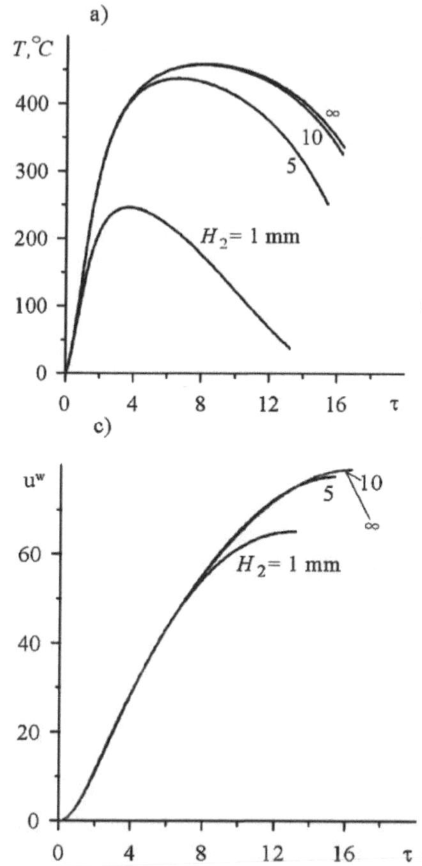

Fig. 4.6: Time histories of contact temperature $T(\tau)$ (a), dimensionless sliding velocity $v(\tau)$ (b) and wear $u^w(\tau)$ (c) during braking for curve 2 (Fig. 4.2a) at $H_1 = 15\,\text{mm}$ and various thicknesses H_2.

4.3. Computation of the contact characteristics of the two semi-space brake models

As is known, for different layers of thickness $H_n > 10\,\text{mm}$, one may assume that the contact layers are semi-spaces. In this case the solution is governed by (4.13)–(4.16) and (4.22).

In this section we consider the case of constant friction coefficient $f(T) = f_0$ and the linear dependence of the wear coefficient versus temperature (in formula (4.9) $d_2 = d_4 = 0$). Furthermore, a lack of external load ($B_1 = 0$) is assumed. The system of nonlinear Equations (4.13)–(4.16) has been solved analytically for the case of two semi-spaces [Yevtushenko and Pyryev (1997)].

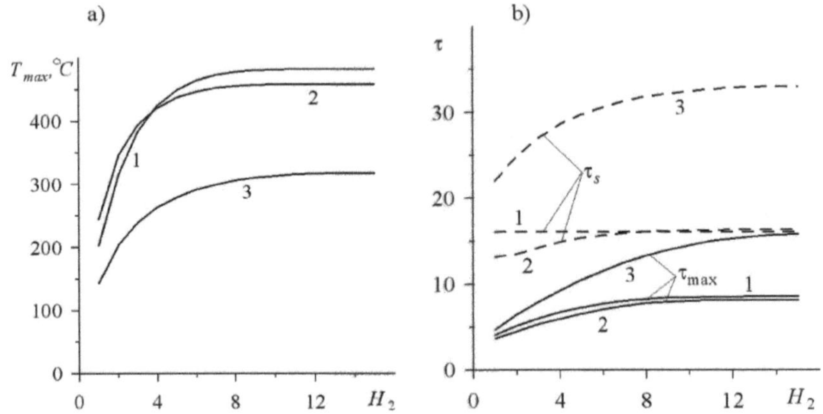

Fig. 4.7: Maximal contact temperature $T_{\max}$, dimensionless time $\tau_{\max}$, and braking time τ_s (b) versus thickness H_2 for $H_1 = 15\,\text{mm}$.

4.3.1 Contact temperature and wear during braking

The frictional wear law governed by (4.8) serves for our computational purpose. Assume that the dependence of K_n^w $(n = 1, 2)$ on the temperature is linear [Aleksandrov and Annakulova (1990), Pyryev, Yevtushenko (2000)]

$$K_n^w(t) = K_{1n}^w(t) + K_{2n}^w(t)\beta_n T(t), \qquad \beta_n = (1+\nu_n)\alpha_n/(1-\nu_n), \tag{4.26}$$

where $T(t) \equiv T_n(0,t)$ is the contact temperature; and K_{jn}^w are the coefficients obtained from experimental investigations [Chichinadze et al. (1979)], $n, j = 1, 2$; ν_n, α_n denote Poisson ratios and heat extension coefficients, respectively. It is assumed that the following load is applied (4.4) when $B_1 = 0$.

Let us introduce, besides parameters (4.10)–(4.12), also the following ones

$$K_0^w = K_{11}^w + K_{12}^w, \qquad \varepsilon = T_0 \frac{K_{21}^w + K_{22}^w}{K_0^w}. \tag{4.27}$$

After integrating the equation of motion (4.13) we obtain the speed-changing law during braking

$$\mathrm{v}(\tau) = \delta^{-1}(1 + \delta - \tau - \exp(-\tau)). \tag{4.28}$$

Upon substitution of Equations (4.14) and (4.28) into Equation (4.15) and integration we find the dimensionless temperature on the contact surface $X = 0$ in the form

$$\theta(\tau) = \delta^{-1}[(1+2\delta)\sqrt{\tau} - 4\tau\sqrt{\tau}/3 + (2\tau - 3 - 2\delta)\, F_D(\sqrt{\tau}) + \sqrt{2}\, F_D(\sqrt{2\tau})], \tag{4.29}$$

where $F_D(\tau)$ is Douson's integral [Abramowitz and Stegun (1965)]

$$F_D(\tau) = \exp(-\tau^2) \int_0^{\tau} \exp(t^2)dt. \tag{4.30}$$

To calculate $F_D(T)$ we use the formulas

$$F_D(\tau) = \sum_{m=0}^{\infty} \frac{(-2\tau^2)^m}{(2m+1)!!}, \qquad 0 \le \tau \le 3;$$

$$F_D(\tau) = \sum_{m=0}^{\infty} \frac{(2m-1)!!}{(2\tau^2)^{m+1}}, \qquad 3 < \tau < \infty.$$

Having an analytical expressions for the speed of v(τ), Equation (4.28) and the contact temperature $\theta(\tau)$ Equation (4.29), from Equation (4.16) we find the wear in the form

$$u^w(\tau) = u_0^w(\tau) + \varepsilon u_1^w(\tau), \tag{4.31}$$

where

$$u_0^w(\tau) = \frac{1}{\delta}\left[\sum_{m=0}^{2} b_m \tau^m + (1+\delta-\tau)\exp(-\tau) - \exp(-2\tau)/2\right],$$

$$u_1^w(\tau) = \frac{2}{\delta^2}\left\{\left[a^{(0)} + \sum_{m=0}^{3} a_{m+1}^{(1)}\tau^{m+1/2} + \exp(-\tau)\sum_{m=0}^{5} a_m^{(2)}\tau^{m/2}\right.\right.$$

$$\left. + \sqrt{\tau}(a_0^{(3)} + a_1^{(3)}\tau)\exp(-2\tau) + a^{(4)}\mathrm{erf}\sqrt{\tau} + a^{(5)}\mathrm{erf}\sqrt{2\tau}\right]$$

$$+ F_D(\sqrt{\tau})\left[\sum_{m=0}^{2} a_m^{(6)}\tau^m + \exp(-\tau)\sum_{m=0}^{5} a_m^{(7)}\tau^m + (a_0^{(8)} + a_1^{(8)}\tau)\exp(-\tau)\right]$$

$$\left. + F_D(\sqrt{2\tau})[a_0^{(9)} + a_1^{(9)}\tau + (a_0^{(10)} + a_1^{(10)}\tau)\exp(-\tau) + a^{(11)}\exp(-2\tau)]\right\},$$

$$b_0 = -1/2 - \delta,\ b_1 = 1+\delta,\ b_2 = -1/2,\ a^{(0)} = -2-\delta,\ a_1^{(1)} = -3/4 - \delta^2,$$

$$a_2^{(1)} = (1+5\delta+2\delta^2)/3,\ a_3^{(1)} = -2(1+\delta)/3,\ a_4^{(1)} = 4/21,\ a_0^{(2)} = -a^{(0)},$$

$$a_1^{(2)} = 1/3 + \delta/4 + \delta^2,\ \ a_2^{(2)} = -a^{(0)},\ \ a_3^{(2)} = -5(1+4\delta)/12,\ \ a_4^{(2)} = 0,$$

$$a_5^{(2)} = 2/3,\ a_0^{(3)} = -(1/6+\delta)/2,\ a_1^{(3)} = 1/3,$$

$$a^{(4)} = \sqrt{\pi}(35/36+11\delta/6-\delta^2)/4,\ a_0^{(7)} = -(11/4+3\delta+\delta^2)/2,\ \ a_1^{(7)} = -a_1^{(6)},$$

$$a_2^{(7)} = -1/2,\ a_0^{(8)} = -(7/6+\delta)/3,\ a_1^{(8)} = -1/3,\ \ a_0^{(9)} = \sqrt{2}(1/2+\delta)/4,$$

$$a_1^{(9)} = \sqrt{2}/4,\ a_0^{(10)} = \sqrt{2}(5/3+\delta)/6,\ a_1^{(10)} = -\sqrt{2}/6,\ \ a^{(11)} = -\sqrt{2}/8,$$

and $\mathrm{erf}\,(x)$ is the probability function [Abramowitz and Stegun (1965)].

A numerical analysis is carried out using analytical solution forms (4.28), (4.29), and (4.31). The following dimensionless quantities δ and ε (4.10), (4.27) serve as initial input parameters.

The parameter $0 < \delta < \infty$ characterises the ratio of braking time t_0 under uniform deceleration to time t_m, when the load reaches maximum value and $0 < \varepsilon < \infty$ characterises the influence of the temperature on wear.

Using the condition $\mathrm{v}(\tau_s) = 0$ from Equation (4.28) we obtain the equation for dimensionless braking time $\tau_s = t_s/t_m$,

$$\tau_s = 1 + \delta - \exp(-\tau_s). \tag{4.32}$$

Numerical analysis has also shown that the following approximations can be used,

$$\tau_s = \sqrt{2\delta} \qquad \text{for} \qquad \delta < 0.1, \qquad \tau_s = 1 + \delta \ \text{ for } \ \delta > 1.2. \tag{4.33}$$

The expressions of Equation (4.33) also imply that for $t_0 > 1.2t_m$ the braking time t_s increases relative to time t_0 at the value of t_m.

An analytical expression of the contact temperature Equation (4.29) allows us to construct such asymptotes

$$\theta(\tau) = 4\tau\sqrt{\tau}/3[1 - 2\tau/5 - 4(3-\delta)\tau^2/(35\delta) + O(\tau^3)], \tag{4.34}$$

$$\theta(\tau) = \sqrt{\tau}/\delta[2(1+\delta) - 4\tau/3 - (1+2\delta)/(2\tau) + O(\tau^{-2})]. \tag{4.35}$$

Comparing approximate values of the temperature, Equations (4.34) and (4.35), with the corresponding accurate values Equation (4.29), we obtain the following engineering expressions for the maximum dimensionless contact temperature $\theta_{\max}$ and for the dimensionless time needed to reach it $\tau_{\max}$.

$$\theta_{\max} = 4\sqrt{\tau_{\max}}\delta^{-1}(1 + \delta - 4\tau_{\max}/3) \quad \text{for } \ \delta > 5.5, \tag{4.36}$$

$$\tau_{\max} = 1 + \delta/2 \quad \text{for } \ \delta > 1.5, \tag{4.37}$$

$$\theta_{\max} = 8\tau_{\max}\sqrt{\tau_{\max}}(1 + 4\tau_{\max}/15)/9 \quad \text{for } \ \delta < 0.1, \tag{4.38}$$

$$\tau_{\max} = 1/4(\sqrt{5\delta(36-7\delta)} - 5\delta)/(3-\delta) \quad \text{for } \ \delta < 0.08. \tag{4.39}$$

Figure 4.9b shows an influence of contact temperature (parameter ε) on wear evolution. When the smallest wear occurs frictional heat is not generated ($\varepsilon = 0$).

An essential influence of temperature on wear of the contacting bodies appears for $\varepsilon = 3.4$.

Numerical results have been computed for the friction couple cast iron/metal–ceramics, for which: $P_0 = 0.98 \cdot 10^6$ Pa, $f = 0.8$, $V_0 = 30\,\mathrm{m\,s^{-1}}$, $t_m = 0.2\,\mathrm{s}$, $W = 3.54 \cdot 10^5\,\mathrm{N\,m^{-1}}$, $\lambda_1 = 50.96\,\mathrm{W\,m^{-1}{}^\circ C^{-1}}$, $a_1 = 14 \cdot 10^{-6}\,\mathrm{m^2\,s^{-1}}$, $\lambda_2 = 34.3\,\mathrm{W\,m^{-1}{}^\circ C^{-1}}$, $a_2 = 15.2 \cdot 10^{-6}\,\mathrm{m^2\,s^{-1}}$, $\eta = 0$, $B_1 = 0$.

Additional computations yield $t_0 = 3\,s$, $\delta = 15$, $T_0 = 264.8°\mathrm{C}$. For $\delta = 15$ and from Figs. 4.8a and 4.9a one obtains $\tau_s = 16$, $\theta_{\max} = 3.64$, $\tau_{\max} = 8.56$. The contact temperature achieves the largest value $T_{\max} = 963.9°\mathrm{C}$ in the second two of the braking process, that is, for time instant $t = 1.7\,\mathrm{s}$.

Numerical analysis of the contact characteristics was carried out and general dependencies of the friction and wear coefficients on contact temperature were obtained (see Fig. 4.2). Computational results are displayed in Fig. 4.10 ($\eta = 0.5$, $T_0 = 132°\mathrm{C}$). Numbers of curves displayed in Fig. 4.10 correspond to the numbers

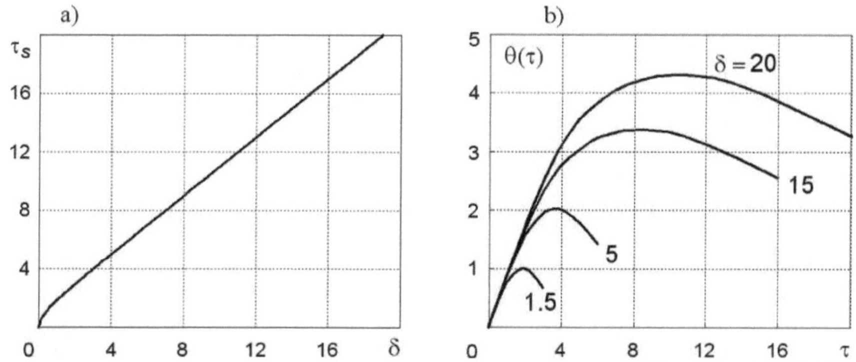

Fig. 4.8: Braking time versus parameter δ (a); time history contact temperature (b).

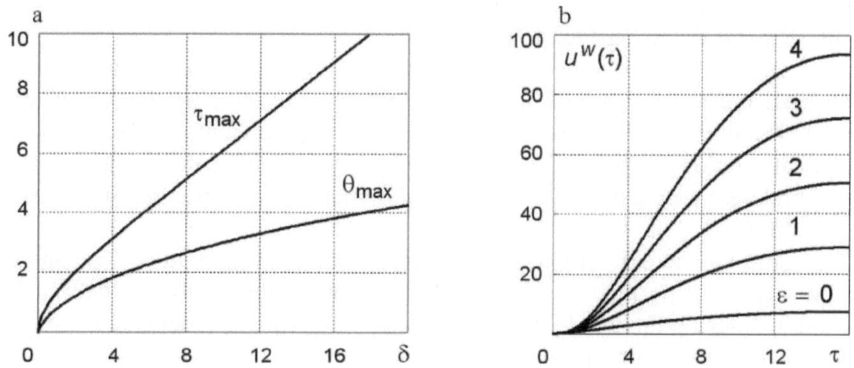

Fig. 4.9: Maximal contact temperature $\theta_{\max}$ and $\tau_{\max}$ versus parameter δ during braking (a); time history wear for various parameter values ε (b).

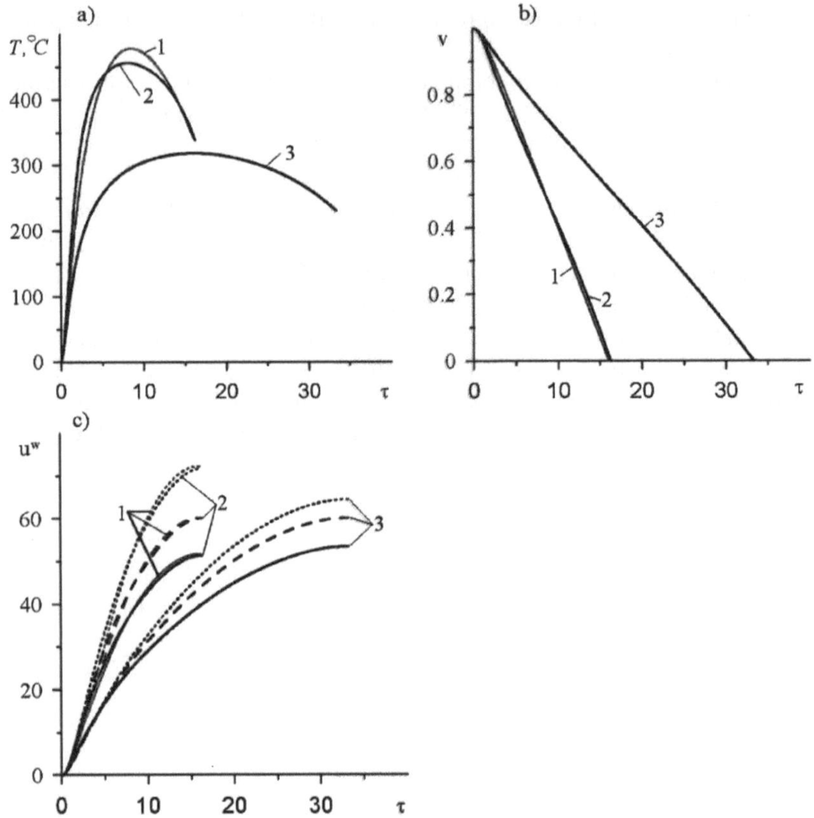

Fig. 4.10: Contact temperature $T(\tau)$ (a), dimensionless velocity $\mathrm{v}(\tau)$ (b) and wear $u^w(\tau)$ versus dimensionless time τ. Numbers and curve shapes correspond to the curve numbers from Fig. 4.2.

of curves given in Fig. 4.2a). Shape of the curves shown in Fig. 4.10c corresponds to the shape of curves in Fig. 4.2b.

The analysis leads to a conclusion that for frictional metallic–plastic (curve 1) and plastic (curve 2) materials, the change in time of the contact characteristics is similar. Therefore, one may assume that the friction coefficient is constant and the analytical solution (4.28), (4.29) can be applied. Note that the contact temperature of the braking process has a maximum in the middle of the braking time interval (Fig. 4.10a).

Owing to the increase of parameter δ both maximum temperature $\theta_{\max}$ and the time to reach it $\tau_{\max}$ increase. As Fig. 4.10b shows, a uniformly retarded motion is exhibited during the whole braking process. The negligible deviation from this motion occurs in the first phase of braking.

Kinetics of wear is illustrated in Fig. 4.10c. Note that in almost the whole braking phase a lapping process occurs when the velocity is constant. Wear achieves its maximum in the last phase of braking.

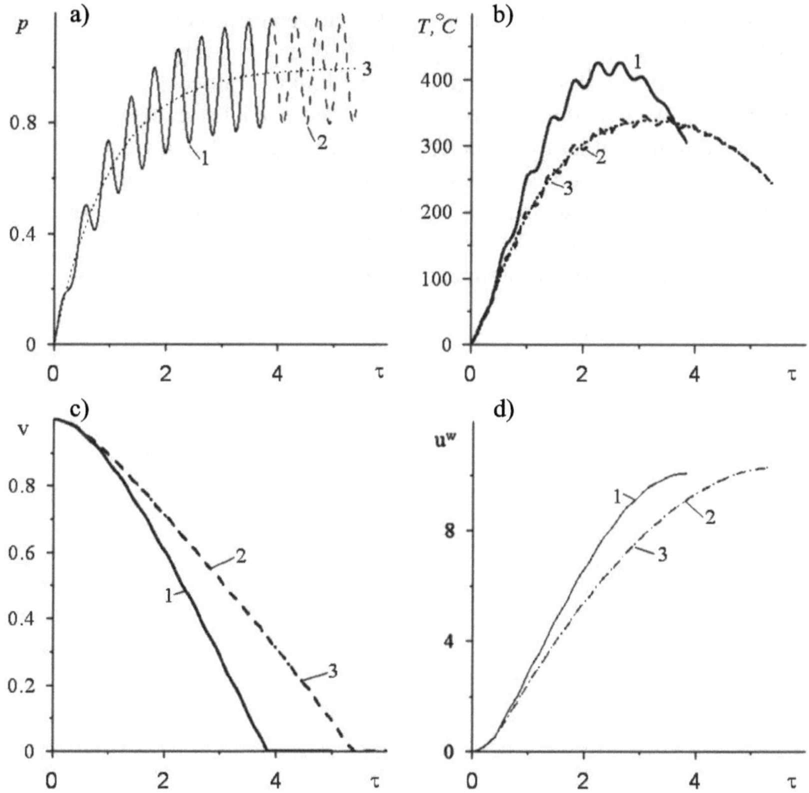

Fig. 4.11: Dimensionless contact pressure $p(\tau)$ (a), contact temperature $T(\tau)$ (b), velocity $\mathrm{v}(\tau)$ (c) and wear $u^w(\tau)$ (d) versus dimensionless time τ. Curve 1: $B_1 = 0.2$, $B_2 = 15$, $\lambda_f = 0$; curve 2: $B_1 = 0.2$, $B_2 = 15$, $\lambda_f = 1.5 \cdot 10^{-3\circ}\mathrm{C}^{-1}$; curve 3: $B_1 = 0$, $\lambda_f = 1.5 \cdot 10^{-3\circ}\mathrm{C}^{-1}$.

4.3.2 Contact temperature and wear during braking and harmonic load excitations

Consider the case of harmonic load input (4.4) and the exponential dependence of the friction coefficient on temperature (4.3). The contact characteristics of a brake for the frictional pair consist of alloy cast iron (disc, body 1) and metallic–ceramic (budding strap element, body 2) [Chichinadze et al. (1979)] have the following input parameters: $P_0 = 0.98\,\mathrm{MPa}$, $t_m = 1.2\,\mathrm{s}$, $V_0 = 30\,\mathrm{m\,s^{-1}}$, $W = 35.4\,\mathrm{MN\,m^{-1}}$, $\eta = 0.5$, $f_0 = 0.7$, $\lambda_f = 1.5 \cdot 10^{-3\circ}\mathrm{C}^{-1}$, $d_0 = 6$, $d_1 = 0$, $d_2 = 3$, $d_3 = 10^{-2\circ}\mathrm{C}^{-1}$, $d_4 = 5$, $d_5 = 10^{-2\circ}\mathrm{C}^{-1}$, $T_1^w = 100°\mathrm{C}$, $T_2^w = 800°\mathrm{C}$ (curve 2 in Fig. 4.2b), $K_0^w = 1\,\mathrm{m^2\,N^{-1}}$, $\lambda_1 = 51\,\mathrm{W\,m^{-1\circ}C^{-1}}$), $a_1 = 14\,\mathrm{m^2\,s^{-1}}$, $\lambda_2 = 34.3\,\mathrm{W\,m^{-1\,\circ}C^{-1}}$, $a_2 = 15.2\,\mathrm{m^2\,s^{-1}}$. Furthermore, carrying out simple computations one gets $t_0 = 3.44\,\mathrm{s}$, $\delta = 2.86$, $T_0 = 284°\mathrm{C}$. Figure 4.11a shows three cases of the external load variations. The functions governing the behaviour of contact temperature, sliding velocity and wear versus dimensionless time τ for various loading types are shown in Figs. 4.11b–d, respectively.

Numerical analysis shows that during the braking process contact pressure oscillations (Fig. 4.11a) influence contact temperature to a lesser extent (Fig. 4.11b), and they do not affect either the velocity behaviour (Fig. 4.11c) or wear (Fig. 4.11d).

The contact temperature T achieves its maximum just before the body's stop (Fig. 4.11b). The numerical analysis allows us to conclude also that a decrease of the friction coefficient caused by temperature increase induces enlargement of t_s and $t_{\max}$, whereas the contact temperature is decreased.

Chapter 5
Thermoelastic Contact of Two Moving Layers with Friction and Wear

Geometry of the contacting bodies is the same as in Chapter 4. However, there are essential differences between the two chapters. In Chapter 4 the so-called heat frictional dynamics (HFD) [Chichinadze et al. (1979)] is studied. One has to study heat transfer equations with conditions of heat occurrence and heat transfer on the sliding surface. We treat the contact pressure making a contribution to the heat stream as a known quantity. The sliding velocity responsible for heat generation on the sliding surface is found from the equations of system dynamics. The contact characteristics are coupled via friction and wear coefficients and they depend on temperature. There is no need to solve equations of the theory of elasticity of the contacting bodies.

In this chapter we study the problem of thermoelastic contact of two sliding layers. Contrary to considerations presented in the previous chapter, here we assume initial body position. In this problem the contact pressure is not known, and it can be defined by solving a thermoelastic problem of the theory of thermal stresses with appropriate boundary conditions. In the case of constant sliding velocity the obtained problems fit with that of thermoelastic frictional instability.

5.1. Analysed system

A one-dimensional model of thermoelastic contact of two plane parallel layers is shown in Fig. 5.1. The upper layer of thickness L_2 and mass m, under the action of force $F = F_0\varphi_F(t)$, parallel to the plane of contact, slides at velocity $\mathrm{v}(t)$ over the layer of thickness L_1. The mass and the force are taken per length unit. The friction forces $F_{fr} = f(T)p(t)$ $(f(T) = f_s F(T))$ over the contact surface $x = L_1$ cause a deceleration of the upper layer. The braking process is accompanied by the generation of heat and wear. The sum of the intensities of heat fluxes directed into layers is equal to the specific power of friction forces $fV(t)P(t)$. The distance between solids $U = U_0\varphi_u(t)$ is known beforehand. Newton's law of heat exchange is assumed for fluxes between external surfaces of the layers and the environment.

DOI: 10.1007/978-0-387-09653-7_5,

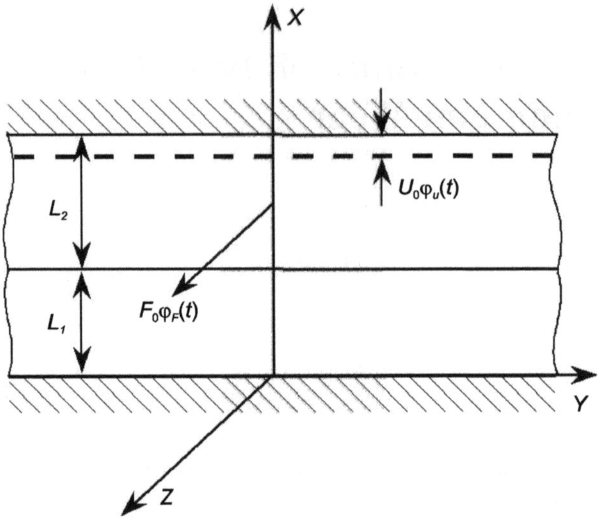

Fig. 5.1: Geometry of the moving solids.

Temperatures $T_n(x,t), n=1,2$ of the layers, temperature of the contact plane $T(t)=T_1(L_1,t)=T_2(L_1,t)$, displacements $U_n(X,t)$ along the X-axis, normal stresses $\sigma_{XX}^{(n)}(X,t)$, wear of the layers $U^w(t)$, and velocity $\mathrm{V}(t)$ are to be found from the solution. From the mathematical standpoint the considered problem consists in solving the system of differential equations of the quasi-static uncoupled thermoelasticity:

$$\frac{\partial}{\partial X}\left[\frac{\partial}{\partial X}U_n(X,t)-\alpha_n\frac{1+\nu_n}{1-\nu_n}T_n(X,t)\right]=0, \tag{5.1}$$

$$\frac{\partial^2}{\partial X^2}T_n(X,\ t)=\frac{1}{a_n}\frac{\partial}{\partial t}T_n(X,\ t), \qquad 0\le t\le t_c, \tag{5.2}$$

$$X\in\begin{cases}(0,L_1), & n=1,\\ (L_1,L_1+L_2), & n=2,\end{cases}$$

and the equation of motion for the upper layer

$$m\frac{d}{dt}V(t)=F_0\varphi_F(t)-f(T)\,P(t), \qquad 0\le t\le t_c. \tag{5.3}$$

We assume the following mechanical boundary conditions:

$$U_1(0,t)=0, \qquad U_2(L_1+L_2,t)=-U_0\varphi_u(t),$$

$$\sigma_{XX}^{(1)}(L_1,t)=\sigma_{XX}^{(2)}(L_1,t), \qquad U_1(L_1,t)-U_1^{\mathrm{w}}(t)=U_2(L_1,t)+U_2^{\mathrm{w}}(t), \tag{5.4}$$

thermal boundary conditions

$$\lambda_1 \frac{\partial T_1(0,t)}{\partial X} - \alpha_1^T T_1(0,t) = 0,$$

$$\lambda_2 \frac{\partial T_2(L_1+L_2,t)}{\partial X} + \alpha_2^T T_2(L_1+L_2,t) = 0,$$

$$\lambda_1 \frac{\partial T_1(L_1,t)}{\partial X} - \lambda_2 \frac{\partial T_2(L_1,t)}{\partial X} = f(T)V(t)P(t),$$

$$T_1(L_1,t) = T_2(L_1,t) = T(t), \tag{5.5}$$

and the initial conditions

$$T_n(X,0) = 0, \qquad n = 1,2, \qquad V(0) = V_0. \tag{5.6}$$

Let us consider the abrasive wear in the form [Aleksandrov, Annakulova (1992)]

$$U^{\mathrm{w}}(t) = U_1^{\mathrm{w}}(t) + U_2^{\mathrm{w}}(t) = \int_0^t K^w(T)V(\tau)P(\tau)d\tau, \tag{5.7}$$

where $K^w(T) = K_a^w K(T)$ is the wear coefficient, which depends on contact temperature.

Normal stresses can be found by means of the following Duhamel relations

$$\sigma_{XX}^{(n)} = \frac{E_n}{1-2\nu_n}\left[\frac{1-\nu_n}{1+\nu_n}\frac{\partial U_n}{\partial X} - \alpha_n T_n\right], \qquad n = 1,2. \tag{5.8}$$

In Equations (5.1)–(5.8) we have used the following symbols: E_n is the Young's modulus, ν_n are Poisson's ratios, a_n are coefficients of thermal diffusivity, λ_n are coefficients of thermal conductivity, α_n are coefficients of linear thermal expansion, K^w is the coefficient of wear, f is the coefficient of friction, $1/\alpha_n^T$ are the thermal resistances, and t_c is the time when the solids are in contact means that $P(t) \geq 0$, and $V(t) \geq 0$ for $t \in (0, t_c)$).

Making use of the system of Equations (5.1), (5.8), and boundary conditions (5.4), the contact pressure $P(t) = -\sigma_{XX}^{(n)}(L_1,t)$ can be presented in the form

$$P(t) = P_0\left[\tilde{\alpha}_1 \int_0^{L_1} T_1(\xi,t)d\xi + \tilde{\alpha}_2 \int_{L_1}^{L_1+L_2} T_2(\xi,t)d\xi - U^w(t) + U_0\varphi_u(t)\right], \tag{5.9}$$

where

$$P_0 = \left(\frac{L_1}{\tilde{E}_1} + \frac{L_2}{\tilde{E}_2}\right)^{-1}, \quad \tilde{\alpha}_n = \alpha_n \frac{1+\nu_n}{1-\nu_n}, \quad \tilde{E}_n = \frac{E_n(1-\nu_n)}{(1+\nu_n)(1-2\nu_n)}. \tag{5.10}$$

We introduce the following dimensionless variables,

$$x = \frac{X}{L_*}, \quad \tau = \frac{t}{t_*}, \quad \tau_c = \frac{t_c}{t_*}, \quad \mathrm{v} = \frac{V}{V_*}, \quad u_n = \frac{U_n}{U_0}, \quad u^w = \frac{U^w}{U_0}, \quad \theta_n = \frac{T_n}{T_*},$$

$$p = \frac{P}{P_*}, \quad \theta = \frac{T}{T_*}, \quad \tilde{L}_n = \frac{L_n}{L_*}, \quad \tilde{a}_n = \frac{a_n}{a_*}, \quad \tilde{\lambda}_n = \frac{\lambda_n}{\lambda_*}, \quad \mathrm{Bi}_n = \frac{\alpha_n^T L_*}{\lambda_n},$$

$$\xi = \frac{K_a^w P_0 L_*}{\Omega}, \quad f_0 = \frac{F_0}{P_*}, \quad \Omega = f_s V_* P_* \frac{L_*}{T_* \lambda_*}, \quad a = \frac{t_* P_*}{V_* m}, \quad \tilde{\tilde{\alpha}}_n = \frac{\tilde{\alpha}_n}{\alpha_*},$$

where

$$t_* = \frac{L_*^2}{a_*}, \quad V_* = \frac{a_*}{L_*}, \quad P_* = P_0 U_0, \quad T_* = \frac{U_0}{\alpha_* L_*}.$$

Dimensionless equations take the form

$$\frac{\partial^2}{\partial x^2}\theta_n(x, \tau) = \frac{1}{\tilde{a}_n}\frac{\partial}{\partial \tau}\theta_n(x, \tau), 0 \le \tau \le \tau_c,$$

$$x \in \begin{cases} (0, \tilde{L}_1), & n = 1, \\ (\tilde{L}_1, \tilde{L}_1 + \tilde{L}_2), & n = 2, \end{cases}$$

$$\frac{\partial \theta_1(0,\tau)}{\partial x} - \mathrm{Bi}_1 \theta_1(0,\tau) = 0,$$

$$\frac{\partial \theta_2(\tilde{L}_1 + \tilde{L}_2, \tau)}{\partial x} + \mathrm{Bi}_2 \theta_2(\tilde{L}_1 + \tilde{L}_2, \tau) = 0,$$

$$\tilde{\lambda}_1 \frac{\partial \theta_1(\tilde{L}_1, \tau)}{\partial x} - \tilde{\lambda}_2 \frac{\partial \theta_2(\tilde{L}_1, \tau)}{\partial x} = \Omega F(\theta T_*) \mathrm{v}(\tau) p(\tau),$$

$$\theta_1(\tilde{L}_1, \tau) = \theta_2(\tilde{L}_1, \tau) = \theta(\tau), \tag{5.11}$$

where the dimensionless contact pressure, velocity, and wear are as follows.

$$p(\tau) = \varphi_u(\tau) + \tilde{\tilde{\alpha}}_1 \int_0^{\tilde{L}_1} \theta_1(\eta, \tau) d\eta + \tilde{\tilde{\alpha}}_2 \int_{\tilde{L}_1}^{\tilde{L}_1 + \tilde{L}_2} \theta_2(\eta, \tau) d\eta - u^w(\tau),$$

$$\mathrm{v}(\tau) = \mathrm{v}_0 + a \left[f_0 \int_0^{\tau} \varphi_F(\eta) d\eta - f_s \int_0^{\tau} F(\theta T_*) p(\eta) d\eta \right],$$

$$u^w(\tau) = \Omega \xi \int_0^{\tau} K(\theta T_*) \mathrm{v}(\eta) p(\eta) d\eta. \tag{5.12}$$

5.2. Laplace transform

A solution of the boundary value problem (Equations (5.11) and (5.12)) can be obtained by the use of the Laplace integral transforms taken with respect to time τ and expressed in terms of the Laplace transforms of $\varphi_F(\tau), \varphi_u(\tau)$, and $q(\tau)$. Because the thicknesses of the moving solids, in fact, are finite, the Laplace transforms of the solution are the analytical functions of the transform parameter s, except a countable set of poles s_m.

Performing the inverse Laplace transformation with the help of residuals and convolution theorems [Carslaw, Jaeger (1959)] one can find the following relations.

$$\theta_n(x,\tau) = \psi_{\theta n}(x,\tau)$$
$$+ \Omega \left[q_1(\tau) * \left(\frac{\partial}{\partial \tau} g_n(x,\tau) + \tilde{\Omega}\tilde{\xi} g_n(x,\tau) \right) - \tilde{\Omega}\xi q_3(\tau) * g_n(x,\tau) \right],$$

$$p(\tau) = \psi_p(\tau) + \Omega \left[q_1(\tau) * \frac{d^2}{d\tau^2} G_2(\tau) - \xi q_3(\tau) * \frac{d}{d\tau} G_1(\tau) \right],$$

$$u^w(\tau) = \psi_w(\tau)$$
$$+ \Omega \left[\xi q_3(\tau) * \frac{d}{d\tau} G_1(\tau) + \tilde{\Omega} \left(\frac{\tilde{K}}{\tilde{F}} q_1(\tau) - q_3(\tau) \right) * \frac{d}{d\tau} G_2(\tau) \right],$$

$$\mathrm{v}(\tau) = \psi_{\mathrm{v}}(\tau)$$
$$- a f_s \left[q_2(\tau) * H(\tau) + \Omega \tilde{F} \left(q_1(\tau) * \frac{d}{d\tau} G_2(\tau) - \xi q_3(\tau) * G_1(\tau) \right) \right],$$

$$q_1(\tau) = r_l(\tau) p(\tau), \qquad l = 1,2,3, \qquad r_1(\tau) = F(\theta T_*)\mathrm{v}(\tau) - \tilde{F}\tilde{\mathrm{v}},$$
$$r_2(\tau) = F(\theta T_*) - \tilde{F}, \qquad r_3(\tau) = K(\theta T_*)\mathrm{v}(\tau) - \tilde{K}\tilde{\mathrm{v}}, \tag{5.13}$$

where

$$\psi_{\theta n}(x,\tau) = \tilde{\Omega}\varphi_u(\tau) * \frac{\partial}{\partial \tau} g_n(x,\tau), \qquad \psi_p(\tau) = \varphi_u(\tau) + \tilde{\Omega}\varphi_u(\tau) * \frac{d}{d\tau} G_3(\tau),$$

$$\psi_u(\tau) = \tilde{\Omega}\tilde{\xi}\varphi_u(\tau) * \frac{d}{d\tau} G_1(\tau),$$

$$\psi_{\mathrm{v}}(\tau) = \mathrm{v}_0 + a[f_0 \varphi_F(\tau) * H(\tau) - f_{\mathrm{a}}\tilde{F}(\varphi_u(\tau) * H(\tau) + \tilde{\Omega}\varphi_u(\tau) * G_3(\tau))],$$

$$g_1(\tilde{L}_1,\tau) = g_2(\tilde{L}_1,\tau) = g(\tau) = \sum_{m=1}^{\infty} \frac{D_1(s_m) D_2(s_m)}{\Delta'(s_m)} \exp(s_m \tau),$$

$$G_j(\tau) = \frac{1}{\tilde{\Omega}\gamma_j} + \sum_{m=1}^{\infty} \frac{\Delta_j(s_m)}{s_m \Delta'(s_m)} \exp(s_m \tau), \qquad j = 1,2,3,$$

$$\Delta'(s_m) = \frac{d}{ds}\Delta(s)\bigg|_{s=s_m}, \qquad \Delta_1(s) = \frac{1}{\tilde{R}_1} M_1(s) D_2(s) + \frac{1}{\tilde{R}_2} M_2(s) D_1(s),$$

$$\Delta_2(s) = \alpha_1^L N_1(s) D_2(s) + \alpha_2^L N_2(s) D_1(s), \qquad \Delta_3(s) = s\Delta_2(s) - \tilde{\xi}\Delta_1(s),$$

$$\Delta(s) = s\Delta_1(s) - \tilde{\Omega}\Delta_3(s), \qquad D_n(s) = C_n + \mathrm{Bi}_n^L S_n,$$

$$N_n(s) = S_n + \mathrm{Bi}_n^L C_{n0}, \qquad M_n(s) = \mathrm{Bi}_n^L C_n + \eta_n^2 S_n,$$

$$C_n = \cosh(\eta_n), \qquad S_n = \frac{\sinh(\eta_n)}{\eta_n}, \qquad C_{n0} = \frac{1 - C_n}{s}, \qquad \eta_n = \sqrt{\frac{s}{\mathrm{Fo}_n}},$$

$$\tilde{R}_n = \frac{\tilde{L}_n \lambda_*}{K_n}, \qquad \alpha_n^L = \frac{\tilde{\alpha}_n \tilde{L}_n}{\alpha_*}, \qquad \mathrm{Bi}_n^L = \mathrm{Bi}_n \tilde{L}_n, \qquad \mathrm{Fo}_n = \frac{a_n t_*}{L_n^2},$$

$$\mathrm{Bi}_n^1 = 1 + \mathrm{Bi}_n^L, \qquad \mathrm{Bi}_n^2 = 1 + \mathrm{Bi}_n^L/2, \qquad n = 1,2,$$

$$\gamma_1 = \tilde{\xi}, \qquad \gamma_2 = \tilde{\xi} b_0 a_0^{-1}, \qquad \gamma_3 = -1, \qquad \tilde{\Omega} = \Omega \tilde{v} \tilde{F}, \qquad \tilde{\xi} = \xi \tilde{K} \tilde{F}^{-1},$$

$$b_0 = \frac{1}{\tilde{R}_1} \mathrm{Bi}_1^L \mathrm{Bi}_2^1 + \frac{1}{\tilde{R}_2} \mathrm{Bi}_2^L \mathrm{Bi}_1^1, \qquad a_0 = \alpha_1^L \mathrm{Bi}_1^2 \mathrm{Bi}_2^1 + \alpha_2^L \mathrm{Bi}_2^2 \mathrm{Bi}_1^1.$$

The asterisk '$*$' denotes the convolution of functions with respect to the time variable; $H(\tau)$ is Heaviside's step distribution; s_m are the roots of the characteristic equation $\Delta(s) = 0$; $\mathrm{Re} s_m \geq \mathrm{Re} s_{m+1}, m = 1,2,\ldots$; and $\tilde{v}$, $\tilde{F}$, $\tilde{K}$ are the arbitrary parameters. They provide a possibility to obtain the limiting cases, and allow for an efficient construction of the numerical algorithm associated with the considered problem.

From the analysis of the roots of the characteristic equation, and for small values of parameter ξ, we find that $\mathrm{Re}(s_m) < 0, \mathrm{Im}(s_m) = 0$, for $m = 3,4,\ldots$, whereas for $m = 1,2$ the roots lie either on the left- or on the right-hand side of the complex plane s depending on the value of the parameters. For the root of the greatest value on the right-hand side of the complex plane s we obtain $s_1 = (\xi_* - \xi)\Omega$ for $\Omega \to \infty$, where

$$\xi_* = r(\alpha_1^L \sqrt{\mathrm{Fo}_1} + \alpha_2^L \sqrt{\mathrm{Fo}_2}), \qquad r = \left(\frac{1}{\tilde{R}_1 \sqrt{\mathrm{Fo}_1}} + \frac{1}{\tilde{R}_2 \sqrt{\mathrm{Fo}_2}} \right)^{-1}. \tag{5.14}$$

If $\xi < \xi_*$ then for $0 < \Omega < \mathrm{v}_2$ roots s_1 and s_2 are negative. For $\mathrm{v}_2 < \Omega < \mathrm{v}_3$ the roots are complex conjugate; moreover when Ω passes v_1 then roots s_1 and s_2 pass from the left-hand side to the right-hand side of the complex plane s. For $\Omega > \mathrm{v}_3$ the roots are real and positive. However, for $\xi > \xi_*$ the roots always lie in the left half of the complex plane. Let us note that in the case of quasi-stationary problems of thermoelasticity the roots of the characteristic equation lie on the real axis only. In the case when wear is taken into account the roots of the characteristic equation become complex.

Some properties of the function used in solving Equation (5.13) are given below:

$$G_1(\tau) = \tau + O(\tau^2), \qquad G_2(\tau) = \xi_* \tau^2/2 + O(\tau^3),$$

$$G_3(\tau) = (\xi_* - \tilde{\xi})\tau + O(\tau^2),$$

$$g(\tau) = 2r\sqrt{\tau/\pi} + O(\sqrt{\tau^3}), \quad G_3(\tau) = \frac{d}{d\tau}G_2(\tau) - \tilde{\xi}G_1(\tau). \tag{5.15}$$

5.3. Algorithm of solutions

Equations (5.13) are integral and nonlinear of the Volterra–Hammerstein type [Verlan, Sizikov (1986)]. A solution is found using an iterative method and applying the trapezoidal rule. The following input parameters are taken: $\tilde{F} = F(0)$, $\tilde{K} = K(0)$, $\tilde{\mathrm{v}} = \mathrm{v}_0$. For the time instants $\tau = \tau_i = i\,h$, $i = 1, 2, \ldots$, where h is the mesh step, one gets

$$\theta_i^{(j)} = \psi_{\theta i} + \Omega h \left[\frac{1}{2} q_{1,i-1} g_1' + (1 - \delta_{i1})(1 - \delta_{i2}) \sum_{m=1}^{i-2} q_{1,m} g_{i-m}' \right.$$
$$\left. + \frac{2}{3}\sqrt{\frac{1}{\pi h}} r (2q_{1,i}^{(j-1)} + q_{1,i-1}) + \tilde{\Omega}(1 - \delta_{i1}) \sum_{m=1}^{i-1} (\tilde{\xi} q_{1,m} - \xi q_{3,m}) g_{i-m} \right],$$

$$p_i^{(j)} = \frac{\psi_{pi} + \Omega h (1 - \delta_{i1}) \left[\sum_{m=1}^{i-1} q_{1,m} G_{2,i-m}'' - \xi \sum_{m=1}^{i-1} q_{3,m} G_{1,i-m}' \right]}{1 - \Omega h (\xi_* r_{1,i}^{(j-1)} - \xi r_{3,i}^{(j-1)})/2},$$

$$\mathrm{v}_i^{(j)} = \psi_{\mathrm{v}i} - a f_s h \left[(1 - \delta_{i1}) \sum_{m=1}^{i-1} q_{2,m} + \frac{1}{2} q_{2,i}^{(j-1)} \right.$$
$$\left. + \Omega \tilde{F} (1 - \delta_{i1}) \left(\sum_{m=1}^{i-1} q_{1,m} G_{2,i-m}' - \xi \sum_{m=1}^{i-1} q_{3,m} G_{1,i-m} \right) \right],$$

$$u_i^w = \psi_{ui} + \Omega h \left[\xi \frac{1}{2} q_{3,i} + \xi (1 - \delta_{i1}) \sum_{m=1}^{i-1} q_{3,m} G_{1,i-m}' \right.$$
$$\left. + \tilde{\Omega}(1 - \delta_{i1}) \sum_{m=1}^{i-1} \left(\frac{\tilde{K}}{\tilde{F}} q_{1,m} - q_{3,m} \right) G_{2,i-m}' \right],$$

$$\theta_i^{(0)} = \theta_{i-1}^{(j_{i-1})}, \qquad p_i^{(0)} = p_{i-1}^{(j_{i-1})}, \qquad \mathrm{v}_i^{(0)} = \mathrm{v}_{i-1}^{(j_{i-1})}, \qquad j = 1, 2, \ldots,$$

$$\theta_0 = 0, \qquad p_0 = 1, \qquad u_0^w = 0, \tag{5.16}$$

where

$$\theta_i = \theta(\tau_i), \qquad p_i = p(\tau_i), \qquad u_i^w = u^w(\tau_i),$$

$$\mathrm{v}_i = \mathrm{v}(\tau_i), \qquad g_i = g(\tau_i),$$

$$q_{l,i} = q_l(\tau_i), \qquad q_{l,i}^{(j-1)} = r_{l,i}^{(j-1)} p_i^{(j-1)}, \qquad G_{l,i} = G_l(\tau_i), \qquad l = 1, 2, 3,$$

$$r_{1,i}^{(j-1)} = F(\theta_i^{(j-1)} T_*)\mathrm{v}_i^{(j-1)} - \tilde{F}\tilde{\mathrm{v}}, \qquad r_{2,i}^{(j-1)} = F(\theta_i^{(j-1)} T_*) - \tilde{F},$$

$$r_{3,i}^{(j-1)} = K(\theta_i^{(j-1)} T_*)\mathrm{v}_i^{(j-1)} - \tilde{K}\tilde{\mathrm{v}},$$

$$\psi_{\theta i} = \psi_\theta(\tau_i), \qquad \psi_{pi} = \psi_p(\tau_i), \qquad \psi_{ui} = \psi_u(\tau_i), \qquad \psi_{\mathrm{v}i} = \psi_\mathrm{v}(\tau_i).$$

The iterative process (5.16) is finished, when for $j = j_i$ the following inequalities are satisfied

$$\left| \frac{p_i^{(j)} - p_i^{(j_i-1)}}{p_i^{(j)}} \right| < \varepsilon, \qquad \left| \frac{\theta_i^{(j)} - \theta_i^{(j_i-1)}}{\theta_i^{(j)}} \right| < \varepsilon, \qquad \left| \frac{\mathrm{v}_i^{(j)} - \mathrm{v}_i^{(j_i-1)}}{\mathrm{v}_i^{(j)}} \right| < \varepsilon.$$

5.4. Solution analysis

Numerical analysis is carried out for the bottom steel layer ($\alpha_2 = 14 \cdot 10^{-6\circ}\mathrm{C}^{-1}$, $\lambda_2 = 21\,\mathrm{W\,m^{-1}} \cdot {}^\circ\mathrm{C}$, $a_2 = 5.9 \cdot 10^{-6}\,\mathrm{m^2\,s^{-1}}$, $\nu_2 = 0.3$, $E_2 = 19 \cdot 10^{10}\,\mathrm{Pa}$) for $L_2 = 0.02\,\mathrm{m}$, $\mathrm{Bi}_2 = 9.52$ and for the upper aluminium layer ($\alpha_1 = 22 \cdot 10^{-6\circ}\mathrm{C}^{-1}$, $\lambda_1 = 173\,\mathrm{W\,m^{-1}} \cdot {}^\circ\mathrm{C}$, $a_2 = 6.72 \cdot 10^{-5}\,\mathrm{m^2\,s^{-1}}$, $\nu_1 = 0.32$, $E_1 = 7.2 \cdot 10^{10}\,\mathrm{Pa}$) for $L_1 = 0.01\,\mathrm{m}$, $\mathrm{Bi}_1 = 9.25$; $V_0 = 1\,\mathrm{m\,s^{-1}}$; $\Omega = 0.36$; $\xi = 0.7 \cdot 10^{-4}$, $f_0 = 0.17 \cdot 10^{-3}$; $f_s = 0.8$; $a = 1.3 \cdot 10^5$. The parameters associated with the steel layer serve as the key ones for derivation of the dimensionless equations. In this case we have $t_* = 67.8\,\mathrm{s}$, $P_* = 5.7 \cdot 10^6\,\mathrm{Pa}$, $T_* = 3.57^\circ\mathrm{C}$. The external load of the tribological system is taken in the form $\varphi_u(\tau) = \varphi_F(\tau) = (1 - \exp(-\delta\tau^2))H(\tau)$ for $\delta = 100$.

Functions $F(T)$, $K(T)$ approximate the experimental results of friction and wear coefficients versus temperature [Chichinadze et al. (1979)], and they are as follows,

$$F(T) = r_1 + \frac{r_2}{[r_3(T - T_m)]^2 + 1},$$

$$K(T) = d_1 + \frac{d_2}{[d_3(T - T_1^\mathrm{w})]^2 + 1} + \frac{d_4}{[d_5(T - T_2^\mathrm{w})]^2 + 1},$$

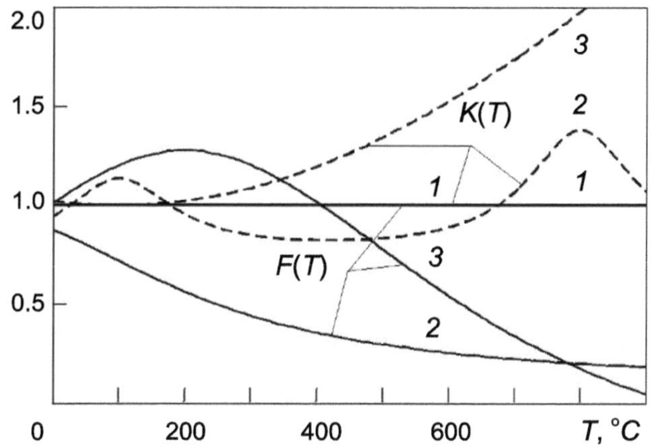

Fig. 5.2: Friction coefficient (solid curves) and wear (dashed curves) versus contact temperature.

where T_m, T_n^w, r_l, d_j are the coefficients obtained in the same way as in the reference [Chichinadze et al. (1979)].

A typical relation of friction coefficient $F(T)$ (solid curves) and wear $K(T)$ (dashed curves) versus contact temperature is shown in Fig. 5.2. Curve 1 corresponds to the case $F(T) = 1$, $K(T) = 1$. Solid curve 2 is associated with the values $r_1 = 0.1$, $r_2 = 0.845$, $r_3 = 3.02 \cdot 10^{-3}{}^\circ\mathrm{C}^{-1}$, $T_m = -100^\circ\mathrm{C}$. Dashed curve 2 corresponds to the values $d_1 = 0.75$, $d_2 = 0.375$, $d_3 = 0.01^\circ\mathrm{C}^{-1}$, $d_4 = 0.625$, $d_5 = 0.01^\circ\mathrm{C}^{-1}$, $T_1^w = 100^\circ\mathrm{C}$, $T_2^w = 800^\circ\mathrm{C}$; curve 3 – $d_1 = 10$, $d_2 = -9$, $d_3 = 5 \cdot 10^{-4}{}^\circ\mathrm{C}^{-1}$, $d_4 = 0$, $T_1^w = 100^\circ\mathrm{C}$.

Time histories of temperature θ, contact pressure p, sliding velocity v, and wear are shown in Fig. 5.3. Curves 1–3 correspond to the coefficients shown in Fig. 5.2. Owing to numerical analysis, the contact temperature achieves its maximum value for the constant values of the friction and wear coefficients. The value of layer wear achieves a maximum for the case of the largest braking time, whereas the friction coefficient decreases with an increase of temperature. The contact pressures are almost the same in the stop instant of the upper layer.

5.5. Frictional thermoelastic instability

In this section a frictional joint model exhibiting a Frictional ThermoElastic Instability (FTEI) is studied. In particular, our attention is focused on the interaction of nonstationary heat generation and wear, and their influence on the occurrence of FTEI.

Owing to a fast contact between the bodies, some instabilities of the contacting parameters are expected yielding a harmful behaviour of the tribological system [Barber (1969), Chichinadze et al. (1979), Barber (1999)]. Recall (see

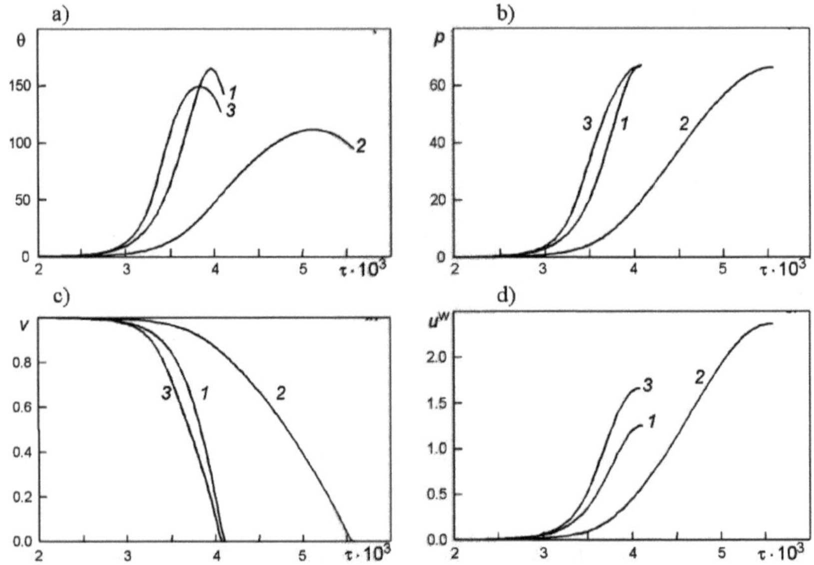

Fig. 5.3: Contact temperature θ (a), contact pressure p (b), sliding velocity v (c), and wear u^w (d) versus time τ during braking.

Section 2.5.1.1) that FTEI is characterized by an exponential increase of the contacting characteristics (i.e., temperature, pressure, and wear) assuming that a relative velocity of sliding bodies overcomes a certain threshold velocity.

Experimental verification of TEI has been presented by [Dow, Stockwell (1977)].

Consider a tribological system which may be modelled by two different layers. One of them moves at a constant velocity on the surface of a second fixed layer (Fig. 5.4).

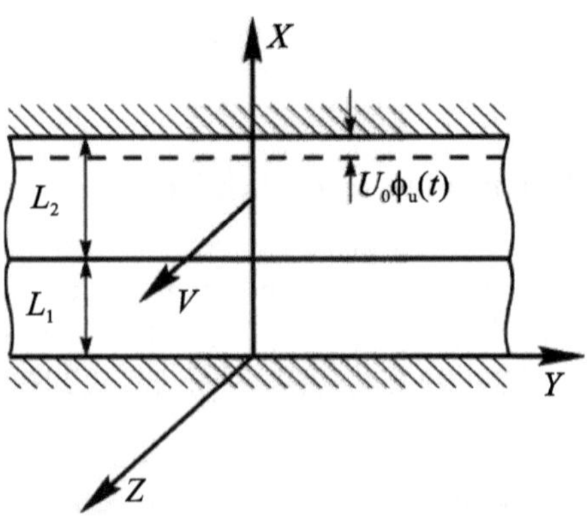

Fig. 5.4: Scheme of the system exhibiting a frictional thermoelastic rubbing instability.

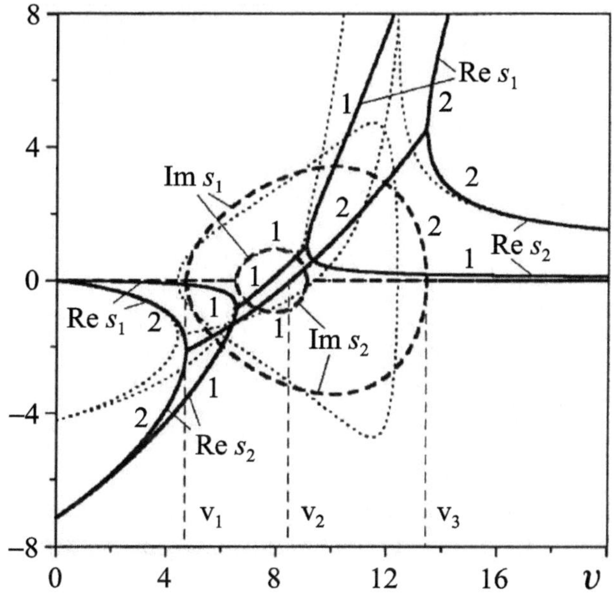

Fig. 5.5: Positions of roots s_1, s_2 of the characteristic equation.

The action of frictional forces produces heat and wear. The introduced simple geometry of contacting bodies enables monitoring of all physical aspects associated with the appearance of FTEI.

Observe that here we assume relatively slow variations of mechanical and thermal loading, which allows us to omit inertial terms and hence enables direct application of quasi-static equations of thermoelasticity in motion equations (1). However, although this approach allows us to study the so-called ThermoElastic Instability (TEI) in our system, it does not enable a study of ThermoElastic Dynamic Instability (TEDI) (see, e.g., [Afferante et al. (2006), Afferrante, Ciavarella (2006), Yi (2006)]).

The solution obtained earlier for constant friction and wear coefficients are used: $F(T) = 1$, $K(T) = 1$ (nonlinear functions take the forms $q_1(\tau) = q_3(\tau) = (\mathrm{v}(\tau) - \tilde{\mathrm{v}})\, p(\tau) \equiv q(\tau)$, $q_2(\tau) = 0$). It is assumed that the layer moves at a constant velocity $\mathrm{v}(\tau) = \tilde{\mathrm{v}} = \mathrm{const}$. This means that $q(\tau) = 0$ and a solution to the started problem is governed by the first term of relation (5.13).

Analysis of stability is limited to the investigation of root locations of the characteristic equation $\Delta(s) = 0$. For the largest root the following estimation holds, $s_1 \approx (\xi_* - \xi)v$, $v \to \infty$ ($v = \tilde{\Omega}$, whereas ξ_* is determined from (5.14)).

Observe that for $\tilde{\xi} < \xi_*$ and $v < \mathrm{v}_2$ roots s_1, s_2 are negative; for $\mathrm{v}_2 < v < \mathrm{v}_3$ the roots are conjugated; and when $v = \mathrm{v}_1$ the real root part changes sign. For $v > \mathrm{v}_3$, the roots are real and positive.

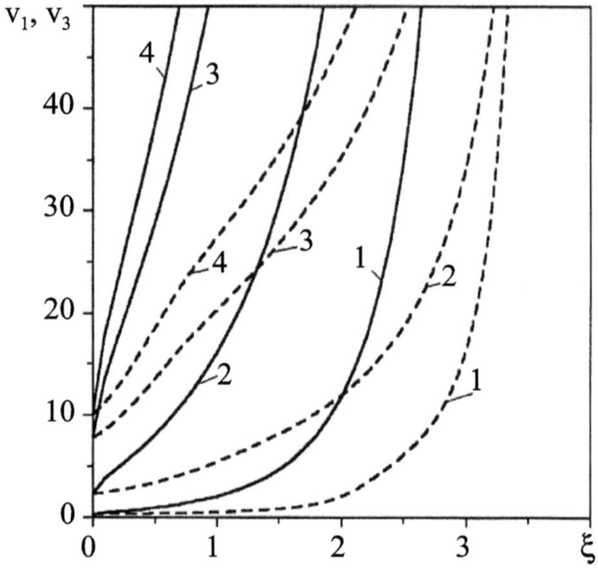

Fig. 5.6: Critical parameters v_3 (solid curves) and v_1 (dashed curves) versus parameter ξ.

For $\tilde{\xi} > \xi_*$ all roots have a negative real part.

In Fig. 5.5, for $\mathrm{Bi}_1 = \mathrm{Bi}_2 = 10$, real (solid curves) and imaginary (dashed curves) parts of the roots versus dimensionless velocity v for $\xi = 0.01$ (curve 1) and $\xi = 0.1$ (curves 2) are shown. For $\xi = 0.1$ the critical values of v_3 (solid curves) and velocity v_1 (dashed curves) ξ for different values Bi are shown in Fig. 5.6. Curves 1–4 correspond to the values of $\mathrm{Bi} = 0.1;\ 1;\ 10;\ \infty$.

The characteristic function $\Delta(s)$ has the form

$$\Delta(s) = \sum_{j=0}^{\infty} s^j d_j, \tag{5.17}$$

where the coefficients d_j follow

$$d_0 = v\xi b_0, \qquad d_1 = b_0 - v(a_0 - \xi(c_0 + b_1)),$$

$$d_j = b_{j-1} - v\,(a_{j-1} - \xi\,(c_{j-1} + b_j)) + c_{j-2}, \qquad j = 2, 3, \ldots$$

$$a_j = \frac{1}{2(2j+1)!}(\mathrm{Fo}_+^{\pm}\alpha_1^L\sqrt{\mathrm{Fo}_1} + \mathrm{Fo}_-^{\pm}\alpha_2^L\sqrt{\mathrm{Fo}_2})$$

$$+\frac{\mathrm{Bi}_2^L\alpha_1^L + \mathrm{Bi}_1^L\alpha_2^L}{2(2j+2)!}\mathrm{Fo}_-^{-}\sqrt{\mathrm{Fo}_1\mathrm{Fo}_2}$$

$$+\sum_{i=0}^{j}\left\{\frac{\mathrm{Bi}_1^L\mathrm{Bi}_2^L}{(2i+1)!(2j-2i+2)!}(\alpha_1^L\mathrm{Fo}_2^{-i}\mathrm{Fo}_1^{i-j}+\alpha_2^L\mathrm{Fo}_1^{-i}\mathrm{Fo}_2^{i-j})\right.$$

$$\left.+\frac{1}{(2i)!(2j-2i+2)!}(\alpha_1^L\mathrm{Bi}_1^L\mathrm{Fo}_2^{-i}\mathrm{Fo}_1^{i-j}+\alpha_2^L\mathrm{Bi}_2^L\mathrm{Fo}_1^{-i}\mathrm{Fo}_2^{i-j})\right\};$$

$$b_j=\frac{\mathrm{Bi}_1^L/\tilde{R}_1+\mathrm{Bi}_2^L/\tilde{R}_2}{2\,(2j)!}\mathrm{Fo}_+^-+\frac{\mathrm{Bi}_1^L\mathrm{Bi}_2^L}{2\,(2j+1)!}\left(\frac{\mathrm{Fo}_+^+\sqrt{\mathrm{Fo}_1}}{\tilde{R}_2}+\frac{\mathrm{Fo}_-^+\sqrt{\mathrm{Fo}_2}}{\tilde{R}_1}\right);$$

$$c_j=\frac{\mathrm{Fo}_-^-}{2\,(2j+2)!}\left(\sqrt{\frac{\mathrm{Fo}_1}{\mathrm{Fo}_2}}\frac{\mathrm{Bi}_1^L}{\tilde{R}_2}+\sqrt{\frac{\mathrm{Fo}_2}{\mathrm{Fo}_1}}\frac{\mathrm{Bi}_2^L}{\tilde{R}_1}\right)$$

$$+\frac{1}{2\,(2j+1)!}\left(\frac{\mathrm{Fo}_+^+}{\tilde{R}_1\sqrt{\mathrm{Fo}_1}}+\frac{\mathrm{Fo}_-^+}{\tilde{R}_2\sqrt{\mathrm{Fo}_2}}\right);\qquad j=0,1,2,\ldots$$

$$Fo^{\pm}=\frac{1}{\sqrt{\mathrm{Fo}_1}}\pm\frac{1}{\sqrt{\mathrm{Fo}_2}},\qquad \mathrm{Fo}_\pm^+=((\mathrm{Fo}^+)^{2j+1}\pm(\mathrm{Fo}^-)^{2j+1}),$$

$$\mathrm{Fo}_-^-=((\mathrm{Fo}^+)^{2j+2}-(\mathrm{Fo}^-)^{2j+2}),\qquad \mathrm{Fo}_+^-=((\mathrm{Fo}^+)^{2j}+(\mathrm{Fo}^-)^{2j}).$$

Owing to the form of the characteristic function, roots s_1, s_2 (in the vicinity of the origin) can be expressed in an analytical form. Namely, the following analytical roots estimation holds

$$\mathrm{v}_1=\frac{b_0}{a_0-\xi(b_1+c_0)},$$

$$\mathrm{v}_{2,3}=b_0\left(\frac{a_0+\xi(b_1+c_0)\mp\sqrt{\xi[a_0(b_1+c_0)-b_0(a_1-\xi(c_1+b_2))]}}{(a_0-\xi(b_1+c_0))^2+4\xi b_0(a_1-\xi(c_1+b_2))}\right).\tag{5.18}$$

In Fig. 5.6 dashed curves correspond to variations of real and imaginary parts of roots s_1, s_2 computed analytically versus dimensionless velocity v. Three first terms of the series (5.17) are included in the computations.

Solution properties. Relations defining v_1, v_2, and v_3 enable prediction of the behaviour of thermoelastic contact characteristics in the conditions of frictional wear and heat expansion. As an example, the characteristics of thermoleastic contact for $\varphi_u(\tau)=H(\tau)$ are considered.

In the initial time instant, a solution is taken in the form

$$p(\tau)=1+v(\xi-\xi_*)\tau,\qquad u^w(\tau)=v\xi\tau,$$

$$\theta(\tau)=\frac{2v}{(\tilde{R}_1\sqrt{\mathrm{Fo}_1})^{-1}+(\tilde{R}_2\sqrt{\mathrm{Fo}_2})^{-1}}\sqrt{\frac{\tau}{\pi}},\qquad \tau\to 0,$$

where $\theta(\tau)$ is the contact temperature ($\theta(\tau) = \theta_1(\tilde{L}_1, \tau) = \theta_2(\tilde{L}_1, \tau)$). For $\xi < \xi_*$ (the wear value is lower than heat expansion), and for $v < \mathrm{v}_2$ time durability of the contact $\tau_c = \infty$, and the contact characteristics tend to stationary value $p_{st} = 0$, $\theta_{st}(x) = 0$, $u^w_{st} = 1$ with time. The time interval to achieve the mentioned points increases when $v \to \mathrm{v}_2$. For $\mathrm{v}_2 < v < \mathrm{v}_3$ the time of the contact is bounded. For $v \to \mathrm{v}_3$ the maximum values of pressure, temperature, and wear increase.

When velocity v is greater than the critical value v_3, both temperature and contact pressure increase exponentially; that is, a frictional thermoelastic instability occurs (the system does not keep pace with cooling down).

For $\xi > \xi_*$ (the wear value is greater than heat expansion) and for $v < \mathrm{v}_2$, the contact characteristics also tend to the values of stable equilibrium positions. For $v > \mathrm{v}_2$ the contact time is bounded, although a formally stable solution does exist.

Numerical analysis. Numerical analysis is carried out for the steel and aluminium layers for $U_0 = 10^{-6}$ m, $\mathrm{Bi}_1 = \mathrm{Bi}_2 = 10$.

In Fig. 5.7 time histories of contact pressure $p(\tau)$ and wear $u^w(\tau)$ for various dimensionless velocity v are reported (1, $v = 6$; 2, $v = 8$; 3, $v = 10$; 4, $v = 14$). Solid curves correspond to the value of $\xi = 0.01$. In this case $\mathrm{v}_1 = 7.77$ ($\mathrm{v}_1 \approx 7.84$), $\mathrm{v}_3 = 9.1$ ($\mathrm{v}_3 \approx 9.04$). Dashed curves correspond to the value of $\xi = 0.1$; that is, in this case $\mathrm{v}_1 = 8.53$ ($\mathrm{v}_1 \approx 9.4$), $\mathrm{v}_3 = 13.5$ ($\mathrm{v}_3 = 12.39$). In the brackets the approximated values for certain characteristic velocities are given, and they are derived from relations (5.18). The numerical analysis confirms that for $v > \mathrm{v}_3$ a frictional thermoelastic instability occurs (see solid curves 3 and dashed curves 4 in Fig. 5.7).

Owing to velocity increase, time duration of the contact decreases, whereas maximum contact pressure values, contact temperature, and wear increase. For the constant velocity, an increase of the wear coefficient ξ causes a decrease of the

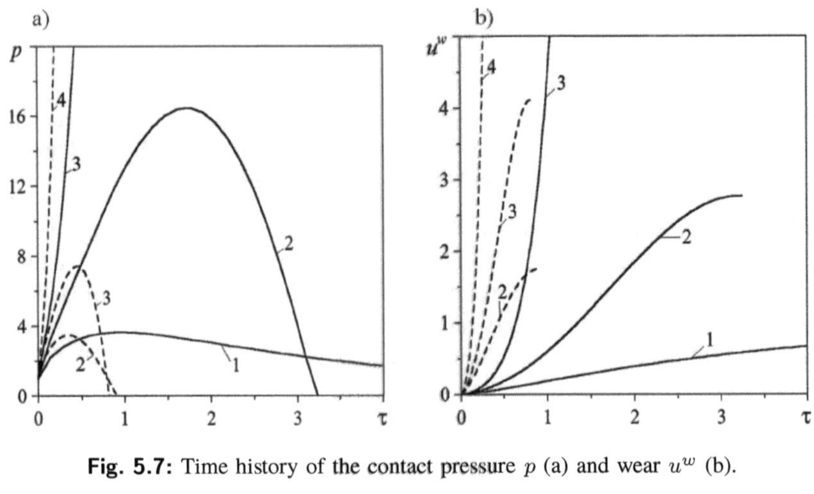

Fig. 5.7: Time history of the contact pressure p (a) and wear u^w (b).

contact time duration; that is, an intensive wear increase at the beginning occurs and yields a decrease of the total amount of wear.

Conclusions. Finally, let us briefly summarize the results obtained in this chapter. The frictional contact of two layers with constant thickness has been investigated. Inertia of the contacting bodies has been included. Both friction and wear dependence on the contact temperature as well as the problem of drawing wear of two layers have been examined. The proposed model can be used for computation of the contact characteristics during acceleration and braking processes.

Note that in the case of constant coefficients of friction and wear $F(T) = 1$, $K(T) = 1$, nonlinear functions take the form $q_1(\tau) = q_3(\tau) = (\mathrm{v}(\tau) - \tilde{\mathrm{v}})p(\tau) \equiv q(\tau)$, $q_2(\tau) = 0$. In the case when the layer moves at the constant velocity $\mathrm{v}(\tau) = \tilde{\mathrm{v}} = \mathrm{const}$, $q(\tau) = 0$ and a solution of the corresponding linear problem is governed by the first term of series (5.13).

One-layer movement with constant velocity and with constant friction and wear has been analysed earlier by [Grilitskiy (1996)], whereas a similar but nonlinear model has been studied in reference [Aleksandrov, Annakulova (1992)]. Also a similar problem devoted to consideration of one layer with a constant friction coefficient and without (with) wear has been examined by [Pyryev (1994), (2000a)] ([Olesiak, Pyryev (1998)]).

In addition, in this chapter the mechanism of occurrence of frictional thermoelastic instability of the contacting bodies that move at a constant relative velocity is investigated. The proposed model can govern the behaviour of movable contact-type seals, braking pads, or other tribological systems. An equation yielding critical velocities is derived. It is illustrated how an account of wear yields an increase of the critical speed, and hence occurrence of the frictional thermoelastic instability. In the case when wear is greater than a heat expansion, frictional thermoelastic instability does not occur. It is worth noting that a frictional thermoelastic instability (in the case of wear account) appears when the roots of the characteristic equation lie not only in the right half-plane of the Laplace transform variable ($\upsilon > \mathrm{v}_1$, $\xi < \xi_*$), but also in the real axis ($\upsilon > \mathrm{v}_3$).

The results obtained in this chapter can be used in computation of material strength of movable joints of machines and mechanisms, to formulate criteria related to optimal choice of materials of the frictional pairs, and to investigate heat transfer and wear in transporting mechanisms during braking or acceleration processes.

References

Abramowitz M., Stegun I., 1965. Handbook of Mathematical Functions: with Formulas, Graphs, and Mathematical Tables. Dover, New York, NBS.

Afferante L., Ciavarella M., Barber J.R., 2006. Sliding ThermoElastoDynamic instability. Proc. Roy. Soc. London, Ser. A, **462**, 2161–2176.

Afferrante L., Ciavarella M., 2006. "Frictionless" and "frictional" ThermoElastic Dynamic Instability (TEDI) of sliding contacts. J. Mech. Phys. of Solids, **54**, 11, 2330–2353.

Afferrante L., Ciavarella M., 2007. Thermo-elastic dynamic instability (TEDI) in frictional sliding of two elastic half-spaces. J. Mech. Phys. of Solids, **55**, 4, 2007, 744–764.

Agelet de Saracibar, C., Chiumenti, M., 1999. On the modeling of frictional wear phenomena. Comput. Methods. Appl. Mech. Engrg., **177**, 401–426.

Aleksandrov V.M., Annakulova G.K., 1990. Contact problem of thermoelasticity under the conditions of wear and heat release induced by friction. Friction and Wear (Treniye i Iznos), **11**, 1, 24–28, in Russian.

Aleksandrov V.M., Annakulova G.K., 1992. Interaction of the coatings bodies allowing for deformability, wear and heat release due to friction. Friction and Wear (Treniye i Iznos), **13**, 1, 154–160, in Russian.

Alexandrov V.M., Pozharskii D.A., 2001. Three-Dimensional Contact Problems. Dordrecht, Kluwer Academic.

Aleksandrov V.M., Romalis G.K., 1986. Contact Problems in Machines Construction. Moscow, Mashinostroyeniye, in Russian.

Andreaus U., Casini P., 2002. Dynamics of friction oscillators excited by a moving base and/or driving force. Math. Comput. Model., 36, 259–273.

Andronov A.A., Witt A.A., Khaikin S.E., 1966. Theory of Oscillations. Oxford, Pergamon Press.

Andrzejewski R., Awrejcewicz J., 2005. Nonlinear Dynamics of a Wheeled Vehicle. Berlin, Springer.

Archard J.F., 1953. Contact and rubbing of flat surface. J. Appl. Physics, **24**, 8, 981–988.

Archard J.F., 1959. The temperature of rubbing surfaces. Wear, **2**, 6, 438–455.

Arnold V.I., Kozlov V.V., Neishtadt A.I., 1997. Mathematical Aspects of Classical and Celestial Mechanics. Berlin, Springer-Verlag.

Awrejcewicz J., 1988. Chaotic motion in a nonlinear oscillator with friction. J. Korean Soc. Mech. Eng., **2**, 22, 104–109.

Awrejcewicz J., 1989. Bifurcation and Chaos in Simple Dynamical Systems. Singapore, World Scientific.

Awrejcewicz J., 1991. Bifurcation and Chaos in Coupled Oscillators. Singapore, World Scientific.

Awrejcewicz J., 1996. Oscillations of Lumped Deterministic Systems. Warsaw, WNT, in Polish.

Awrejcewicz J., Andrianov I.V., Manievich L.I., 1998. Asymptotic Approaches in Nonlinear Dynamics: New Trends and Applications. Berlin, Springer-Verlag.

Awrejcewicz J., Grzelczyk D., Pyryev Y., 2007. On the stick-slip vibrations continuous friction model. Proceedings of the 9th Conference on Dynamical Systems – Theory and Applications, Eds: J. Awrejcewicz, P. Olejnik, J. Mrozowski, Lódz, Poland, December 17–20, 2007, 113–120.

Awrejcewicz J., Holicke M., 1999. Melnikov's method and stick-slip chaotic oscillations in very weakly forced mechanical systems. Int. J. Bifurcation Chaos, **9**, 3, 505–518.

Awrejcewicz J., Holicke M.M., 2007. Smooth and Nonsmooth High Dimensional Chaos and the Melnikov-Type Methods. Singapore, World Scientific.

Awrejcewicz J., Krysko V.A., 2003. Nonclassical Thermoelastic Problems in Nonlinear Dynamics of Shells. Berlin, Springer-Verlag.

Awrejcewicz J., Krysko V.A., Vakakis A.F., 2004. Nonlinear Dynamics of Continuous Elastic Systems. Berlin, Springer-Verlag.

Awrejcewicz J., Kudra G., Lamarque C.H., 2002. Nonlinear dynamics of triple pendulum with impacts. J. Tech. Physics, **43**, 2, 97–112.

Awrejcewicz J., Lamarque C. H., 2003. Bifurcation and Chaos in Nonsmooth Mechanical Systems. New Jersey, London, Singapore, World Scientific.

Awrejcewicz J., Mrozowski J., 1989. Bifurcations and chaos of a particular Van der Pol-Duffing's oscillator. J. Sound Vibration, **132**, 1, 89–100.

Awrejcewicz J., Olejnik P., 2005. Analysis of dynamics systems with various friction laws. Appl. Mech. Rev., **58**, 6, 389–411.

Awrejcewicz J., Pyryev Yu., 2002. Thermoelastic contact of a rotating shaft with a rigid bush in conditions of bush wear and stick-slip movements. Int. J. Engng. Sci., **40**, 1113–1130.

Awrejcewicz J., Pyryev Yu., 2003a. De Saint-Venant principle and an impact load acting on an elastic half- space. J. Sound Vibration, **264**, 1, 245–251.

Awrejcewicz J., Pyryev Yu., 2003b. Determination of deflection of elastic half space yielded by impact loading. X-th Scientific Conference on Vibrotechnics and Vibroacoustics in Technical Systems, Wibrotech 2003. Krakov, CD-ROM, 1–6, in Polish.

Awrejcewicz J., Pyryev Yu., 2003c. Influence of Tribological Processes on a Chaotic Motion of a Bush in a Cylinder-Bush System. Meccanica, **38**, 6, 749–761.

Awrejcewicz J., Pyryev Yu., 2004a. Tribological periodic processes exhibited by acceleration or braking of a shaft-pad system. Commun. Nonlinear Sci. Numer. Simul., **9**, 6, 603–614.

Awrejcewicz J., Pyryev Yu., 2004b. Contact phenomena in braking and acceleration of bush-shaft system. J. Thermal Stresses, **27**, 5, 433–454.

Awrejcewicz J., Pyryev Yu., 2005. Thermo-mechanical model of frictional self-excited vibrations. Int. J. Mech. Sci., **47**, 9, 1393–1408.

Awrejcewicz J., Pyryev Yu., 2006a. Chaos prediction in the Duffing-type system with friction using Melnikov's functions. Nonlinear Anal. Real World Appl., **7**, 1, 12–24.

Awrejcewicz J., Pyryev Yu., 2006b. Dynamics of a two-degrees-of-freedom system with friction and heat generation. Special Issue of Commun. Nonlinear Sci. Numer. Simul., **11**, 5, 635–645.

Awrejcewicz J., Pyryev Yu., 2006c. Regular and chaotic motion of a bush-shaft system with tribological processes. Mathematical Problems in Engineering, 2006, 5, 13 pages (DOI: 10.1155/MPE/2006/86594).

Awrejcewicz J., Pyryev Yu., 2007. Dynamical damper of vibration with thermoelastic contact. Arch. Appl. Mech., **77**, 281–291.

Baker G.L., Gollub J.P., 1996. Chaotic Dynamics: An Introduction, 2nd ed. Cambridge, Cambridge University Press.

Balandin D., 1993. Frictional vibrations in a gap. Izv. Ross. Akad. Nauk. Mekh. Tverd. Tela, 1, 54–60, in Russian.

Balandin D., Bolotnik N., Pilkey W., 2001. Optimal Protection from Impact, Shock and Vibration. Toronto, Gordon and Breach Science.

Banerjee A.,K., 1968. Influence of kinetic friction on the critical velocity of stick-slip motion. Wear, **12**, 107–116.

Barber J.R., 1969. Thermoelastic instabilities in the sliding of conforming solids. Proc. Roy. Soc. London Ser. A, **312**, 381–394.

Barber J.R., 1973. Indentation of the semi-infinite elastic solid by a hot sphere. Int. J. Mech. Sci., **15**, 813–819.

Barber J.R., 1975. Thermoelastic contact of a rotating sphere and a half-space. Wear, **35**, 2, 283–289.

Barber J.R., 1976. Some thermoelastic contact problems involving frictional heating. Q. J. Mech. Appl. Math., **29**, 1–13.

Barber J.R., 1999. Thermoelasticity and contact. J. Thermal Stresses, **22**, 513–525.

Barber J.R., Beamond T. W., Waring J. R., Pritchard C., 1985. Implications of thermoelastic instability for the design of brakes. J. Tribol., **107**, 206–210.

Barber J.R., Ciavarella M., 2000. Contact mechanics. Int. J. Solids Struct., **37**, 29–43.

Barber J.R., Comninou M., 1989. Thermoelastic contact problems. Thermal Stresses III (ed. R. Hetnarski), North Holland, Amsterdam, 1–106.

Barber J.R., Zhang R., 1988. Transient behaviour and stability for the thermoelastic contact of two rods of dissimilar materials. Int. J. Mech. Sci., **30**, 9, 691–704.

Batako A. D., Babitsky V.I., Halliwell N.A., 2003. A self-excited system for percussive-rotary drilling. J. Sound Vibration, **259**, 1, 97–118.

Belajev N.M., 1945. Strength of Materials. Moscow-Leningrad, Gos. Izd. Tech.-Teor. Lit., in Russian.

Bell R., Burdekin M., 1969. A study of stick-slip motion of machine tool feed drives. Proc. Inst. Mech. Engrs., **184**, 1, 543–557.

Bielski W. R., Telega J.J., 2001. Modelling contact problems with friction in fault mechanics. J. Theor. Appl. Mech. Wars., **39**, 3, 475–505.

Blau P.J., 1996. Friction Science and Technology, New York, Decker.

Blok H., 1937. Teoretical study of temperature rise at surfaces of actual contact under oiliness lubricating conditions. Proceedings of the General Discussion on lubrication and lubricants. London, Institute of Mechanical Engineers, 1–14.

Blok H., 1937. Les Températures de Surface Dans Des Conditions de Graissage Sous Extrême Pression, Proceedings of 2nd World Congress Paris, **3**, 471–486.

Blok H., 1940. Fundamental mechanical aspects of boundary lubrication. S.A.E. J., **40**, 2, 54–68.

Bo L.C., Pavelesku D., 1976. Stability criterion for stick-slip motion using a discontinuous dynamic friction model. Wear, **40**, 113–120.

Bogolubov N.N., Mitropolski Yu.A., 1961. Asymptotic Methods in the Theory of Nonlinear Oscillations. New York, Gordon and Breach.

Boley B.A., Weiner J.H., 1960. Theory of Thermal Stresses. New York, London, John Wiley and Sons.

Bowden F.P., Ridler K.E.W., 1935. A note on the surface temperature of sliding metals. Proc. Camb. Philos. Soc., **31**, 431–432.

Bowden F.P., Ridler K.E.W., 1936. Physical properties of surfaces. Part III: The surface temperature of sliding metals. The temperature of lubricated surfaces. Proc. R. Soc. Lond., Ser. A, **154**, 640–656.

Brockley C.A., Cameron R., Potter A.F., 1967. Friction-induced vibration. ASME J. Lubric. Technol., **89**, 101–108.

Brogliato B., 1999. Nonsmooth Mechanics. London, Springer.

Burton R.A., Nerlikar V., Kilaparti S.R., 1973. Thermoelastic instability in a seal-like configuration. Wear, **24**, 189–198.

Capone G., D'Agostino V., Della Valle S.,. Guida D., 1993. Influence of the variation between static and kinetic friction on stick-slip instability. Wear, **161**, 121–126.

Carslaw H.S., Jaeger J.C., 1959. Conduction of Heat in Solids. Oxford, Clarendon Press.

Chichinadze A.V., 1995. Processes in heat dynamics and modelling of Friction and Wear (dry and boundary friction). Tribology International, **8**, 1, 55–58.

Chichinadze A.V., Braun E.D., Ginsburg A.G., Ignat'eva Z.V., 1979. Calculation, testing, and selection of friction couples. Moscow, Nauka, in Russian.

Chin J. H., Chen C. C., 1993. A study of stick-slip motion and its influence on the cutting process. Int. J. Mech. Sci., **35**, 5, 353–370.

Ciavarella M., Barber J. R., 2005. Stability of thermoelastic contact for a rectangular elastic block sliding against a rigid wall. Euro. J. Mech. A (Solids), **24**, 371–376.

Cockerham G., Cole M., 1976. Stick-slip stability by analogue simulation. Wear, **36**, 189–198.

Cockerham G., Symmons G.R., 1976. Stability criterion for stick-slip motion using a discontinuous dynamic friction model. Wear, **40**, 113–120.

Conway H.D., Farnham K.A., Ku T.C., 1967. The indentation of a transversely isotropic half-space by a rigid sphere. Trans. ASME, J. Appl. Mech., **34**, 2, 491–492.

Dao N.V., Dinh N.V., 1999. Interaction between nonlinear oscillating systems. Hanoi, Vietnam National University Publishing House.

Den Hartog, J.P., 1952. Mechanical Vibrations. Berlin, Springer-Verlag.

Derjagin B.V., Push V. E., Tolstoj D.M., 1957. A theory of stick-slip sliding of solids. Proc. Conf. on Lubrication and Wear, Inst. Mech. Engrg., London, 257–268.

Dow T.A., Stockwell R.D., 1977. Experimental verification of thermoelascic instabilities in sliding contact. Trans. ASME, J. Lubr. Technol., **99**, 3, 359–364.

Dragon-Louiset M., 2001. On a predictive macroscopic contact-sliding wear model based on micromechanical considerations. Int. J. Solids Struct., **38**, 1625–1639.

Engel Z., 1993. Environment Protection Against Vibration and Noise. Warsaw, PWN, in Polish.

Fazekas G.A.G., 1953. Temperature gradients and heat stresses in brake drums, SAE Trans., **61**, 279–284.

Feeny B.F., Moon F.C., 1993. Bifurcation sequences of a Coulomb friction oscillator. Nonlinear Dynamics, **4**, 25–37.

Filippov A.F., 1988. Differential equations with discontinuous right-hand sides. Mathematics and Its Applications. Dordrecht, Kluwer Academic.

Finigenko I.A., 2001. On right-hand solution for a class discontinuous systems. Automatica i Telemechanika, 9, 149–158, in Russian.

Fremond M., 2002. Non-Smooth Thermomechanics. Berlin, Springer Verlag.

Galin L.A., 1961. Contact problems in the classical theory of elasticity (Engl. Trans. Edited by Sneddon, I.N.). Tech. Rep. G16447. North Carolina State College, Raleigh, NC.

Ganghoffer J.F., Schultz J., 1995. A deductive theory of friction. Wear, **188**, 88–96.

Generalov M.B., Kudryavtsev B.A., Parton V.Z., 1976. The contact problem of thermoelasticity for rotating bodies. Izv. Akad. Nauk SSSR. Mekh. Tverd. Tela, 3, 46–52, in Russian.

Giergiel J., 1990. Damping of Mechanical Vibrations. Warsaw, PWN, in Polish.

Gladwell G.M.L., 1980. Contact Problems in the Classical Theory of Elasticity. Alphen aan den Rijn, Sijthoff and Noordhoff.

Gladwell G.M.L., Barber J.R. Olesiak Z.S.,1983. Thermal problems with radiation boundary conditions, Q. J. Mech. Appl. Math., **36**, 387–401.

Goryacheva I.G., 1998. Contact mechanics in tribology, Dordrecht, Kluwer.

Grinchenko V.T., Ulitko A.F., 1999. Local singularities in mathematical models of physical fields. J. Math. Sci. (New York), **97**, 1, 3777–3795.

Grudziński K., Warda B., Zapłata M., 1995. Influence of mass and elastic system parameters on friction and self-excited frictional vibrations. Archive Technol. Mach. Automat., **14**, 209–224.

Grudziński K., Wedman S., 1998. Simulating of stick-slip motion during kinematic external excitations, XXXVII Symposium "Modeling in Mechanics", 135–142.

Grilitskiy D., 1996. Thermoelastic Contact Problems in Tribology. Kyiv, In-t zmistu i metodiv navchannia MO Ukrainy, in Ukrainian.

Grilitskiy D., Pyryev Yu., Mandzyk Yu., 1997. Quasistatic thermoelastic contact problem for infinite two-layer circular cylinder under friction heating. J. Thermal Stresses, **20**, 47–65.

Grilitskiy D., Pyryev Yu., Mandzyk Yu., 1998. The quasistatic contact problem of thermoelasticity for a two-layer cylinder under frictional heating and nonideal thermal contact. J. Math. Sci., New York, **88**, 3, 413–418.

Gu R.J., Shillor M., 2001. Thermal and wear analysis of an elastic beam in sliding contact, Int. J. Solids Struct., **38**, 2323–2333.

Guckenheimer J., Holmes P., 1983. Nonlinear Oscillations, Dynamical Systems, and Bifurcations of Vector Fields. New York-Berlin-Heidelberg-Tokyo, Springer-Verlag.

Halling J., 1975. Principles of Tribology. London, Macmillan Press.

Hamilton G.M., Goodman L.E., 1966. The stress field created by a circular sliding contact. Trans. ASME, J. Appl. Mech., **33**, 371–376.

Haraldsson A., Wriggers P., 2000. A strategy for numerical testing of frictional laws with application to contact between soil and concrete. Comput. Meth. Appl. Mech. Engrg., **190**, 963–977.

Hasselgruber H., 1963. Der schalfvorgang einer trockenreibungs Kupplung bei kleinster Erwärmung. Konstruktion, **15**, 41–45.

Hayashi Ch., 1968. Nonlinear Oscillations in Physical Systems. New York, McGraw-Hill.

Hebda M., Chichinadze A.V., 1989, Vol. 1; 1990, Vol. 2; 1992, Vol. 3. Reference Manual on Tribology. Moscow, Mashinostrojenie, in Russian.

Hebda M., Wachal A., 1980. Tribology. Warsaw, WNT, in Polish.

Hertz H., 1882. Über die Berührung fester elastischer Körper, J. reine und angewandte Mathematik, **92**, 156–171.

Hertz H., 1895. Gesammelte Werke. Bd. I. Johann Ambrosius Barth, Leipzig, 179–195.

Hess D., Soom A., 1991. Normal vibration and friction under harmonic loads: Part I – Hertzian contacts, part II – rough planar contacts. J. Tribol., **113**, 80–86, 87–92.

Hills D.A., Nowell D., Sackfield A.,1993. Mechanics of Elastic Contacts, Oxford, Butterworth.

Ibrahim R. A., 1994. Friction-induced vibration, chatter, squeal and chaos. Part 1: Mechanics of contact and friction. Part 2: Dynamics and modeling. Appl. Mech. Rev., **47,** 7, 209–253.

Jaeger J., C., 1942. Moving sources of heat and the temperature at sliding contacts. Proc. R. Soc., N.S.W., **76**, 203–224.

Jaeger J., 2005. New Solutions in Contact Mechanics. Southampton: WIT Press.

Jarzebowski A. Mróz Z., 1994. On slip and memory rules in elastic, friction contact problems. Acta Mechanica, **102**, 199–216.

Johansson L., Klarbring A., 1993. Thermoelastic friction contact problems: modeling, finite element approximation and numerical realization. Comp. Mech. Appl. Mech. Engng., **105**, 181–210.

Jonson K., 1985. Contact Mechanics. Cambridge, Cambridge University Press.

Kaczyński A., Matysiak S.J., 1988. Plane contact problems for a periodic two-layered elastic composites. Ingenieur Archiv., **58**, 137–147.

Kaczyński A., Matysiak S.J., 1993. Rigid sliding punch on a periodic two-layered elastic half-space. J. Theor. Appl. Mech. Wars., **31**, 2, 295–305.

Kalker J.J., 1990. Three-Dimensional Elastic Bodies in Rolling Contact. Dordrecht, Boston, London, Kluwer Academic.

Kapica S.P., Kurdiumov S.P., Malineckij S.P., 2003. Synergetics and prognosis of a Future. Series Synergetics from Past to Future. Moscow, in Russian.

Kappus R., 1939. Zur Elastizitätstheorie endlicher Verschiebungen. ZAMM, **19**, 5, 271–285, and **19**, 6, 344–361.

Karnopp D., 1985. Computer simulation of stick-slip friction in mechanical dynamic systems. ASME J. Dynamic Sys. Measure. Control, 107, 100–103.

Kauderer H., 1958. Nichtlineare Mechanik. Berlin, Springer-Verlag.

Kennedy F.E., 1981. Surface temperatures in sliding systems - a finite element analysis. ASME J. Lubric. Technol., **103**, 1, 90–96.

Kennedy F.E., Ling F.F., 1974. A thermal, thermoelastic, and wear simulation of a high-energy sliding contact problem. ASME J. Lubric. Technol., **96**, 497–505.

Klarbring A., 1986. General contact boundary conditions and the analysis of frictional system. Int. J. Solids Struct., **22**, 12, 1377–1398.

Klarbring A., 1990. Derivation and analysis of rate boundary-value problem of frictional contact. Eur. J. Mech., **9**, 53–85.

Kononienko V.O., 1964. Vibrating Systems with a Limited Power Supply. Moscow, Nauka, in Russian.

Kovalenko A.D., 1975. Thermoelasticity. Kiev, Vyshaya Shkola, in Russian.

Kragelsky I.V., Dobychin M.N., Kombalov V.S., 1982. Friction and Wear: Calculation Methods. Oxford, UK, Pergamon Press.

Kragelsky I.V., Gitis N.V., 1987. Frictional self-oscillations. Moscow, Mashinostrojenije, in Russian.

Kragelsky I.V., Shchedrov V. S., 1956. Development of the Science of Friction. Moscow, Izd. Akad. Nauk SSSR, in Russian.

Kudinov V. A., 1967. Dynamics of Tool-Lathe. Moscow, Mashinostroenie, in Russian.

Kulchytsky-Zhyhailo R., Olesiak Z., 2000. When can we avoid the paradoxes in the solution to the problems of two thermoelastic cylinders in contact. J. Theor. Appl. Mech. Wars., **38**, 2, 297–314.

Kunze M., 2000. Non-Smooth Dynamical Systems. Berlin, Springer.

Kurnik W., 1997. Divergent and Oscillatory Bifurcation. Warsaw, PWN, in Polish.

Kuzmenko A. G., 1981. Contact problems with wear and sliding resistance. Friction and Wear (Treniye i Iznos), **2**, 3, 502–512, in Russian.

Laursen T.A, Oancea V. G., 1997. On the constitutive modelling and finite element computation of rate-dependent frictional sliding in large deformations. Comput. Meth. Appl. Mech. Engrg., **143**, 197–227.

Laursen T.A., 1999. On the development of thermodynamically consistent algorithms for thermomechanical friction contact. Comput. Meth. Appl. Mech. Engrg., **177**, 973–287.

Laursen T.A., Simo J.C., 1993. A continuum-based finite element formulation for the implicit solution of multi-body, large deformation frictional contact problems. Int. J. Numer. Methods Engrg., **36**, 3451–3485.

Lawrowski Z., 1993. Tribology, Friction, Wear and Lubrication. Warsaw, PWN, in Polish.

Leine R.I., Campen D.H., 2002a. Discontinuous bifurcations of periodic solutions. Math. Comput. Model., 36, 259–273.

Leine R.I., Campen D.H., 2002b. Discontinuous fold bifurcations in mechanical systems. Arch. Appl. Mech., 72, 138–146.

Leine R.I., Campen D.H., 2006. Bifurcation phenomena in non-smooth dynamical systems. Euro. J. Mech. A/Solids, 25, 595–616.

Leine R.I., Campen D.H., De Kraker A., Van den Steen L., 1998. Stick-slip vibrations induced by alternate friction models. Nonlinear Dynamics, 16, 1, 41–54.

Leine R.I., Campen D.H., Vrande B.L., 2000. Bifurcation in nonlinear discontinuous systems. Nonlinear Dynamics, 23, 105–164.

Ling F.F., 1959. A quasi-iterative method for computing interface temperature distribution. ZAMP, **10**, 461–474.

Ling F.F., 1973. Surface Mechanics. New York, John Wiley.

Ling F.F., Rice J.S., 1966. Surface temperature with temperature-dependent thermal properties. ASLE Trans., **9**, 195–201.

Litak G., Spuz-Szpos G., Szabelski K., Warmiński J., 1999. Vibration of externally forced Froude pendulum. Int. J. Bifurcation Chaos, **9**, 3, 561–570.

Maksimov I.L., 1988. Thermal instability of sliding and oscillations due to frictional heating effect. J. Tribol., **110**, 69–72.

Marchelek K., 1991. Dynamics of Machine Tools. Warsaw, WNT, in Polish.

Martins J. A. C., Oden J.T., Simoes F.M.F., 1990. A study of static and kinetic friction. Int. J. Engng. Sci., **28**, 1, 29–92.

Matysiak S., Yevtushenko A., Kulchytsky-Zhyhailo R., 2000. Contact problems of thermoelasticity for a half-space of functionally gradient material. J. Math. Sci. (New York), **99**, 5, 1569–1583.

Matysiak S., Yevtushenko A., 2001. On heating problems of friction. J. Theor. Appl. Mech. Wars., **39**, 3, 577–584.

Maugis D., 2000. Contact, Adhesion and Rupture of Elastic Solids. Berlin, Springer-Verlag.

Melan E., Parkus H., 1959. Wärmespannungen infolge stationärer temperaturfelder. Wien, Springer-Verlag.

Melnikov V. K., 1963. On the stability of the center for time periodic perturbations. Trans. Moscow Math. Soc., **12**, 1, 3-52, in Russian.

Michalowski R., Mróz Z., 1978. Associated and non-associated sliding rules in contact friction problems. Arch. Mech., **30**, 259–276.

Mindlin R.D., 1949. Compliance of elastic bodies in contact. J. Appl. Mech., **16**, 259–268.

Minorsky N., 1962. Nonlinear Oscillations. Van Nostrand, Princeton.

Mokrik R., Pyryev Yu., 1993. The state of stress in a semiinfinite cylinder upon ion exchange. Glass Physics Chem., **19**, 5, 384–386.

Moon F.C., 1987. Chaotic Vibrations. New York, John Wiley.

Moore D. F., 1975. Principles and Applications of Tribology. Oxford, Pergamon Press.

Moreau J.J., Panagiotopoulos P.D.(Eds), 1988. Non-smooth Mechanics and Applications, CISM Courses and Lectures No 302. Wien, New York, Springer-Verlag.

Morov V.A., 1985. Analysis of Thermoelastic Instability in Tribosystems of the Type of Moving Compressors with an Axially Symmetric Perturbation. In Investigation of Tribotechnical Systems under Cold Climate Conditions. Yakutsk, Izd. Yakutsk. Fil. Sib. Otd. Akad. Nauk SSSR, 21–33, in Russian.

Mostaghel N., 1999. Analytical Description of Pinching, Degrading Hysteretic Systems. J. Eng. Mech. **125**, 2, 216–224.

Mostaghel N., 2005. A non-standard analysis approach to systems involving friction. J. Sound Vibration. **284**, 3–5, 583–595.

Mróz Z., 2000. On the stability of friction contact. J. Theor. Appl. Mech. Wars., **38**, 2, 315–329.

Mróz Z., Stupkiewicz S., 1994. An anisotropic friction and wear model. Int. J. Solids Structures, **31**, 1113–1131.

Nakai M., 1998. Railway wheel squeal (squeal of disk subjected to periodic excitation). ASME J. Vibration Acoustics, **120**, 614–622.

Nayfeh A.H., 1981. Introduction to Perturbation Techniques. New York, John Wiley & Sons.

Nayfeh A.H., Balachandran B., 1995. Applied Nonlinear Dynamics. New York, John Wiley.

Nayfeh A.H., Mook D,T., 1979. Nonlinear Oscillations. New York, John Wiley.

Neimark Yu.I., 1978. Dynamical Systems and Controlled Processes. Moscow, Nauka, in Russian.

Nowacki W., 1962. Thermoelasticity. Oxford, Pergamon Press.

Nowacki W., 1970. Theory of Elasticity. Warsaw, PWN, in Polish.

Nowacki W., Olesiak Z., 1991. Thermodiffusion in Solid Bodies. Warsaw, PWN, in Polish.

Novogilov V. V., 1958. Theory of Elasticity. Leningrad, Sudpromgiz, in Russian.

Nusse H.E., Yorke J.A., 1994. Dynamics: Numerical Explorations. New York, Springer-Verlag.

Oancea V. G., Laursen T.A., 1996. Dynamics of a state variable frictional law in finite element analysis. Finite Elements in Analysis and Design, 22, 25–40.

Oden J.T., Martins J.A.C., 1985. Models and computational methods for dynamic friction phenomena. Comput. Meth. Appl. Mech. Engrg., **52**, 1–3, 527–634.

Olędzki A, Siwicki I., 1997. Modelling of vibrations in a certain self-locking system. Proceedings XXXVI Symposium Modelling in Mechanics, 265–268.

Olesiak Z., Pyryev Yu., 1996a. On nonuniqueness and stability in Barber's model of thermoelastic contact. Trans. ASME J. Appl. Mech., **63**, 3, 582–586.

Olesiak Z., Pyryev Yu., 1996b. Transient response in a one-dimensional model of thermoelastic contact. Trans. ASME J. Appl. Mech, **63**, 3, 575–581.

Olesiak Z., Pyryev Yu., 1997. A coupled quasi-stationary problem of thermodiffusion for an elastic cylinder. Int. J. Engng. Sci., **33**, 6, 773–780.

Olesiak Z., Pyryev Yu., 1998. A model of thermoelastic dynamic contact in conditions of frictional heat and wear. J. Theor. Appl. Mech. Wars., **36**, 2, 305–320.

Olesiak Z., Pyryev Yu., 2000. Determination of temperature and wear during braking. Acta Mech., **143**, 1–2, 67–78.

Olesiak Z., Pyryev Yu., Yevtushenko A., 1997. Determination of temperature and wear during braking. Wear, **210**, 120–126.

Onsager L., 1931. Reciprocal relations in irreversible processes. Phys. Rev., **37**, 405–427.

Osiński Z., 1979. Theory of Vibrations. Warsaw, PWN, in Polish.

Ott E., 1993. Chaos in Dynamical Systems. New York, Cambridge University Press.

Panovko J.G., 1980. Introduction in Theory of Mechanical Vibrations. Moscow, Nauka, in Russian.

Parkus H., 1959. Instationäre Wärmespannungen. Wien, Springer-Verlag.

de Pater A.D., Kalker J.J. (Eds.), 1975. Mechanics of Contact Between Deformable Media. Proc. IUTAM Symposium, Delft University Press, Enschede.

Pauk V.J., 1994. Plane contact problem involving heat generation and radiation. J. Theor. Appl. Mech. Wars., **32**, 829–839.

Pauk V.J., Woźniak Cz., 1999. Plane contact problem for a half-space with boundary imperfections. Int. J. Solids Structures, **36**, 3569–3579.

Pfeiffer F., 1984. Mechanische Systemen mit unstetigen Übergängen. Ingenieur-Achive Band **54**, 3, 232–240.

Pomeau Y., Manneville P., 1980. Intermittent transition to turbulence in dissipative dynamical systems. Commun. Math. Phys., **74**, 189–197.

Pyryev Yu., 1994. Dynamical model of thermoelastic contact in the conditions of the frictional heat and the restrained thermal expansion. Friction and Wear (Trenije i iznos), **15**, 6, 941–948, in Russian.

Pyryev Yu., 1999. Propagation of Waves in Elastic Solids with Taking Into Account Coupling of Physical-Mechanical Fields. Moscow, SIP RIA, in Russian.

Pyryev Yu., 2000a. Investigation of stick-slip motion of body with frictional heat generation and wear. Visn. L'viv. Univ. Ser. Mekh. Mat., 57, 128–132, in Ukrainian.

Pyryev Yu., 2000b. Frictional contact of the cylinder and the bush with inertia, heat generation and wear taking into account. Physico-Chem. Mech. Mater. **36**, 3, 53–58, in Ukrainian.

Pyryev Yu., 2001. Investigation of peculiarities of thermoelastic contact of two bodies with nonlinear dependence of thermal resistance on loading taken intro account. Math. Meth. Physicomech. Fields, **44**, 2, 100–106, in Ukrainian.

Pyryev Yu., 2002. Investigation of the contact interaction between two layers with regard for the temperature dependence of friction and wear coefficients. J. Math. Sci., New York, **109**, 1, 1257–1265.

Pyryev Yu., 2004. Dynamics of Contacting Systems with an Account of Heat Transfer, Friction and Wear. Łódź, Scientific Bulletin of Łódź Technical University, No 936, in Polish.

Pyryev Yu., Grilitskiy D., Mokrik R., 1995. Nonsteady-state temperature field and thermoelastic state of a bilayer hollow circular cylinder upon frictional heating. Int. Appl. Mech., **31**, 1, 38–43.

Pyryev Yu., Grilitskiy D., 1995. Nonlinear nonstationary problem on the frictional contact for cylinder with inertia and heat generation. Reports of NAS of Ukraine, 9, 34-37, in Ukrainian.

Pyryev Yu., Grzelczyk D., Awrejcewicz J., 2007. On a novel friction model suitable for simulation of the stick-slip vibration. Khmelnitskiy State University's Bulletin, **1**, 4, 86–92.

Pyryev Yu., Grilitskiy D., 1996. Nonstationary problem of frictional contact for a cylinder under the conditions of heat release and wear. J. Appl. Mech. Tech. Phys., **37**, 6, 857–861.

Pyryev Yu., Mandzyk Yu., 1996. An analysis of stability of thermoelastic contact in tribosystems consisted of radial seal for a cylinder. Friction and Wear, **17**, 5, 621-628, in Russian.

Pyryev Yu., Mokryk R., 1996. Nonlinear Transient Problem about Frictional Contact with Heat Resistance. Visn. L'viv. Univ. Ser. Mekh. Mat., **43**, 51–55, in Ukrainian.

Pyryev Yu., Yevtushenko A., 2000. The influence of the brakes friction elements thickness on the contact temperature and wear. Heat and Mass Transfer, **36**, 319–323.

Rabinowicz E., 1965. Friction and Wear of Materials. New York, John Wiley.

Radi E., Bigoni D., Tralli A., 1999. On uniqueness for frictional contact rate. J. Mech. Phys. Solids, 47, 275–296.

Richard T., Detournay E., 2000. Stick-slip motion in a friction oscillator with normal and tangential mode coupling. C. R. Acad. Sci. Paris, Friction, Adhesion, Lubrification, 328, 671–678.

Riznichenko G.Yu., 2002. Lectures on Mathematical Models in Biology. Moscow, NITs RKhD, in Russian.

Rojek J., Telega J.J., Stupkiewicz S., 2001. Contact problems with friction, adhesion and wear in orthopaedic biomechanics. Part II - Numerical implementation and application to implanted knee joints. J. Theor. Appl. Mech. Wars., **39**, 3, 679–706.

Rozman M.G., Urbakh M., Klafter J., 1996. Stick-slip motion and force fluctuations in driven two-wave potential. Phys. Rev. Lett., **77**, 4, 683–686.

Sadowski J., 1999. Termodynamic Interpretation of Friction and Wear. Radom, Radom Technical University, in Polish.

Saux C., Leine R.I., Grocker C., 2005. Dynamics of a rolling disk in the presence of dry friction. J. Nonlinear Sci., **15**, 27–61.

Schmidt G., Tondl A., 1986. Non-Linear Vibrations. Berlin, Akademie Verlag.

Schuster H.G., 1995. Deterministic Chaos: An Introduction, 3rd ed. New York, Wiley.

Shtaerman I.Y., 1949. Contact Problem in the Theory of Elasticity. Moscow, Gostekhizdat, in Russian.

Sneddon I. N., 1966. Mixed Boundary Value Problems in Potential Theory. Amsterdam, Holland, North-Holland.

Solski P., Ziemba S., 1965. Problems of Dry Friction. Warsaw, PWN, in Polish.

Stefański A., Wojewoda J., Wiercigroch M, Kapitaniak T., 2003. Chaos caused by non-reversible dry friction. Chaos, Solitons Fractals, **16**, 661–664.

Strömberg N., 1999. Finite element treatment of two-dimensional thermoelastic wear problems. Comput. Methods Appl. Mech. Engrg., **177**, 441–455.

Strömberg N., Johansson L., Klarbring A., 1996. Derivation and analysis of a generalized standard model for contact, Friction and Wear. Int. J. Solids Structures, **33**, 13, 1817–1836.

Stupkiewicz S., Mróz Z., 2001. Modelling of friction and dilatancy effects at brittle interfaces for monotonic and cyclic loading. J. Theor. Appl. Mech. Wars. **39**, 3, 707–739.

Szabelski K., 1984. Self-Excited Vibrations of a system excited parametrically and with nonilinear elasticity. Theor. Appl. Mech., **1/2**, 22, 171–183.

Szabelski K., 1991. The vibrations of self-excited system with parametric excitation and non-symmetric elasticity characteristic. J. Theor. Appl. Mech., **29**, 1, 57–81.

Szefer G., 1997. Dynamic contact of bodies experiencing large deformation. Acta Mechanica, Wien, Springer, **125**, 217–233.

Tarng Y.S., Cheng H.E., 1995. An investigation of stick-slip friction on the contouring accuracy of CNC machine tools. Int. J. Mach. Manufact., **35**, 4, 565–576.

Telega J., 1988. Topics on unilateral contact problems of elasticity and inelasticity. In: Non-smooth mechanics and applications (ed. J.J. Moreau and P.D. Panagiotopoulos). Berlin, Springer CISM Courses and Lectures, No. 302, 341–462.

Tichy J. A., Meyer D. A., 2000. Review of solid mechanics in tribology. Int. J. Solids Structures, **37**, 391–400.

Timoshenko S., Goodier J.N., 1951. Theory of Elasticity. New York, McGraw-Hill.

Tondl A., 1970. Self-excited Vibration. Monographs and Memoranda, No.9, Běchovice, National Research Institute for Machine Design.

Tondl A., 1978. On the Interaction Between Self-excited and Parametric Vibrations. Monographs and Memoranda, No.25, Běchovice, National Research Institute for Machine Design.

Ulitko A. F., 1990. Three-dimensional motion of elastic solid. Izv. Akad. Nauk. SSSR, Mekh. Tverd. Tela, **6**, 55–66, in Russian.

Van De Velde F., De Beats P., 1998. A new approach of stick-slip based on quasi-harmonic tangential oscillation. Wear, **216**, 15–26.

Van de Wouw N., Leine R.I., 2004. Attractivity of equilibrium sets of systems with dry friction. Nonlinear Dynamics, **35**, 19–39.

Váradi K., Néder Z., Friedrich K., Flöck J., 2000. Contact and thermal analysis of transfer film covered real composite-steel surfaces in sliding contact. Tribol. Int., 33, 11, 789–802.

Verlan A.F., Sizikov V.S., 1986. Integral Equations: Methods, Algorithms, and Codes. Kyiv, Naukova Dumka, in Russian.

Wagg D.J., 2003. Periodic and chaotic dynamics in asymmetric elastoplastic oscillator. Chaos, Solitons Fractals, 16, 779–786.

Warmiński J., Litak G., Szabelski K., 2000, Synchronisation and chaos in a parametrically and self-excited system with two degrees of freedom. Int. J. Nonlinear Dynamics, **22**, 135–153.

Wikieł B, Hill J.M., 2000. Stick-slip motion for two coupled masses with side friction. Int. J. Non-Linear Mech., **35**, 953–62.

Willis J.R., 1966. Hertzian contact of anisotropic bodies. J. Mech. Phys. Solids, **14**, 163–176.

Wriggers P., 1995. Finite element algorithms for contact problems. Arch. Comput. Methods Engrg. **2**, 4, 1–49.

Wriggers P., 2002. Computational Contact Mechanics. New York, John Wiley and Sons.

Yevtushenko A., Kulchytsky-Zhyhailo R., 1995. Determination of limiting radii of the contact area in axi-symmetric contact problems with frictional heat generation. J. Mech. Phys. Solids, **43**, 4, 599–604.

Yevtushenko A., Pyryev Yu., 1997. The influence of wear on the development of thermoelastic instability of the frictional contact. Mech. Solids, **32**, 1, 91–97.

Yevtushenko A., Pyryev Yu., 1998. Calculation of the contact temperature and wear of frictional elements of brakes. Mater. Sci., **34**, 2, 249–254.

Yevtushenko A., Pyryev Yu., 1999a. The applicability of the hereditary model of wear with an exponential kernel in the one-dimensional contact problem taking frictional heat generation into account. J. Appl. Math. Mechs., **63**, 5, 795–801.

Yevtushenko A., Pyryev Yu., 1999b. Calculation of the contact temperature and wear during braking. J. Math. Sci., New York, **96**, 1, 2892–2896.

Yi Y.B., 2006. Finite element analysis of thermoelastodynamic instability involving frictional heating. Trans. the ASME. J. Tribol., **128**, 718–724.

You H. I., Hsia J. H., 1995. The influence of friction-speed relation on the occurrence of stick-slip motion. J. Tribol., **117**, 450–455.

Zajtsev V.I., Shchavelin V.M., 1989. Method of solutions to contact problems taking into account real properties of interacting surface bodies. Izv. Akad. Nauk. SSSR, Mekh. Tverd. Tela, 1, 88–94, in Russian.

Zboiński G., Ostachowicz W., 2001. Three-dimensional elastic and elasto-plastic frictional contact analysis of turbomachinery blade attachments. J. Theor. Appl. Mech., **39**, 3, 769–790.

Ziegler H., 1963. Some extremum principle in irreversible thermodynamics with application to continuum mechanics. Prog. Solid Mech., **4**, 93–193.

Zmitrowicz A., 1987. A thermodynamical model of contact, friction and wear: I governing equations; II constitutive equations for materials and linearized theories; III constitutive equations for friction, wear and frictional heat. Wear, **114**, 2, 135–168, 169–197, 199–221.

Zmitrowicz A., 1999. Illustrative examples of anisotropic friction with sliding path curvature effects. Int. J. Solids. Structures, **36**, 2849–2863.

Zmitrowicz A., 2001. Variation descriptions of wearing out solids and wear particles in contact mechanics. J. Theor. Appl. Mech. Wars., **39**, 3, 791–808.

Index

MIX
Papier aus verantwortungsvollen Quellen
Paper from responsible sources
FSC® C105338

If you have any concerns about our products,
you can contact us on
ProductSafety@springernature.com

In case Publisher is established outside the EU,
the EU authorized representative is:
Springer Nature Customer Service Center GmbH
Europaplatz 3, 69115 Heidelberg, Germany

Printed by Libri Plureos GmbH
in Hamburg, Germany